RADIOACTIVE WASTE MANAGEMENT

RADIOACTIVE WASTE MANAGEMENT

2nd Edition

Edited by

**James H. Saling &
Audeen W. Fentiman**

TAYLOR & FRANCIS

New York

Denise T. Schanck, *Vice President*
Robert H. Bedford, *Editor*
Catherine M. Caputo, *Assistant Editor*
Tom Hastings, *Marketing Director*
Mariluz Segarra, *Marketing Associate*

Published in 2001 by
Taylor & Francis
29 West 35th Street
New York, NY 10001

Published in Great Britain by
Taylor & Francis
11 New Fetter Lane
London EC4P 4EE

Copyright © 2001 by Taylor & Francis

Printed in the United States of America on acid-free paper.

Library of Congress Cataloging-in-Publication Data

Radioactive waste management / edited by James H. Saling & Audeen W. Fentiman.–2nd ed.
    p.   cm.
    Rev. ed. of: Radioactive waste management / Y.S. Tang, James H. Saling.
    Includes bibliographical references and index.
    ISBN 1-56032-842-8 (alk. paper)
    1. Radioactive waste disposal.   2. Radioactive wastes–Management.
I. Saling, James H.   II. Fentiman, Audeen W.   III. Tang, Y.S. (Yu S.), 1922-Radioactive waste management.   IV. Title.
TD812.R33   2001
363.72'89—dc21

                    2001023959

CONTENTS

Disposal of nuclear wastes has been studied for more than five decades under the auspices of the U.S. Atomic Energy Commission (AEC) and its successors, the U.S. Energy Research and Development Administration (ERDA) and the U.S. Department of Energy (DOE). In 1982, the U.S. Congress passed the Nuclear Waste Policy Act, which, among other things, directed the DOE to establish the Office of Civilian Radioactive Waste Management. The purpose of this organization was to design and construct facilities that would accept and start disposing of spent nuclear fuel and high-level wastes by January 31, 1998. Almost two decades have passed since that act was signed into law, and after the expenditure of several billion dollars, we still have no firm date for disposal of high-level nuclear wastes.

Similar delays have occurred in disposing of other types of radioactive waste. Construction of the Waste Isolation Pilot Plant, designed for the permanent disposal of transuranic waste, was essentially complete in 1989, but the first shipment of waste was not accepted until 1999. The Low-Level Waste Policy Act of 1980 made states responsible for providing disposal capacity for their own low-level waste (LLW) and allowed them to form compacts for that purpose, but in the intervening 20 years, no state or compact has opened a new LLW disposal facility.

One wonders why a country with the technological capability to put a man on the moon in a single decade has been unable to accomplish the seemingly much less challenging task of placing nuclear wastes safely underground in five decades. There may be several reasons for this. The Department of Energy supported the development of many technologies for packaging, storing, handling, transporting, and disposal of nuclear wastes, including spent fuel. As a result, the technical community developed a competitive atmosphere with each developer insisting that its method was best, when in fact, many of the technologies were quite adequate to safely handle the wastes. Most scientists agreed that many of the technologies were adequate, but they still pushed for their particular technology to be selected. This failure of the technical community to agree on a single concept, particularly for high-level waste disposal, often confused the public and led them to question whether any disposal system would be safe. Over time, the waste management issue has become politicized, and many believe decisions are often made for political reasons rather than technical ones, further eroding public confidence.

We believe it is important for the federal government, industry, educators, and other stakeholders to provide complete, factual, research-based information to the public on nuclear power and nuclear waste and to refute any misinformation that has been provided over the past several years. The information provided should include facts on the volume of nonrenewable resources consumed for electricity production and the many other uses of those resources for which there are currently no other economical raw materials. In a democratic society, the public needs this factual information in order to make decisions. Today's choices regarding fuel for electricity generation can have a significant impact on the quality of life and national security for many decades to come.

The original authors were motivated to write the first edition of this book by the lack of suitable textbooks or reference books in the radioactive waste management field when they were organizing a short training course in 1984, a nuclear waste management overview for Westinghouse Electric Corporation. In the revised edition, we retain much of the original material because of its historical value and add more recent information. In addition, we broaden the scope to include chapters on mixtures of hazardous and radioactive wastes and on environmental restoration of Department of Energy sites. This should allow the book to be used in environmental engineering courses as well as courses in radioactive waste management. We also hope that the wider use of this book will provide future decision makers with information that will allow them to consider nuclear power as one of several methods to meet our country's energy needs.

The purposes of the new edition of this book are as follows:

1. To broaden the scope to make the book useful for environmental engineers as well as for radioactive waste managers.
2. To create a general awareness of technologies developed for radioactive waste management and environmental restoration.
3. To summarize the current status of such technologies.
4. To prepare practicing scientists, engineers, administrative personnel, and students for participation in working teams applying such technologies.

Thus this book is aimed at serving as a textbook for students in nuclear engineering and students in environmental engineering and as a reference book for those who have decision-making roles at various levels in government and private industry.

We are indebted to a number of people, too numerous to list, for their help in the preparation of this book. We do wish, however, to acknowledge the contributions of

Amber Climer, a graduate student in Nuclear Engineering and in Environmental Science at The Ohio State University, who gathered data, prepared tables and figures, typed several of the chapters, and provided valuable insights from the student's perspective. Finally, we thank our spouses and families for their encouragement and understanding, without which we could not have completed this undertaking.

James H. Saling
Audeen W. Fentiman

CHAPTER ONE

INTRODUCTION

Radioactive waste management is not a new issue, for it arose with the advent of nuclear energy. During the first conference on the Peaceful Uses of Atomic Energy, held in Geneva in 1955, management concepts were introduced and discussed. Subsequently, several management techniques have been developed and put into practice. Only in the recent past has the management of radioactive waste (or radwaste*) become a matter of great public concern. As the geochemist Konrad B. Krauskopf, Stanford University Professor Emeritus and an advisor to federal agencies concerned with waste disposal,† said,[1] nuclear waste disposal can be done safely. The problem is political, not technological. This point of view has been shared by many scientists and engineers, including Edward Teller. The National Research Council, through the Waste Isolation Systems Panel, made a study of the isolation system for geologic disposal of radwaste for the U.S. Department of Energy (DOE). Among other overall conclusions, the panel's report indicated that the waste technology identified in the study should be more than adequate for isolating radwastes from the biosphere and for protecting public health and safety.[2] This view has also been supported by most of the organized technical community.[3] In 1982 President Reagan was eager for legislation that would establish a deadline for each step of the waste disposal procedure. After resolving disagreement by compromises, Congress passed the Nuclear Waste Policy Act (NWPA), which directed the DOE to accomplish specific goals at specific times. This was the push

---

*The term radwaste will be used for radioactive waste throughout this book.

†Dr. Krauskopf has served on committees and subcommittees of the Department of Energy, the Nuclear Regulatory Commission (NRC), and the Environmental Protection Agency (EPA) and is a past chairman of the National Academy of Sciences' Board on Radwaste Management (BRWM).

needed for the implementation of waste management. However, public acceptance and cooperation are indispensable for the success of this program and the public is apprehensive about the potential hazard of radwaste material. The importance of understanding the management technologies, design philosophy, and performance evaluation related to processing, packaging, storing, transporting, and disposing of radwaste cannot be overemphasized and such information must be disseminated. Thus, the purposes of this book are as follows:

1. To provide broad information useful for environmental engineers as well as for radioactive waste managers.
2. To create a general awareness of technologies and programs for both fields.
3. To summarize the current status of such technologies.
4. To prepare practicing scientists, engineers, administrative personnel, and students to apply such technologies.

The present chapter covers the nuclear fuel cycle, types and sources of both radioactive (rad) waste and nonradioactive hazardous wastes, waste management activities and responsibilities for both types of waste, regulatory agencies, and legislative involvement.

1.1 NUCLEAR FUEL CYCLE AND POWER GENERATION

The nuclear fuel cycle is the generic term for various support activities essential to the operation of power reactors. Different types of operations exist within the fuel cycle (Figure 1.1), and Figure 1.2 depicts the types of waste generated with each of these operations.[4]

Uranium mining. Uranium, one of the fuels used in nuclear power plants, is a naturally occurring element in the earth's crust. Like many other ores, it is mined, with an average yield of 1–5 tons of uranium recovered from 1000 tons of ore. Normally the uranium is extracted from ore with an organic solvent. Natural uranium is composed of two essential isotopes: ^{235}U (0.72%) and ^{238}U (99.27%).

Mill processing. Uranium mills, where uranium is extracted and converted to oxide forms, are usually built near the mines. The crude uranium normally consists of oxides such as pitchblende (U_3O_8) and salts such as sodium diuranate ($Na_2U_2O_7$) or ammonium diuranate [$(NH_4)_2U_2O_7$]. The purification process removes the remaining contaminants and produces as a product one of the oxides UO_3, U_3O_8, or UO_2. Dilute and dispersed effluents are discharged from the mills, while the concentrate and fines from liquid sludge are tailings to be disposed of.

Fuel conversion. The final step is to convert the uranium into a form suitable for enrichment, such as UF_6, through either a dry process or a wet process. The nature of the radioactive effluents from the two processes differs, as the dry process releases gaseous and solid radwastes and the wet process releases liquid effluents.

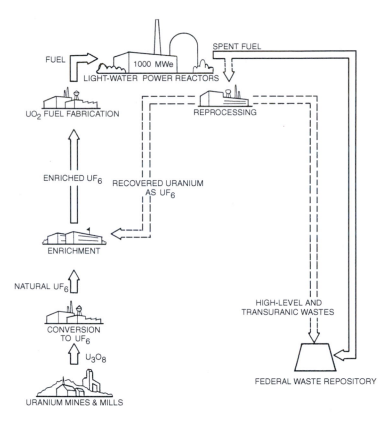

Figure 1.1 Nuclear fuel cycle. From An identification of the waste generated within the nuclear fuel cycle by R. A. Wolfe, *AIChE Symp. Ser.*, no. 154, vol. 72, p. 1, 1976. Copyright 1976 by American Institute of Chemical Engineers. Reprinted by permission.

Enrichment. Gaseous diffusion has been the method most commonly used to enrich the uranium used in nuclear power plants. This process makes use of the phenomenon of molecular effusion to effect separation. In a vessel containing a mixture of two gases, molecules of the gas with the lower molecular weight travel faster and strike the walls of the vessel more frequently, relative to their concentration, than do the molecules of the gas with the higher molecular weight. If the walls of the vessel have perforations just large enough to allow passage of individual molecules without permitting bulk flow of the gas as a whole (diffusion barrier), more of the light molecules flow through the walls relative to their concentration than do the heavy molecules.

Figure 1.3 illustrates a simple type of gaseous diffusion cascade. As the mixture entering the diffuser on a stage flows past a diffusion barrier, a portion of the gas flows through the barrier into the region of lower pressure on its downstream side. The gas flowing through the barrier is enriched in the component of lower molecular weight (^{235}U in this case) and constitutes the light fraction, which is fed to the stage nearer the top of the cascade. Similarly, the gas that does not flow through the barrier is enriched in the component of higher molecular weight and constitutes the heavy fraction, which

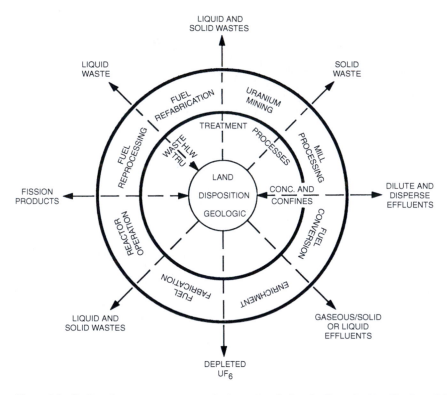

Figure 1.2 Radioactive waste management in the nuclear fuel cycle. From An identification of the waste generated within the nuclear fuel cycle by R. A. Wolfe. *AIChE Symp. Ser.*, no. 154, vol. 72, p. 1, 1976. Copyright 1976 by American Institute of Chemical Engineers. Reprinted by permission.

is fed to the stage nearer the bottom of the cascade. On each stage a pump compresses the gas from the pressure prevailing on the downstream side of the diffusion barrier to the pressure on the upstream side, and a cooler removes the heat of compression from the stage feed. As only a limited degree of separation is attainable in a single stage, it is necessary to repeat the process a number of times (typically many hundreds of times) to obtain the degree of separation required. The UF_6 is withdrawn from the diffusion process with desired ^{235}U, and the depleted UF_6 in the other stream is stored at the diffusion plant for possible later use.

Gaseous diffusion requires high-compression-ratio pumps, special fine-grained diffusion membranes, and a very large number of stages, which makes it a very expensive separation method. The search for other methods of enrichment, including gas centrifuge and laser separation, has continued.

Fuel fabrication. The nuclear reaction is sustained by properly arranged fuel assemblies consisting of the fuel elements maintained in the desired configuration by structural support material. Normally the fuel is received at the fuel fabrication plant in the form of either UO_2 powder or uranyl nitrate solution. In the latter case it is converted to UO_2

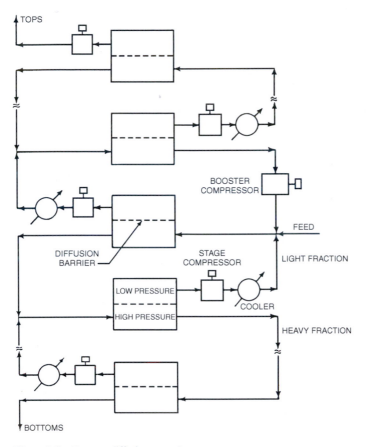

Figure 1.3 Gaseous diffusion cascade.

powder before fabrication. The UO$_2$ powder is cold pressed into pellets that are typically about 0.89 cm (0.35 in.) in diameter by about 0.89 cm (0.35 in.) long. These pellets are then sintered in a reducing atmosphere at temperatures ranging from 1300°C to 2000°C (2400–3600°F), arranged in 10-ft-long stacks, and loaded into zircalloy seamless tubes to form the fuel rods. Waste products from the fabrication plants are liquids and solids contaminated with uranium and fluorides.

The fuel rods are then loaded into fuel assemblies with spacers or grids to keep them separated so that the reactor coolant can flow around the rods and remove the heat. The reactor core is made up of the correct number of assemblies to provide criticality and produce the heat needed to drive the turbines and produce electricity. The number of assemblies and the number of fuel rods differ for different types of reactor. Each boiling water reactor (BWR) assembly contains 63 fuel rods each of which is about 1.25 cm. (1/2 inch) in diameter and each pressurized water reactor (PWR) assembly contains 264 fuel rods, each of which is 0.95 cm. (1/3 in.) in diameter. Figure 1.4 shows BWR and PWR fuel assemblies. A PWR reactor core is made up of approximately 180 fuel assemblies. Until the late 1980s approximately 1/3 of the core was replaced each year when the fuel had

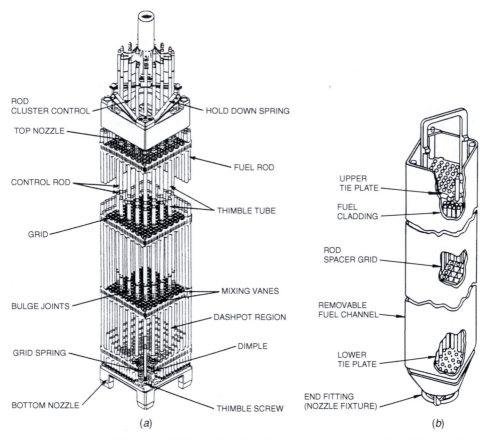

Figure 1.4 (a) Typical light water PWR core. The 17 × 17 assembly has a total weight of 658 kg, contains 461 kg U, has a fuel element outer diameter of 0.95 cm, and has 264 fuel elements per assembly. (b) Schematic drawing of a typical BWR fuel assembly. The 8 × 8 assembly has a total weight of 320 kg, contains 189 kg U, has a fuel element outer diameter of 1.25 cm, and has 63 fuel elements per assembly.

achieved a burnup of about 33,000 megawatt days (MWD) per ton of heavy metal (MTH) for PWRs (less for BWRs). At this time utilities started looking at extended burnups. They are now achieving burnups in the order of 45,000–50,000 MWD/MTU and they replace 1/3 of the fuel every 18 months. This operating period without refueling is sometimes pushed to 24 months and 55,000 MWD per ton burnup for PWR reactors. A list of light water reactor (LWR) nuclear fuel assembly characteristics is provided in Table 1.1 (Ref. 5,6,7).

Reactor operation. Most commercial reactors can be classified as either boiling water reactors (BWRs) or pressurized water reactors (PWRs). The BWR heats the cooling water flowing around the fuel rods to boiling, thus producing steam, which is then used to drive the turbines to produce electricity. The PWR operates at high pressure and thus produces higher temperature steam, which is then fed to an intermediate heat exchanger, which transfers the heat to a secondary system to produce steam to drive the turbines and produce electricity. Because of this difference in design, the PWR produces fewer low-level wastes than does the BWR, as seen in Table 1.2.

Table 1.1 Integrated Data Base reference characteristics of LWR nuclear fuel assemblies

Characteristics	BWR[a]	PWR[b]
Overall assembly length (m)	4.470	4.059
Cross section (cm)	13.9 × 13.9	21.4 × 21.4
Fuel rod length (m)	4.064	3.851
Active fuel height (m)	3.759	3.658
Fuel rod outer diameter (cm)	1.252	0.950
Fuel rod array	8 × 8	17 × 17
Fuel rods per assembly	63	264
Assembly total weight (kg)	319.9	657.9
Uranium/assembly (kg)	183.3	461.4
UO$_2$/assembly (kg)	208.0	523.4
Zircaloy/assembly (kg)	103.3[c]	108.4[d]
Hardware/assembly (kg)	8.6[e]	26.1[f]
Total metal/assembly (kg)	111.9	134.5
Nominal volume/assembly (m^3)	0.0864[g]	0.186[g]

[a] Ref. 6
[b] Ref. 7
[c] Includes zircaloy fuel-rod spacers and fuel channel.
[d] Includes zircaloy control-rod guide thimbles.
[e] Includes stainless steel tie-plates, Inconel springs, and plenum springs.
[f] Includes stainless steel nozzles and Inconel-718 grids.
[g] Based on overall outside dimension. Includes spacing between the stacked fuel rods of an assembly.

From Integrated Data Base Report-1996: U.S. Spent Fuel and Radioactive Waste Inventories, Projections, and Characteristics, DOE/RW-0006, Rev. 13, U.S. Department of Defense, Washington, D.C., December 1997. (Reference 5) Reprinted by permission.

The fuel elements in the reactor core are surrounded by ordinary water to remove the heat from the fuel pins and also to act as a moderator to slow the fission neutrons down to thermal energies. Neutrons of thermal energies are far more likely to cause more fissions in a light water reactor than fast neutrons. In order to control the reaction, spaces are provided within the reactor core for control rods made of a material that has a very high absorption rate for the neutrons being produced during reactor operation. At reactor startup, the control rods are slowly withdrawn and a neutron source is inserted to initiate the chain reaction. The control rods can be withdrawn further or inserted further to increase or decrease the reactor power generation rate. The heat generated by the fission process produces steam to drive the turbines to produce electricity. The fission products retained in the fuel elements are a source of radwaste when the fuel elements are discharged from the reactor as spent fuel. The spent fuel may be stored or the fuel may be reprocessed to recover the unused uranium for future use. The United States does not reprocess commercially generated spent fuel. Other countries, such as France and Japan, do reprocess their spent fuel and recover the unused uranium and the plutonium that is created during the fission process. Radwaste is also produced by neutron absorption in the structural members of the reactor. The radwaste produced by neutron absorption in reactor structural members and contaminated materials generated during the reactor operations is treated and packaged at the reactor site.

Table 1.2 Low-level waste from nuclear energy plants[a]

Waste type	1978–1981 PWR	1978–1981 BWR	1982–1985 PWR	1982–1985 BWR	1985–1986 PWR	1985–1986 BWR
	Average volume (ft³/unit yr)					
Dry waste						
Compacted	4,800	14,300	6,350	10,450	4,300	7,850
Noncompacted	5,500	8,050	3,700	8,050	2,100	4,900
Filters	250	50	250	100	200	50
Subtotal	10,550	22,400	10,300	18,600	6,600	12,800
Wet waste						
Resins	1,050	2,100	1,500	2,200	1,100	2,400
Sludges	0	5,550	250	6,000	400	4,350
Concentrates	4,000	4,600	1,250	1,750	800	1,700
Oils			300	900	300	1,100
Miscellaneous			100	50	150	200
Subtotal	5,050	12,250	3,400	10,900	2,750	9,750
Total average volume	15,600	34,650	13,700	29,500	9,350	22,550
	Median volume (ft³/unit yr)					
Dry waste	8,600	21,600	8,550	16,850	6,150	12,400
Wet waste	2,300	9,700	2,700	10,550	1,900	9,250
Total waste	12,800	31,650	11,400	25,900	8,800	19,700

[a]Units in data base: 1978–1981, 18 BWRs and 28 PWRs; 1982–1985, 27 BWRs and 53 PWRs; 1985–1986, 27 BWRs and 56 PWRs. Note that 1 ft³ = 0.0283 m³.

From Radioactive Waste Generation Survey Update by G. S. Daloisio and C. P. Deltete, Vols. 1 and 2, NP-5526, Final Report, Electric Power Research Institute, Palo Alto, Calif., 1988. (Reference 8) Copyright 1988 by Electric Power Research Institute. Reprinted by permission.

Fuel reprocessing. Tributyl phosphate, an organic solvent, is commonly used to extract plutonium and uranium from the acidic solution in which the spent fuel is dissolved. Additional solvent extraction or ion exchange separates plutonium from uranium. The remainder of the acidic solution is the origin of the high-level waste. Fuel reprocessing operations also produce the largest volume of transuranic-contaminated waste. The recovered uranium is then fed to the enrichment plant for fuel fabrication.

1.2 CLASSIFICATION OF RADWASTES

1.2.1 Physical States of Radwaste

Radwaste material can exist in all three physical states:

1. *Gaseous waste.* The gaseous waste comes from gaseous effluents of reprocessing plants and nuclear power plants. This waste usually contains airborne radionuclides such as ^{85}Kr, ^{3}H, and ^{131}I. Treatment of gaseous waste is of particular importance in protecting the environment from such airborne radionuclides.

2. *Liquid waste.* The liquid waste comes mostly from spent fuel reprocessing plants, such as the aqueous waste from the first-cycle extraction system in irradiated fuel processing. Liquid discharge from different parts of the reactor plant or steam generators is also liquid waste.
3. *Solid waste.* The solid waste comes from the mining and milling of uranium and thorium ores, from sludges in storage tanks containing waste solutions, and from contaminated equipment and structures.

1.2.2 Classification by Type and Level of Radioactivity

Radwastes can be classified according to their origin (i.e., defense wastes or commercial wastes), the type of material present (e.g., transuranic waste or spent nuclear fuel), and their level of radioactivity (e.g., high- or low-level wastes). The general classification of radwastes used in this book is as follows.

1. *High-level waste (HLW)* results from the reprocessing of spent fuel from a defense or commercial reactor. The amount of plutonium and other heavy isotopes remaining in the solutions is small (0.5% of original Pu and U); the residue consists mainly of fission products.
2. *Spent nuclear fuel (SNF)* discharged from the reactor may be stored at the reactor site and eventually placed in a waste repository without reprocessing, as has been planned in the United States. In this case, SNF is treated as a high-level waste unless and until it is retrieved and reprocessed at a future time.
3. *Transuranic (TRU) waste* is defined as waste material that is contaminated with alpha-emitting radionuclides of sufficiently long life (>20 years) of elements with atomic number 92 or larger and concentrations greater than 100 nanocuries per gram (nCi/g). The TRU waste classification includes all transuranic nuclides except ^{238}Pu and ^{241}Pu and also includes ^{233}U and its daughter products.[9] Some wastes containing ^{238}Pu and ^{241}Pu are handled as TRU waste in accordance with local requirements.[10]
4. *Low-level waste (LLW)* often has relatively little radioactivity and contains practically no TRU elements. Most LLW requires little or no shielding, may be handled by direct contact, and may be buried in near-surface facilities. Some LLW, however, has high enough radioactivity that it must be given special treatment and disposal.[11] Wastes of the latter type are classified in Europe as intermediate-level waste (ILW).
5. *Mill tailings* from uranium mills constitute another type of waste with a low level of radioactivity, but they are not usually classified as LLW. The tailings contain elements such as thorium and radium, which are by-products of the decay of ^{238}U and are not removed in the extraction of uranium. These tailings leave the mill as a liquid sludge and are allowed to dry. They are collected in piles within enclosures. It is necessary to take precautions to prevent the tailings from contaminating ground water or getting into the air as dust.[12] The passage of the National Environmental Policy Act (NEPA) prompted the National Research Council to initiate a generic environmental impact statement (EIS) that provides some specific guidelines about ways to handle tailings.

1.3 SOURCES OF RADWASTE

Table 1.3 shows the general sources of different kinds of waste, as described in more detail below.

1.3.1 Sources of HLW and SNF

Almost all of the HLW generated from reprocessing of spent fuel in the United States has originated from defense activities and almost all of the SNF accumulated in the United States has originated from commercial reactors. Most of the radioactivity, however, is associated with the commercially generated SNF.

Although a great amount of money and effort has been spent on studies and attempts to provide storage and/or disposal facilities for all types of wastes, the U.S. government has yet to provide any facilities for disposal of HLW or spent fuel, and there are only three LLW disposal sites in operation all of which are commercially operated. Therefore, all HLW is currently being stored at the DOE facilities that are generating it; with few exceptions, all spent nuclear fuel (SNF) is being stored at the reactor sites where it was generated; and a fairly large portion of the LLW is being stored at the generator sites. Table 1.4 provides the basic assumptions that were used to project future quantities of HLW and SNF. These assumptions are based on the DOE Energy Information Administration (EIA) 1997 Reference Case projections.[13] These data show a major decrease in the production of nuclear power in the United States over the next several years. There are no projections that show what energy source will be used to replace the energy currently supplied by nuclear sources.

1.3.2 Sources of TRU Waste

Typically, TRU waste is in the form of metal, glassware, process equipment, soil, and laboratory waste such as ion exchanger resins, filters, clothing, and paper products. The estimated physical characteristics of future commercial TRU waste are summarized in Table 1.5.[14,15]

Table 1.3 Nuclear waste types and sources

Source	Spent nuclear fuel (SNF)	High-level waste (HLW)	Transuranic waste (TRU)	Low-level waste (LLW)
Commercial nuclear fuel cycle operations	X	X		X
Institutions (hospitals, universities, etc.)	X		X	X
Industrial users	X		X	X
Decontamination and decommissioning of fuel cycle			X	X
Defense-related activities	X	X	X	X

Table 1.4 Major assumptions used in projecting future quantities of HLW and SNF

Inventory/projection basis

Inventories (except where indicated) are reported as of the end of FY1996 (September 30, 1996)

Projections are generally reported for the FY1997–2030

HLW solidification activities

For Hanford, HLW solidification (borosilicate glass production) starts in 2002 and concludes in 2028

For INEEL (Idaho National Engineering and Environmental Laboratory), HLW solidification (immobilization) starts in 2019 and continues through 2034

For SRS (Savannah River Site), HLW solidification (glass production) at the Defense Waste Processing Facility (DWPF) started in 1996 and continues through 2019

For WVDP (West Valley Demonstration Project), HLW solidification (glass production) started in 1996 and will be completed in 2001

Commercial activities[a,b]

DOE/EIA projections of installed net LWR electrical capacity:

Calendar year	1997	2000	2005	2010	2015	2020	2025	2030
Capacity (GWe)	101	99	95	89	63	49	22	2

DOE/EIA assumptions for LWR fuel enrichment and design burnup:

LWR fuel	Calendar year fuel is loaded	Fuel enrichment ($\%^{235}U$)	Design burnup (MWD/MTIHM)
BWR	1993	3.14	36,000
	1996	3.12	40,000
	2000	3.47	43,000
	2010	3.58	46,000
PWR	1993	3.84	42,000
	1997	4.11	46,000
	2001	4.38	50,000
	2008	4.74	55,000

[a] SNF from commercial reactors is not reprocessed. Thus, a fuel cycle without reprocessing is assumed for all commercial projections.

[b] This case assumes that each reactor will be retired when the expiration date specified in its operating license is reached.

From Integrated Data Base Report-1996, DOE/RW-0006, Rev. 13, U.S. Department of Energy, Washington, D.C. Reprinted by permission.

1.3.3 Sources of LLW

Cleanup operations which generate LLW are necessary in all steps of the fuel cycle and in various aspects of nuclear energy use. As shown in Table 1.3, LLW comes from hospitals, research laboratories, and industrial users. In the United States, all radwaste material other than HLW, SNF, and TRU waste belongs to this classification, except for mill tailings, which are treated somewhat differently (see Section 1.2.2). The LLW generated during fuel cycle activities is summarized by Godbee et al.[15] Table 1.6 shows the projected volume and radioactivity of commercial fuel cycle LLW, based on the 1982 mid-case forecast for nuclear power growth as developed by the EIA,[16] with all the electricity

Table 1.5 Estimated physical characteristics of future commercial TRU waste

Waste material[a]	Uncompacted waste fraction (vol %)	Combustible waste fraction (vol %)
Low-level general-purpose trash	38.5	80
Fuel element hulls	28.2	0
High-level general-purpose trash	18.8	80
Process filters (low activity)	3.9	0
Fluorinator bed solids	3.2	0
Failed equipment	3.0	0
Process filters (high activity)	2.8	40
Sample/analytical items	1.4	40
Ventilation filters (high activity)	0.1	0
Ventilation filters (low activity)	0.1	0

[a] High activity refers to dose rates greater than 50 mR/hr at the package surface and low activity to dose rates less than 50 mR/hr.

From Estimation of Nuclear Waste Types, Characteristics and Quantities from the Barnwell Nuclear Fuel Plant by W. H. Carr et al., ONWI/3092/Top-01, Rev. 1 (E512-09600R), October 1982. Reprinted by permission.

being generated by LWRs. The conversion of the uranium mill product, yellowcake, to UF_6 gives rise to wastes that contain essentially natural uranium. Approximately 0.05 m^3 (1.8 ft^3) of solid and chemical waste and 0.06 m^3 (2.1 ft^3) of liquid LLW result from processing ore containing 1 metric ton of initial heavy metal (MTIHM). As shown in Table 1.6, only small amounts of LLW are generated during the fuel enrichment process as an estimated 3.11×10^{-5} m^3 (1.1×10^{-3} ft^3) of waste per separative work unit (SWU) is assumed. These wastes contain low-enriched uranium (LEU) of 2–3% ^{235}U.

The fuel fabrication wastes also contain only slightly enriched uranium. Solid trash and ash are assumed to account for only about 2.5 m^3 (88 ft^3) per MTIHM and liquid wastes for 86 m^3 (3040 ft^3) per MTIHM. The LLW associated with reprocessing spent LWR fuel is also shown in Table 1.6 to indicate the potential radioactivity in the waste; the values shown are based on the flow sheet for the Allied General Nuclear Service (AGNS) Barnwell Nuclear Fuel Plant, which was assumed hypothetically to have started in 1989. Reactor operations account for more than half of the LLW shipped to commercial disposal sites. A survey of the volume of solid LLW shipped from nuclear power operations in three different time periods has been reported.[17] As shown in Table 1.2, the dry waste consists of compacted and uncompacted waste and filter cartridges.

Wet waste consists of bead resins (used in processing liquid streams), resins and sludges used for precoat materials, evaporator concentrates, and contaminated oils. Two volumes are shown in Table 1.2: average volumes, representing the average or mean of the data reported in a specific category, and median volumes, representing the middle or midpoint of the data reported in a specific category. Thus the median values for dry waste volume and wet waste volume do not necessarily come from the same plant. Notably, significant volume reductions were realized from 1978–1982 to 1982–1985 and from 1982–1985 to 1985–1986. For instance, the median total volumes for 1982–1985 and 1985–1986 show a 23% reduction for PWRs and a 24% reduction for BWRs. The types of radioactive wastes produced by nuclear reactors are fission products, which emit beta

Table 1.6 Projected volume and radioactivity of commercial fuel cycle LLW in 5-year increments by type of operation

End of calendar year	Volume (10^3 m^3)		Radioactivity (Ci)	
	Annual rate	Accumulation	Annual rate	Accumulation[a]
	UF$_6$ conversion[b]			
1983	1.5	1.5	9.6	9.6
1985	1.8	5.2	11.4	34.3
1990	1.7	14.3	11.2	96.6
1995	1.9	23.7	12.3	159.9
2000	2.4	34.1	15.6	230.2
2005	2.8	47.0	18.2	317.5
2010	2.9	61.9	18.8	418.3
2015	3.3	78.2	21.3	528.3
2020	3.5	94.7	22.4	640.2
	Uranium enrichment			
1983	0.7	1.4	1.4	2.0
1985	0.8	3.1	1.6	3.4
1990	0.8	7.2	1.6	6.7
1995	0.9	11.4	1.8	10.1
2000	1.1	16.0	2.2	14.1
2005	1.3	21.8	2.6	19.0
2010	1.3	28.5	2.7	24.3
2015	1.5	35.8	3.1	30.4
2020	1.6	43.2	3.2	36.4
	Fuel fabrication[b]			
1983	7.9	7.9	29.6	29.6
1985	8.8	23.6	33.2	89.8
1990	8.8	66.8	33.2	254.6
1995	9.1	109.7	34.3	418.8
2000	9.7	155.8	36.5	595.0
2005	11.6	213.6	43.6	815.8
2010	14.2	281.6	53.3	1,075.7
2015	14.2	354.1	53.4	1,353.1
2020	14.7	428.4	55.1	1,637.1
	Reprocessing			
1990	0.9	1.3	883.4	1,215.7
1995	1.4	8.1	2,402.7	10,957.0
2000	1.8	15.3	2,939.0	20,636.7
2005	2.7	28.4	4,803.2	39,500.6
2010	4.1	44.7	7,373.1	63,058.1
2015	4.1	65.0	7,850.3	92,338.3
2020	4.1	85.3	7,749.4	117,136.9

[a] Radioactive decay as a function of time is accounted for. No historical data are included.
[b] Includes nitrate or other chemical wastes and settling-pond sludges.

From Nuclear fuel cycle: An introductory overview by H. W. Godbee, A. H. Kibbey, C. W. Forsberg, W. L. Carter, and K. J. Notz, in *Radioactive Waste Technology*, edited by A. A. Moghissi, H. W. Godbee, and S. Hobart, chapter 1, American Society of Mechanical Engineers/American Nuclear Society, Engineering Center, New York, 1986. Copyright 1986 by the American Society of Mechanical Engineers. Reprinted by permission.

Table 1.7 Activation products in reactor materials

Nuclide	Percent of activity	Half-life (years)	Radiation[a]	Comment
Iron-55	49	2.7	γ, x	—
Cobalt-60	36	5.3	γ, β	Main contributor for almost 100 years
Nickel-63	5	100	β	High activity but low dose
Manganese-54 and	10	0.85	γ	—
cobalt-58		0.19	β^+, γ	—
Nickel-59 and	<1	75,000	x	Eventual dominant isotope
niobium-94		20,300	γ, β, x	—

[a] γ, gamma ray; x, x ray; β, beta particle; β^+, positron.

From *Understanding Radioactive Waste* by R. L. Murray, Battelle Memorial Institute Press, Columbus, Ohio, 1989, 3rd Ed., pp. 77, 78. Copyright 1989 by Battelle Memorial Institute Press. Reprinted by permission.

and gamma rays are mostly retained in fuel claddings; TRU, which are principally alpha emitters and are present as activation products in structural materials; and corrosion products and impurities in the cooling water. The activation products in structural materials are shown in Table 1.7.[11] Note that most fission fragments and TRU waste are retained in the reactor until the plant is in the decontamination and decommissioning stage. Thus such wastes are listed in Table 1.3 as generated by decontamination and decommissioning.

1.3.4 Volume of Uranium Mill Tailings

Projections for mill tailings are shown in Table 1.8.[14] The values are estimated on the basis that 93.5% of the uranium is extracted from ores that assay 0.113% U_3O_8.[15] The unrecovered 6.5% of the uranium and all the radioactive daughters remain in the tailings, which average in volume 558.7 m^3 (19.7 $\times$ 10^3 ft^3) per MTIHM. All of the U.S. fuel demands are assumed to be met by domestic mining and milling operations.

Small increases in mill tailings were experienced after 1990 and no increases are expected in the future.

1.3.5 Accumulated Waste

Table 1.9 shows the Commercial and DOE wastes accumulated in the United States through 1996 and projected at 10-year intervals through 2030. As mentioned earlier, LLW has the highest volume and SNF the lowest volume. However, because it is highly radioactive, SNF contains the highest radioactivity.

1.4 MIXED WASTES

Mixed wastes are radioactive wastes that also contain nonradioactive hazardous wastes. HLW is considered mixed waste because it is highly radioactive and it is either highly acidic or highly alkaline, depending on how it is being stored. It also contains some heavy metals, which are considered to be hazardous. The hazardous

Table 1.8 Uranium ore processed, U_3O_8 recovery rate, and tailings generated through 1996[a,b]

End of calendar year	U_3O_8 recovery rate		Ore processed		Tailings generated	
	Mass[c] (10^6 t)	Grade (% U_3O_8)	Recovery from ore (%)	Product[d] (10^3 t)	Mass[e] (10^6 t)	Volume[f] (10^6 m^3)
Prior to 1978	NA	NA	NA	NA	108.8	68.0
1978	12.5	0.134	91	15.6	12.6	7.9
1979	14.6	0.113	91	15.3	14.5	9.1
1980	15.3	0.118	93	17.2	15.2	9.5
1981	13.2	0.115	94	14.5	13.2	8.2
1982	7.9	0.119	96	9.9	8.1	5.0
1983	5.4	0.128	97	7.0	5.4	3.4
1984	3.9	0.112	95	4.4	4.0	2.5
1985	1.6	0.161	96	2.8	1.6	1.0
1986	1.2	0.338	97	4.0	1.2	0.7
1987	1.3	0.284	96	3.8	1.3	0.8
1988	1.1	0.288	95	3.2	1.1	0.7
1989	1.1	0.323	95	3.7	1.0	0.7
1990	0.7	0.293	94	2.1	0.7	0.4
1991	0.6	0.188	92	1.2	0.6	0.4
1992	0.2	0.229	96	0.6	0.2	0.2
1993	0.0	0.000	0	0.0	0.0	0.0
1994	0.0	0.000	0	0.0	0.0	0.0
1995	0.1[g]	0.531	93	0.8	0.1	0.1
1996	<0.1[g]	0.524	87	0.7	<0.1	<0.1
Total[h]					189.7	118.7

[a] Sources: Prior to 1984, U.S. Department of Energy, Grand Junction Area Office data files. 1984–1996, Energy Information Administration, "Uranium Industry Annual Survey," Form EIA-858.

[b] This table has been revised based on a detailed study of milling data from the Grand Junction Project Office and EIA files. The values shown include all tailings. NA, Not available.

[c] Before in-process inventory adjustments.

[d] Conventional U_3O_8 concentrate production.

[e] Includes adjustments to ore-fed amounts for annual mill circuit inventory changes and uranium concentrate production.

[f] Calculated assuming that the average density of tailings is 1.6 t/m^3.

[g] Stockpiled ore mined before 1993.

[h] Because of independent rounding, totals may not equal the sum of components.

From Integrated Data Base Report-1996; DOE/RW-0006, Rev. 13, U.S. Department of Energy, Washington, D.C., December 1997. Reprinted by permission.

substances of HLW are defined by the Resource Conservation and Recovery Act (RCRA). All mixed wastes must be managed according to RCRA and Atomic Energy Act (AEA) requirements.

Most mixed wastes are categorized as mixed low-level waste (MLLW), a significant fraction of which is associated with environmental restoration activities. Wastes are currently being generated by ongoing remedial action activities and certain stored wastes are being treated either on- or off-site before their disposal. Because of these activities, waste volumes can change significantly in very short periods of time. In 1992, the NRC and

Table 1.9 Current and projected total quantities of radioactive waste and SNF

Source and type of material	Amount of material at end of given fiscal year[a]				
	1996	2000	2010	2020	2030
DOE sites					
SNF (mass, MTHM)[b]	2,483	NA	NA	NA	NA
HLW					
Interim storage	347.3	310	244	96	3
Glass or glass/ceramic[c]	0.06	0.7	2.9	11.1	18.5
TRU waste					
Buried	141[d]	141	141	141	141
Stored (as generated from site operations)	96.6[d]	NA	NA	NA	NA
Stored (environmental restoration activities)	0.042	NA	NA	NA	NA
LLW					
Buried[e]	3,068	3,277	3,791	4,361	4,577
Stored (site operations)	NA	NA	NA	NA	NA
Stored (environmental restoration activities)	290	NA	NA	NA	NA
MLLW					
Stored (site operations)	76.2	NA	NA	NA	214
Stored (environmental restoration activities)	40	NA	NA	NA	NA
11e(2) by-product material[i]					
Stored (environmental restoration activities)	28,000	NA	NA	NA	NA
Commercial sites					
LWR SNF (no reprocessing) (mass, MTIHM[b,f,g])					
Reference Case	34,252	43,300	63,400	78,500	86,700
HLW (WVDP)					
Interim storage	2.0	0.2	0.0	0.0	0.0
Glass	0.02	0.22	0.24	0.24	0.24
LLW buried[e] (no reprocessing)	1,551	1,588	NA	NA	NA
UMT[g,j]	118,70	NA	NA	NA	NA
MLLW	NA	NA	NA	NA	NA
Other commercial disposal facilities[h]					
LLW	200.0	NA	NA	NA	NA
MLLW	31.0	NA	NA	NA	NA
NARM[k]	296.7	NA	NA	NA	NA
11e(2) by-product	168.6	NA	NA	NA	NA

[a] Quantities are expressed as volume (10^3 m^3) unless otherwise indicated. NA, Not available.

[b] Historically, spent nuclear fuel has been measured in units of mass rather than units of volume.

[c] Includes projections for glass at SRS and glass/ceramic at INEEL.

[d] Includes mixed and nonmixed wastes.

[e] Projections include contributions of LLW from HLW immobilization activities.

[f] The 1996 discharged spent nuclear fuel mass is a BWR and PWR mass sum rounded to the nearest metric ton. Such rounding may result in slight differences between the spent nuclear fuel inventories and projections reported here and those reported by DOE/EIA.

[g] End-of-calendar year data.

[h] Includes wastes from DOE-, commercial-, DOD-, and EPA-sponsored activities.

[i] Material defined in Section 11e(2) of the Atomic Energy Act of 1954.

[j] Uranium Mill Tailings.

[k] Naturally occurring and accelerator-produced radioactive material.

From Integrated Data Base Report-1996, DOE/RW-0006, Rev. 13, U.S. Department of Energy, Washington, D.C., December 1997. Reprinted by permission.

Table 1.10 Summary of estimated total MLLW inventories and FY1996 generation

| Category | Volume (m³) | |
	Total inventory	FY1996 generation[a]
DOE sites		
RCRA and RCRA PCB	71,710[b]	608
(poly chlorinated biphenyl) MLLW		
Non-RCRA PCB MLLW	4,530[b]	73
DOE MLLW total	76,240	681
Major commercial sites[c]	2,116	3,949
Other commercial sites[d]	31,014	0

[a]Except where indicated.

[b]Based on ref. 8. The data for the various DOE sites range from September 1995 to July 1997.

[c]Reported for calendar year 1990.

[d]Wastes from commercial- and government-sponsored (DOE, EPA, DOD) activities that are disposed of at other commercially operated disposal facilities.

From Integrated Data Base Report-1996, DOE/RW-0006, Rev.13, U.S. Department of Energy, Washington, D.C., December 1997. Reprinted by permission.

Environmental Protection Agency (EPA) published a survey study to compile a national profile of the volumes, characteristics, and treatability of commercially generated MLLW. These data were collected to provide a basis for possible federal actions that would effectively manage and regulate the treatment and disposal of mixed wastes.

Table 1.10 is a summary of the currently estimated total MLLW inventories and the amounts generated in FY 1996.

1.5 ENVIRONMENTAL RESTORATION

There are thousands of environmentally impaired facilities and areas in the United States that require some type of cleanup activity. In order to properly address this issue, one must establish which sites should be cleaned up first. Most sites that are considered dangerously contaminated have been characterized to some extent. The data obtained from these sites need to be analyzed and evaluated. Any additional data needed to properly characterize the site should then be collected before remediation efforts are initiated. It is astounding how many remedial activities are conducted by groups without ever clearly understanding the problem. This leads to very inefficient and incomplete cleanup operations. It is also generally true that it is easy to remove a large portion of the contaminant, but that it is more difficult to remove the final traces.

Most large environmental cleanup projects are funded by federal agencies and a large percentage of programs to develop new and improved cleanup technologies are funded and performed in the National Laboratories. Many new technologies have not yet been demonstrated in the field. The old methods continue to be used, such as "dig and haul," when there are potentially many better ways of accomplishing the task. National laboratories such as Fernald, in Ohio, and Oak Ridge National Laboratory, in Tennessee, have conducted some field tests of cleanup technologies.

1.6 WASTE MANAGEMENT ACTIVITIES AND RESPONSIBILITIES

Radwaste management is an integrated system that involves a number of activities:

1. Accumulation
2. Processing
3. Handling
4. Packaging
5. Transportation
6. Storage
7. Disposal
8. Decontamination and decommissioning

The responsibilities for these activities depend on the types of radwaste involved, as discussed in the following.

Management of HLW. Because there is no commercial reprocessing in the United States, all HLW in this country is generated by DOE facilities. All activities related to HLW are therefore DOE's responsibility. Although at present no disposal facility exists, the DOE Office of Civilian Radioactive Waste Management (OCRWM) has the responsibility of providing for disposal of commercial HLW, as specified in the Nuclear Waste Policy Act of 1982.

Management of spent fuel. Commercial SNF is generated mostly by electric utilities, so they are responsible for the accumulation, processing/handling, and packaging activities (if shipping is required). Most SNF is currently stored under several feet of water in spent fuel storage pools in at the reactor site. The OCRWM will provide for transportation, away-from-reactor storage, and eventually disposal of the SNF.

Management of TRU waste. In the United States most TRU waste is generated by DOE; only small amounts are generated by industry. Together, DOE and industry are responsible for all activities involving TRU waste except storage and disposal, which are DOE's responsibility. For instance, DOE stores most TRU waste from fuel fabrication facilities. The Waste Isolation Pilot Plant (WIPP) is the repository for defense TRU waste only.

Management of LLW. LLW is generated by different sources (e.g., utilities, institutions; see Table 1.3), which are responsible for its processing/handling, packaging, and even storage. Low-level waste can also be handled by service vendors or operators under contract to the DOE for its waste. However, the responsibility for disposal has been assigned to the states through regional compacts, as discussed in Chapter 7.

1.7 REGULATION OF ACTIVITIES AND REGULATORY AGENCIES

Activities involving radwaste are regulated primarily by the federal government. These regulatory responsibilities are shared by the Nuclear Regulatory Commission, the Environmental Protection Agency, and the Department of Transportation (DOT), with

NRC being the licensing agency. Thus, NRC regulates and licenses all waste handling/processing, transportation, and disposal activities. EPA promulgates regulations that set standards for exposure of the general public to radiation and reviews environmental impact statements for major projects. DOT enforces the NRC packaging standards, sets qualifications for carrier personnel and handling procedures, and monitors transportation. This division of responsibilities is reflected in the contributions of the different agencies to titles in the Code of Federal Regulations (CFR): Title 10—energy—Parts 1–199 by NRC and Parts 200–end by DOE; Title 40—protection of environment—Parts 190–192 by EPA; and Title 49—transportation—Parts 171–179 by DOT. In defense-related waste management activities, DOE is generally self-regulated. States have come to play an increasing role in regulating LLW management activities through the NRC "agreement states" arrangement, by which the NRC delegates licensing authority to the states. In repository sitings, the public, state and local governments, and Indian tribes are all involved, in compliance with the Nuclear Waste Policy Act (NWPA) of 1982.

Internationally two organizations have such jurisdiction: the International Atomic Energy Agency (IAEA) and the Nuclear Energy Agency (NEA) of the Organization for Economic Co-operation and Development (OECD).

1.8 LEGISLATIVE INVOLVEMENT

1.8.1 Federal Legislation

The NWPA was not Congress's first attempt to establish a national nuclear waste program, as Congress attached a single provision to the Atomic Energy Commission Authorization Act of 1972 (Public Law 92-84). That provision created the National Radwaste Repository Program (Project 72-3-b) under the auspices of the then Atomic Energy Commission (AEC) and actually specified the site for the first repository (Lyons, Kansas). Although the first attempt was short-lived, as it was terminated by 1974, it had a profound influence on the determinations that shaped the NWPA.[20] This influence is shown in the complexity and specificity of the NWPA, which was designed to provide a stable framework for the DOE nuclear waste management program. Before the passage of the NWPA, nuclear waste management policy and programs seemed to change with each incoming administration or with each agency reorganization or change in an agency's official responsibility for such a program. The NWPA provides four cornerstones that form the foundation of a comprehensive program:

1. Detailed schedule milestones for all major decisions related to the federal storage and disposal facilities.
2. Extensive procedures for state and public participation in all major decisions for such facilities.
3. Establishment of the Nuclear Waste Fund.
4. Establishment of a separate office within the DOE (OCRWM) to direct the program.

The NWPA has three titles.[21] Title I deals with repositories for disposal of HLW and SNF, an interim storage program, monitored retrievable storage (MRS), and LLW. Title II deals with research, development, and demonstration related to disposal of HLW and SNF. Title III deals with other provisions related to radwaste. In 1987 this law was amended by the Nuclear Waste Policy Amendment Act,[22] which redirected the program to phase out site-specific activities at all candidate sites other than the Yucca Mountain site and essentially added 5 years to the schedule for the first geologic repository. This Amendment Act also nullified the Oak Ridge siting proposal for MRS and established an MRS Review Commission to evaluate the need for such a facility as part of the nation's nuclear waste management system.

In addition to the NWPAs, two other public laws control the management of radwastes: the Low-Level Radioactive Waste Policy Act (Public Law 96-573) and the Hazardous Materials Transportation Act (Public Law 93-633).

Low-Level Radioactive Waste Policy Act. This act specifies the policy of the federal government that each state is responsible for providing for the disposal of LLW generated within its borders and that LLW can be managed most safely and effectively on a regional basis. To carry out this policy, states may enter into compacts that do not take effect until Congress has consented to the compact by law. When this law was passed (1982) it was stated that after January 1, 1986, any such compact may restrict the use of the regional disposal facilities to the disposal of LLW generated within its compact region. However, because of the delay in forming state compacts, it was necessary to amend the LLW Policy Act to move this date to January 1, 1993, as discussed in Chapter 7.

Hazardous Materials Transportation Act. This Act declares that it is the policy of Congress to improve the regulatory and enforcement authority of DOT in protecting the nation adequately against the risks to life and property that are inherent in the transportation of hazardous materials in commerce. It entitles DOT to issue regulations for the safe transport of, and establish criteria for handling, hazardous materials; it requires transporters and manufacturers of transport devices to register with DOT not less than once every 2 years; it prohibits the transportation on passenger-carrying aircraft in air commerce of radioactive materials (radmaterials) unless they are intended for research, medical diagnosis, or treatment; and it preempts state transportation requirements. In addition, as a member state of IAEA, the United States adopts the IAEA Transportation Regulations, which set forth minimum safety requirements based on performance standards that would be universally applicable and could serve as a basis for national and international regulations. The 1984 revisions of the IAEA regulations include the addition of a dynamic crush test, new values for activity limits for type A packages, and a proposal for a new test of type B packages* to show that each fuel package containing more than 10^6 Ci will not rupture in water at a depth of 200 m. The IAEA Basic Safety Standards for Radiation prescribe maximum permissible

---

*These are waste classifications as defined in Code 10CFR61, Licensing Requirements for Land Disposal of Radioactive Waste.

levels for radiation exposure and fundamental operating principles and provide an appropriate regulatory basis for the protection of health and safety of employees and the public.

1.8.2 State Legislation

A number of states have legislation to control radiation and atomic energy. Examples are the Atomic Energy and Radiation Control Act of the State of Washington, the State of Nevada Rules and Regulations for Radiation Control, and the State of Washington Radiation Control Regulations.

REFERENCES

1. King, C. L., Nuclear waste disposal can be done safely, *Stanford Observer*, p. 5, October 1985.
2. Pigford, T. H., The National Research Council study of the isolation system for geologic disposal of radioactive wastes, in *Scientific Basis for Nuclear Waste Management VII*, p. 461, Elsevier, New York, 1984.
3. American Nuclear Society, Waste management, *Nuclear News*, p. 84, March 1986.
4. Wolfe, R. A., An identification of the waste generated within the nuclear fuel cycle, *AIChE Symp. Ser.*, no. 154, vol. 72, p. 1, 1976.
5. U.S. Department of Energy, Integrated Data Base Report-1996: U.S. Spent Fuel and Radioactive Waste Inventories, Projections, and Characteristics, DOE/RW-0006, Rev. 13, Washington, D.C., December 1997.
6. General Electric Company, General Electric Standard Safety Analysis Report, BWR/6, Docket STN 50-447, San Jose, Calif., 1973.
7. Westinghouse Nuclear Energy Systems, Reference Safety Analysis Report, RESAR-3, Docket STN 50-480 Pittsburgh, Penn., 1972.
8. Daloisio, G. S., and C. P. Deltete, Radioactive Waste Generation Survey Update, vols. 1 and 2, NP-5526, Final Report, Electric Power Research Institute, Palo Alto, Calif., 1988.
9. Jensen, R. T., Inventories and characteristics of transuranic waste, *Nucl. Chem. Waste Manage.*, vol. 4, no. 1, p. 19, 1983.
10. Harmon, K. M., and J. A. Kelman, Summary of Non-U.S. National and International Radioactive Waste Management Programs 1982, PNL-4405, Pacific Northwest Laboratory, Richland, Wash., 1982.
11. Murray, R. L., *Understanding Radioactive Waste*, Battelle Memorial Institute Press, Columbus, Ohio, 1989, 3rd ed., pp. 77, 78.
12. U.S. Department of Energy, The Defense Waste Management Plan, DOE/DP-0015, June 1983.
13. U.S. Department of Energy, Energy Information Administration, Nuclear Power Generation and Fuel Cycle Report 1997, DOE/EIA-0436(97), Washington, D.C., September 1997.
14. Carr, W. H., et al., Estimation of Nuclear Waste Types, Characteristics and Quantities from the Barnwell Nuclear Fuel Plant, ONWI/3092/Top-01, Rev. 1 (E512-09600R), October 1982.
15. Godbee, H. W., A. H. Kibbey, C. W. Forsberg, W. L. Carter, and K. J. Notz, Nuclear fuel cycle: An introductory overview, in *Radioactive Waste Technology*, A. A. Moghissi, H. W. Godbee, and S. Hobart, eds., chapter 1, American Society of Mechanical Engineers/American Nuclear Society, Engineering Center, New York, 1986.
16. U.S. Department of Energy, Estimate of Future U.S. Nuclear Power Growth, SR-NAFD-83-01, Energy Information Administration Service Report, 1983.
17. Daloisio, G. S., and C. P. Deltete, Radioactive Waste Generation Survey Update, vols. 1 and 2. NP-5526, Final Report, Electric Power Research Institute, Palo Alto, Calif., 1988.
18. Manion, W. J., and D. R. Perkins, Disposition of Spent Fuel, Atomic Industrial Forum Workshop. Washington, D.C., September 1979.

19. ORNL, Spent Fuel and Radwaste Inventories, Projections and Characteristics, DOE/NE-0017/2, Oak Ridge National Laboratory, Oak Ridge, Tenn., 1983.
20. Davis, E. M., A critique of the U.S. Department of Energy nuclear waste management program, Presented at the American Nuclear Society Summer National Meeting, New Orleans, June 1984.
21. U.S. Congress, 97th, *Nuclear Waste Policy Act of 1982*, Public Law 97-425, 1983.
22. U.S. Congress, 100th, *Nuclear Waste Policy Amendment Act*, Omnibus Budget Reconciliation Act for FY 1988. Public Law 100–203, December 1987.

CHAPTER TWO

RADIATION SOURCES, EXPOSURE, AND HEALTH EFFECTS

2.1 INTRODUCTION

Nuclear and environmental scientists need to have a basic understanding of radiation science in order to enable them to perform their jobs effectively and to protect themselves and coworkers. This chapter covers the basic types of ionizing radiation and methods for calculating various radiation doses. Also covered are introductory health physics principles, such as somatic and genetic effects of radiation, as well as radiation protection standards and concepts.

2.2 BASIC RADIATION SCIENCE

2.2.1 Types of Ionizing Radiation

Four types of radiation are important when dealing with radioactive waste. They are alpha (α) and beta (β) particles, neutrons (n), and gamma (γ) rays. The first three are particulate radiations consisting of subatomic particles of various masses and charges and the last one is electromagnetic. Radioactivity is a process by which a nucleus spontaneously disintegrates or decays. The nucleus is transformed into another isotope, and one or more types of ionizing radiation are released. The following nuclear reactions are examples of such radioactive decay process:

$$\text{uranium-238} \rightarrow \text{thorium-234} + \text{alpha particle}$$
$$\text{cobalt-60} \rightarrow \text{nickel-60} + \text{beta particle} + \text{gamma ray}$$
$$\text{iodine-131} \rightarrow \text{xenon-131} + \text{beta particle}$$

Alpha particles are positively charged particles (helium nuclei). They can be stopped by a sheet of paper or even the outer layer of skin.

Beta particles are high-speed electrons. They are more penetrating than alpha particles, and can pass through 1 in. of water or human flesh, but they can often be stopped by a thin sheet of aluminum. The energy of a beta particle depends on the isotope from which it was emitted and determines its ability to penetrate matter and cause radiation damage.

Neutrons can be produced when a nucleus is struck by an alpha particle. They are also emitted during spontaneous fission.

Gamma rays are high-energy electromagnetic waves that can pass through the human body like x rays. Dense materials such as concrete and lead can provide shielding against gamma radiation. Many radioisotopes are gamma emitters. Cobalt-60 is a well-known isotope that emits gamma rays that can be used in medicine for diagnosis or cancer therapy.

These radiations are called "ionizing" because they have sufficient energy to ionize atoms with which they interact.

2.2.2 Radioactive Decay and Natural Decay Chains

During radioactive decay, an atom is transformed into an isotope of another element. Sometimes that new isotope is stable (not radioactive) and sometimes the new isotope is radioactive. If it is radioactive, it will decay. When one radioactive isotope decays or is transformed into another radioactive isotope which in turn decays, a "decay chain" is formed. A decay chain can contain two or more radioactive isotopes and always ends with a stable isotope.

The rate of radioactive decay is called *activity* and is expressed as the number of disintegrations per unit time (seconds). The activity decreases with time as the number of radioactive atoms present decreases. The time it takes for half of any sample of identical isotopes to decay is called the *half-life* ($t_{1/2}$). Thus, the smaller the amount of material present and the longer its half-life, the smaller is the activity and consequently the safer is the sample (or isotope).[1] The half-lives of some often-used radionuclides are listed in Table 2.1.

Some natural radioisotopes (radioactive isotopes) have long decay chains. Table 2.2 shows the chain of natural radioactivity starting with uranium-238.[2] The last isotope in the chain is lead-206, which is stable. Uranium as a mineral found in nature is not dangerous, as the half-lives of both its most abundant isotopes are very long ($t_{1/2}$ for ^{238}U is 4.47 billion years and for ^{235}U is 704 million years). However, the presence of decay products increases the hazard associated with mining uranium ore.

2.2.3 Decay Calculations

Radioactive decay occurs when the nucleus of an atom is in an unstable state. There are two types of nuclear instability: when the neutron-to-proton ratio is too high or when it is too low. In the case in which the neutron-to-proton ratio is too low, an alpha particle is emitted. For example, in the reaction $^{235}_{92}U \rightarrow {}^{4}_{2}He + {}^{231}_{90}Th$, the neutron-to-proton ratio is initially 1.55, but after an alpha particle is emitted, the ratio becomes 1.57. However, if the neutron-to-proton ratio is too high, then a beta particle is emitted. In the reaction $^{35}_{16}S \rightarrow {}^{35}_{17}Cl + {}^{0}_{-1}e$,

Table 2.1 Selected radionuclides and their half-lives

Element	Symbol	Half-life	Atomic mass (amu)	Mode of decay
Cobalt-60	Co-60	5.27 yr	59.934	1.33- and 1.17-MeV γ
Cesium-137	Cs-137	30.07 yr	136.907	0.661-MeV g
Radium-226	Ra-226	1600 yr	226.025	4.871-MeV α
Strontium-90	Sr-90	28.78 yr	89.908	0.546-MeV β
Sulfur-35	S-35	87.51 days	34.969	0.167-MeV β
Phosphorous-32	P-32	14.262 days	31.974	1.711-MeV β
Carbon-14	C-14	5730 yr	14.003	0.156-MeV β
Hydrogen-3	H-3	12.33 yr	3.016	0.019-MeV β
Iodine-131	I-131	8.02070 days	130.906	0.971-MeV β
Iodine-125	I-125	59.408 days	124.905	0.186-MeV $\epsilon\chi$
Technetium-99m	Tc-99m	6.01 hr	98.906	0.143-MeV $\iota\tau$
Polonium-210	Po-210	138.376 days	209.983	5.407-MeV α
Radon-222	Rn-222	3.8235 days	222.018	5.590-MeV α
Argon-39	Ar-39	269 yr	38.964	0.565-MeV β
Americium-241	Am-241	432.2 yr	241.057	5.638-MeV α
Uranium-235	U-235	703,800,000 yr	235.044	4.679-MeV α
Uranium-238	U-238	4.468E+9 yr	238.051	4.270-MeV α
Neptunium-237	Np-237	2.144E+6 yr	237.048	4.959-MeV α
Thorium-232	Th-232	1.405E+10 yr	222.038	4.083-MeV α

a beta particle is emitted, which lowers the neutron-to-proton ratio from 1.19 to 1.06. The method by which the nucleus decays depends on the neutron-to-proton ratio as well as the mass-energy of the parent nucleus.[3]

There are six major modes of decay: alpha emission, beta emission, gamma emission, positron emission, electron capture, and internal conversion. The three that are important in the study of radioactive waste are described below.

Table 2.2 Radioactive decay properties of the ^{238}U series[a]

	^{238}U	$\rightarrow$ ^{234}Th	$\rightarrow$ ^{234m}Pa	^{234}Pa	$\rightarrow$ ^{234}U	$\rightarrow$ ^{230}Th
Property	Uranium	Thorium		Protactinium	Uranium	Thorium
Half-life	4.51×10^9 yr	24.1 days		6.75 hr	2.47×10^5 yr	8.0×10^4 yr
Major radiation	α	β, γ		β, γ	α, γ	α, γ

	$\rightarrow$ ^{226}Ra	$\rightarrow$ ^{222}Rn -----$\rightarrow$	^{210}Bi	^{206}Tl ^{210}Po	$\rightarrow$ ^{206}Pb
Property	Radium	Radon	Bismuth	Polonium	Thallium	Lead
Half-life	1602 yr	3.82 days	5.01 days	138.4 days	4.19 min	Stable
Major radiation	α, γ	α, γ	α, β	α, γ	β	—

[a]Radioisotopes with a half-life shorter than several hours are not shown.

From *Table of Isotopes*, by C. M. Lederer and V. S. Shirley, 7th ed., Wiley, New York, 1976. Copyright 1976 by John Wiley & Sons. Reprinted by permission.

Alpha emission. An alpha particle is a helium nucleus: two protons and two neutrons. When an alpha particle is emitted, an element with a lower mass is formed. An example of alpha decay is $^{235}_{92}U \rightarrow {}^{231}_{90}Th + {}^{4}_{2}He$.

Beta emission. A beta particle is an electron. In beta decay, a neutron in the nucleus splits to form a proton and an electron. The electron is emitted, and the proton remains in the nucleus. The total number of nucleons stays the same, and the atomic number increases by one. An example is $^{3}_{1}H \rightarrow {}_{-1}^{0}e + {}^{3}_{2}He$.

Gamma emission. Gamma decay is the release of energy from the nucleus. There is no change in the number of protons or neutrons. However, gamma decay usually occurs together with an alpha or beta decay.

Decay calculations are often done to determine the number of radioactive atoms remaining in a sample at any given time. They can be done using the following equation:

$$N = N_0 e^{-\lambda t} \tag{1}$$

where N is the number of atoms present at time t, N_0 is the number of atoms present at an earlier time t_0, and λ is the decay constant. The decay constant λ is the probability that a decay will occur in a given period of time. It is calculated by dividing the natural log of 2 by the half-life. Note that the units for the time t and the half-life must be the same.

Example 1. If ^{222}Rn has a half-life $T_{1/2}$ of 3.82 days, what fraction of the radioiodine atoms remain after (a) 1 sec? (b) 1 day? See Table 2.1 for selected ^{222}Rn properties.

Solution Start with the equation $N = N_0 e^{-\lambda t}$. Now, N/N_0 equals the fraction of atoms remaining. The variable λ is equal to $(\ln 2)/T_{1/2}$ and $T_{1/2}$ is given as 3.82 days, which is 330,048 secs. In Example 1a, the value for t is 1 sec. Thus

$$\frac{N}{N_0} = e^{-\lambda t} = e^{-(\ln 2)t/T_{1/2}} = e^{-(\ln 2)1 \text{ sec}/330,048 \text{ sec}} = 0.9999979$$

of the atoms remain after 1 sec.

In Example 1b, t equals 1 day. Thus

$$\frac{N}{N_0} = e^{-\lambda t} = e^{-(\ln 2)t/T_{1/2}} = e^{-(\ln 2) 1 \text{ day}/3.82 \text{ days}} \approx .8340903$$

of the atoms remain after 1 day.

Activity. Activity is a measure of how many decays occur in a specific time. The two most common units used are becquerels (Bq) and curies (Ci). One Bq is equal to one decay per second (dps) and 1 Ci is equal to 3.7×10^{10} Bq. Activity A is related to the number of atoms N by the decay constant λ:

$$A = \lambda N \tag{2}$$

Example 2. If we start with 10 mg of ^{3}H, what is the activity of this sample 20 years later?

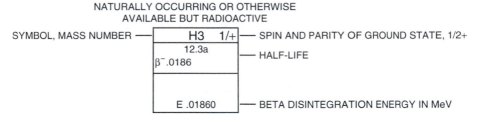

NATURALLY OCCURRING OR OTHERWISE
AVAILABLE BUT RADIOACTIVE

SYMBOL, MASS NUMBER —————— H3   1/+ —— SPIN AND PARITY OF GROUND STATE, 1/2+

12.3a

β⁻.0186 —— HALF-LIFE

E .01860 —— BETA DISINTEGRATION ENERGY IN MeV

Figure 2.1   Information given in an entry in the chart of nuclides.

Solution For this problem it is necessary to determine the number of atoms per milligram of ^{3}H. This quantity is found using Avogadro's number, which is approximately 6.023×10^{23} atoms/mole. The number of grams per mole is equal to the number of amu/atom. For ^{3}H, there is approximately 3.016 amu/atom, so there is approximately 3.016 g of ^{3}H/mole.

Next we calculate the initial activity of the sample, $A_0 = \lambda N_0$. We have

$$N_0 = (10 \text{ mg}) \frac{1 \text{ mole}}{3.016 \text{ g}} \frac{1 \text{ g}}{1000 \text{ mg}} \frac{6.023 \times 10^{23} \text{ atoms of } ^3\text{H}}{\text{mole}}$$

$$= 1.997 \times 10^{21} \text{ atoms of } ^3\text{H}$$

and

$$A_0 = \lambda N_0 = \frac{\ln 2}{12.33 \text{ years}} \frac{1 \text{ year}}{365 \text{ days}} \frac{1 \text{ day}}{24 \text{ hr}} \frac{1 \text{ hr}}{3600 \text{ sec}} (1.997 \times 10^{21})$$

$$= 3.6 \times 10^{12} \text{ dps}$$

Thus we have

$$A = A_0 e^{-\lambda t} = (3.6 \times 10^{12} \text{ dps}) e^{-(\ln 2)20 \text{ years}/12.33 \text{ years}} = 1.16 \times 10^{12} \text{ dps}$$

Chart of the nuclides. Radioactive decay has been studied for nearly 100 years. In 1902, Ernest Rutherford and Frederick Soddy theorized that radioactivity was a process involving changes within the atom.[4] Since then scientists have been compiling vast amounts of information about the elements and their isotopes. This information is summarized in the chart of the nuclides, which includes the nuclide symbol and mass number, natural isotopic composition, half-life, types of radiation emitted and their energies, and isotopic mass. An example of an entry is given in Figure 2.1.

Several organizations produce charts of the nuclides. The most commonly used chart is created by GE Nuclear Energy and can be ordered for a small fee through the mail or on-line.

2.3 RADIATION DOSES

2.3.1 Measurement Units of Radiation

Quantities and doses of ionizing radiation are measured in several units. The oldest unit is the roentgen, which is a measure of the quantity of ionization induced in air, that is, the amount of ionizing radiation required to deposit 1 electrostatic unit (esu) of energy in 1 cm^3 of air at standard temperature and pressure or to deposit 84 ergs per gram of air. The principal units used to express doses absorbed by living matter are the rad (1 rad equals 100 ergs per gram of tissue) and the gray (1 gray equals 1 J per kilogram of tissue, or 100 rad). Since for a particular dose different types of radiation may have different biological effects, a unit called the rem is used to indicate the level of biological hazard. The rem is defined as the product of the dose absorbed by living matter multiplied by the relative biological effectiveness (RBE). The RBE factor is 1 for gamma rays, 1 for beta particles, 10–20 for alpha particles, and 5–10 for neutrons. Thus, 1 rem is the amount of radiation that produces a biological effect equivalent to that resulting from 1 rad of gamma rays. Similarly, 1 sievert (Sv) is the amount of radiation with effect equivalent to that of 1 gray of gamma rays, or 100 rem.

The units used to express collective doses are the person-rem and the person-sievert, which are obtained by multiplying the average dose per person by the number of people exposed. Thus 1 rem to each of 100 people equals 100 person-rem or 1 person-Sv.

Measurement devices. Some of the devices commonly used to measure ionizing radiation include ionization chambers, proportional counters, Geiger–Mueller counters, scintillation detectors, and semiconductor detectors. The most important application of ionization chambers is in the measurement of gamma-ray exposure. Proportional counters are very important in the detection and spectroscopy of low-energy x radiation and the detection of neutrons. A Geiger tube can only function as a counter because all energy information about the incident radiation is lost. Scintillation detectors can be composed of organic or inorganic crystals. Organic crystals are best used in beta spectroscopy and fast neutron detection, whereas inorganic crystals are better for gamma-ray spectroscopy. There are also two main types of semiconductor detectors. The silicon semiconductor is used mainly for charged particle spectroscopy and the germanium semiconductor is used more often in gamma-ray measurements.[5]

2.3.2 Background Radiation

According to the Biological Effects of Ionizing Radiation (BEIR) Committee of the National Academy of Sciences, the typical American receives a background dose of 360 mrem (1 rem = 1000 mrem) or 0.0036 Sv per year. This background dose comes from both natural and man-made sources. Eighty-two percent of the radiation exposure to the U.S. population comes from natural sources. The natural background radiation is composed principally of (1) radon (55% of total), (2) cosmic rays (8% of total), (3) emissions from the disintegration of uranium, thorium, radium, and other radioactive elements in the earth's crust (8% of total), and (4) emissions from ^{40}K, ^{14}C, and other radioisotopes that occur naturally in the body (11% of total). Seventeen percent of the average annual effective dose is composed of radiation from medical x rays (11%), nuclear medicine (4%), and consumer products (3%). Less than 1% comes from occupational exposures, fallout, and the nuclear fuel cycle.[6]

Doses from the various sources of background radiation quoted above are an average for the United States. The actual dose a person receives depends on where he or she

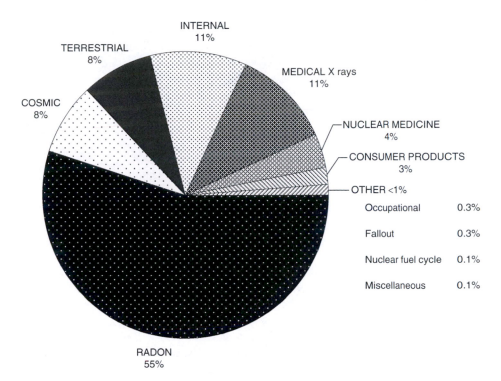

Figure 2.2 Average radiation dose for an individual in the United States from natural background and human activities. Created from data found in BEIR V.

lives and on his or her activities. For example, a person who lives at a high elevation will receive a higher dose from cosmic radiation than a person who lives at sea level. Figure 2.2 shows the percentages of effective dose from different categories.

It is important to note that there are populations elsewhere in the world whose ancestors lived for generations in conditions of much higher radiation exposure, such as the Kerala region of India. Readings taken inside homes in 10 villages in the Kerala region gave a mean dose rate that ranged from 131 to 2814 mrem/year (0.00131–0.02814 Sv/year), which is more than seven times the U.S. average. Some areas in Brazil exhibit high background readings.[7] The absorbed radiation levels in air in three towns built over the monazite sands along the Atlantic coast of Brazil were found to range from 0.100 to 0.200 mrem/hr (0.9–1.8 rem/year), and at spots on the beach where black monazite sand is concentrated, the dose can be up to 2 mrem/hr.[7] The natural background varies widely over the face of the planet, and people who live in areas with high natural radiation levels do not exhibit easily observed abnormalities. Differences one might look for in various populations can be addressed only if there is a control population against which to measure effects.

The largest contribution from man-made sources comes from radiation used for medical diagnosis and treatment. Other man-made sources include radiation-emitting components of television sets, smoke detectors, and other consumer products; fallout from atomic weapons; and accidental leakage from nuclear reactors and other nuclear facilities. X rays became enormously popular in the early 1900s, and the equipment was so cheap that the medical profession adopted it wholeheartedly for diagnostic use. Few precautions were taken, and the early decades of "roentgenology" saw doctors and even more patients succumb to x-ray damage. Radium was also used carelessly then and was freely prescribed in dangerous amounts by doctors. Radiation doses from medical uses and the nuclear industry are discussed in the next sections.

2.3.3 Radiation Doses from Medical Uses

The application of radioactive materials in clinical diagnosis and therapy is a specialty called nuclear medicine. Such applications in medicine began with the availability of inexpensive radioisotopes prepared in the government's nuclear reactor in Oak Ridge, Tennessee, although radioactive materials had previously been used in medicine. In the 1890s, soon after its discovery by Pierre and Marie Curie, radium began to be used in the treatment of cancer. The development of the cyclotron (by Earnest Lawrence), the discovery of the neutron (by James Chadwick), and the discovery of artificial radioactivity (by Frederic Joliot and Irene Curie-Joliot) permitted extension of radioisotope methodology to improve our understanding of human physiology and clinical diagnosis.[8] It may be worth repeating the words of the veteran doctor and clinical educator, Dr. Rosalyn Yalow: "The benefits of radioactivity in the service of men are real and significant. Let us not be kept from these benefits by irrational fears generated by well intentioned or ill intentioned but often uninformed Cassandras."[8]

Table 2.3 shows the mean active bone marrow doses to adults from various medical x rays.

2.3.4 Radiation from Nuclear Weapons Testing Fallout

In the past, nuclear weapons were tested in the atmosphere. The radioactive material from those weapons fell to earth and was called "fallout." The EPA estimates of the annual whole-body dose equivalent for the U.S. population from global fallout from weapons tests through the year 2000 are shown in Table 2.4. The projected annual average whole-body dose equivalent rate for the U.S. population from these tests is <3 mrem/year.

2.3.5 Radiation from the Production of Nuclear Power

As of March 1999, 104 commercial nuclear power reactors were licensed for operation in the United States (433 reactors in the world).[9] To these numbers, we must add non-power reactors used for tests and research and reactors for military use such as propulsion units for ships. Estimated annual whole-body dose equivalents for the U.S. population from the radionuclides of significance (from the standpoint of routine operation of nuclear power plants, i.e., tritium, carbon-14, and krypton-85) are summarized in Table 2.5. Overall estimates show that the dose equivalent rate for the average person in the United States due to environmental release of all radionuclides from nuclear operations is currently less than 1 mrem/year.[10,11]

Doses to workers at the reactor sites are generally greater than doses to the public. To estimate occupational exposure of workers in the nuclear industry, each reactor unit

Table 2.3 Radiation doses to adults from various medical x rays

Examination	Active bone marrow[12] (mrem)	Skin[10] (mrem per examination)
Head and neck		
Skull	78	1,500
Cervical spine	52	1,500
Thorax		
Chest—photofluorography	44	1,500
Chest—radiographic	10	140
Thoracic spine	247	800
Ribs	143	1,200
Upper abdomen		
Upper gastrointestinal series	535	1,700
(Radiographic)	(294)	1,400
(Fluoroscopic)	(241)	8,500
Small bowel series	422	20,000[a]
Lower abdomen		
Barium enema (total)	875	
(Radiographic)	(497)	1,500
(Fluoroscopic)	(378)	20,000
Lumbosacral spine	450	5,000
Abdomen (kidney, ureter, and bladder)	147	1,200
Pelvis		
Pelvimetry	595	8,000
Pelvis	93	3,300
Hip	72	1,400

[a] Fluoroscopy.

From *The Effects on Populations of Exposure to Low Level Ionizing Radiation: 1980* by Committee on Biological Effects of Ionizing Radiation (BEIR), National Research Council, National Academy Press, Washington, D.C., 1980. Copyright 1980 by National Academy Press. Reprinted by permission.

Table 2.4 Projections of annual whole-body dose equivalent irradiation to U.S. population from global weapons testing fallout[a]

Year	Per capita dose equivalent (mrem)
1963	13
1965	6.9
1969	4.0
1980	4.4
1990	4.6
2000	4.9

[a] Data from U.S. Office of Radiation Programs.[13]

From *The Effects on Populations of Exposure to Low Level Ionizing Radiation: 1980,* by Committee on Biological Effects of Ionizing Radiation (BEIR), National Research Council, National Academy Press, Washington, D.C., 1980. Copyright 1980 by National Academy Press. Reprinted by permission.

Table 2.5 Projected annual dose equivalent to the U.S. population from specific nuclides[a]

Radionuclide	Body organ	Per capita dose equivalent (mrem)				
		1960	1970	1980	1990	2000
Hydrogen-3	Whole body	0.02	0.04	0.03	0.02	0.03
Carbon-14	Whole body	0.3	0.6	0.6	0.6	0.6
	Bone	0.5	1.0	1.0	1.0	1.0
Krypton-85	Whole body	0.0001	0.0004	0.003	0.01	0.04
	Skin	0.005	0.02	0.1	0.6	1.6
	Lung	0.0002	0.0006	0.005	0.02	0.06

[a] Data from U.S. Environmental Protection Agency.[14]

From *The Effects on Populations of Exposure to Low Level Ionizing Radiation: 1980* by Committee on Biological Effects of Ionizing Radiation (BEIR), National Research Council, National Academy Press, Washington, D.C., 1980. Copyright 1980 by National Academy Press. Reprinted by permission.

is considered to be operated by approximately 150 people, not counting those in engineering support, maintenance, and inspection and others who may be at a site during the year. Such occupational doses to personnel associated with commercial nuclear power plants and supporting activities (fuel cycle) are tabulated and published annually by the NRC. These data are summarized in Table 2.6.[10]

In addition, there are radioactive wastes from nuclear power production, as stated in Section 1.3. Of the three types of HLW produced by nuclear reactors—fission products, transuranics, and activation products—the first two are much more important than the third. Fission products and transuranics reside in the fuel, which consists of uranium oxide ceramic and remains in the reactor for about 3 years until it is replaced by fresh fuel. If this spent fuel is buried deep underground, as planned, there is a chance that some of the radioactive waste will seep into groundwater and be carried to the accessible environment. The potential hazard of ingesting some of this waste was calculated to be 0.11 eventual fatalities per GWe per year.[15] The LLW generated by cleanup operations is buried in shallow trenches at an average depth of about 4 m in commercial burial grounds (located in South Carolina and Washington State). The important contributions to the LLW from nuclear power plant operations are listed in Table 1.2 (Section 1.3.3). The number of eventual fatalities associated with buried LLW was found to be 2.2×10^{-3} per GWe per year.[15]

2.3.6 Radiation from Consumer and Industrial Products

Radiation exposure of the general population can be caused by a variety of consumer and industrial products, such as television sets, watches with luminous dials, tobacco products, fossil fuels, and building materials. A summary of dose equivalent rates of these products is presented in Table 2.7.[16] The estimated average whole-body dose equivalent rate for the average person in the United States from these sources is 10 mrem/year.[10] Most of this exposure is due to naturally occurring radionuclides in building materials.

The annual dose rates from important sources of radiation exposure in the United States are summarized in Table 2.8.[10]

Table 2.6 Person-rems accumulated, by category of covered licensees, 1973–1976[a]

Covered categories of NRC licensees	Calendar year	Number of licensees reporting	Number of persons monitored	Number of persons with measurable exposure	Total number of person-rems	Average exposure per person (based on total monitored) (rem)	Average exposure per person (based on measurable exposures) (rem)
Commercial power reactors	1976	62	66,800	36,715	26,555	0.40	0.72
	1975	54	54,763	28,034	21,270	0.39	0.76
	1974	53	62,044	21,904	14,083	0.23	0.64
	1973	41	44,795	16,558	14,337	0.32	0.87
Industrial radiography	1976	321	11,245	6,222	3,629	0.32	0.58
	1975	291	9,178	4,693	2,796	0.30	0.60
	1974	319	8,792	4,943	2,938	0.33	0.59
	1973	341	8,206	5,328	3,354	0.41	0.63
Fuel processing and fabrication	1976	21	11,227	5,285	1,830	0.16	0.35
	1975	23	11,405	5,495	3,125	0.27	0.57
	1974	25	10,921	4,617	2,739	0.25	0.59
	1973	27	10,610	5,056	2,400	0.23	0.47
Processing and distribution of by-product material	1976	24	3,501	1,976	1,226	0.35	0.62
	1975	19	3,367	1,859	1,188	0.35	0.64
	1974	24	3,340	1,827	1,050	0.31	0.57
	1973	34	4,251	1,925	1,177	0.28	0.61
Totals	1976	428	92,773	50,198	33,240	0.36	0.66
	1975	387	78,713	40,081	28,379	0.36	0.71
	1974	421	85,097	33,291	20,810	0.24	0.63
	1973	443	67,862	28,867	21,268	0.31	0.74

[a] Data from U.S. Nuclear Regulatory Commission.[17]

From *The Effects on Populations of Exposure to Low Level Ionizing Radiation: 1980*, by Committee on Biological Effects of Ionizing Radiation (BEIR), National Research Council, National Academy Press, Washington, D.C., 1980. Copyright 1980 by National Academy Press. Reprinted by permission.

Table 2.7 Dose equivalent rates from selected consumer products[a]

Product	Body portion considered	Average annual dose equivalent rate (mrem/yr)	
		For persons using product	For average person in U.S. population
Luminous compounds			
Wristwatches	Gonads	1–3	0.2
Clocks	Whole body	9	0.5
Television sets	Gonads	0.3 (females)	0.1 (females)
		1 (males)	0.5 (males)
Construction materials	Whole body	7	3.5
Combustion of fossil fuels			
Coal	Lungs	0.25–4	0.05–10
Oil	Lungs	0.002–0.04	0.004
Natural gas			
Cooking ranges	Bronchial epithelium	6–9	5
Unvented heaters	Bronchial epithelium	22	2
Tobacco products	Bronchial epithelium	8.000[b]	2.000[b]

[a] Data from National Council on Radiation Protection.[16]

[b] Hypothetical maximum at highly localized points.[16]

    From *The Effects on Populations of Exposure to Low Level Ionizing Radiation: 1980*, by Committee on Biological Effects of Ionizing Radiation (BEIR), National Research Council, National Academy Press, Washington, D.C., 1980. Copyright 1980 by National Academy Press. Reprinted by permission.

2.3.7 Risks to a Total Population

Induction of cancer by penetrating radiation has been the subject of many scientific studies. However, it is hard to identify which cancers are caused by radiation since there is no widely accepted way to distinguish them from cancers that occur naturally. There is some indication that a few forms of cancers (e.g., cancer of the pancreas) may be more easily induced by ionizing radiation than other cancers. The BEIR estimated that the natural background causes about 1% of the 563,100 cancer deaths that occur each year.[7] A later appraisal by the U.S. General Accounting Office (GAO), in consultation with the EPA,[17,18] assigned an incidence of 12,000 "health effects" to the annual background, where human health effects were defined as "cancers (including leukemia), serious genetic effects, and increases in diseases that are specifically genetic, e.g., certain forms of mental defects, dwarfism, diabetes, schizophrenia, epilepsy and anemia."

2.4 HEALTH EFFECTS

Radiation effects have been classified traditionally as *somatic* if manifested in the exposed subject and *hereditary* or *genetic* if manifested in the descendants of the exposed subject. The term *stochastic* is used to describe effects whose probability of occurrence in an exposed population (rather than their severity in an affected individual) is a direct

Table 2.8 Annual dose rates from significant sources of radiation exposure in the United States

Source	Exposed group		Body portion exposed	Average dose rate (mrem/yr)	
	Description	Number exposed		Exposed group	Prorated over total population
Natural background					
Cosmic radiation	Total population	220×10^6	Whole body	28	28
Terrestrial radiation	Total population	220×10^6	Whole body	26	26
Internal sources	Total population	220×10^6	Gonads; Bone marrow	28	28
				24	24
Medical x rays					
Medical diagnosis	Adult patients	105×10^6/yr	Bone marrow	103	77
Medical personnel	Occupational	195,000	Whole body	300–350[a]	0.3
Dental diagnosis	Adult patients	105×10^6/yr	Bone marrow	3	1.4
Dental personnel	Occupational	171,000	Whole body	50–125[a]	0.05
Radiopharmaceuticals					
Medical diagnosis	Patients	10×10^6 to 12×10^6/yr	Bone marrow	300	13.6
Medical personnel	Occupational	100,000	Whole body	260–350	0.1
Atmospheric weapons tests, nuclear industry	Total population	220×10^6	Whole body	4–5	4–5
Commercial nuclear power plants (effluent releases)	Population within 10 mi	$<10 \times 10^6$	Whole body	≪10	≪1
Commercial nuclear power plants (occupational)	Workers	67,000	Whole body	400[b]	0.1
Industrial radiography (occupational)	Workers	11,250	Whole body	320	0.02
Fuel processing and fabrication (occupational)	Workers	11,250	Whole body	160	0.01
Handling by-product materials (occupational)	Workers	3,500	Whole body	350	0.01
Federal contractors (occupational)	Workers	88,500	Whole body	~250	0.1

(continued)

Table 2.8 (continued)

Source	Exposed group		Average dose rate (mrem/yr)		
	Description	Number exposed	Body portion exposed	Exposed group	Prorated over total population
Naval nuclear propulsion program (occupational)	Workers	36,000	Whole body	220	0.04
Research activities					
Particle accelerators (occupational)	Workers	10,000	Whole body	Unknown	≪1
X-ray diffraction units (occupational)	Workers	10,000–20,000	Extremities and whole body	Unknown	≪1
Electron microscopes (occupational)	Workers	4,400	Whole body	50–200	0.003
Neutron generators (occupational)	Workers	1,000–2,000	Whole body	Unknown	≪1
Consumer products					
Building materials	Population in brick and masonry buildings	110×10^6	Whole body	7	3–4
Television receivers	Viewing population	100×10^6	Gonads	0.2–1.5	0.5
Miscellaneous					
Airline travel (cosmic radiation)	Passengers	$35 \times 10^{6\,c}$	Whole body	3	0.5
Airline travel (cosmic radiation)	Crew members and flight attendants	40,000	Whole body	160	0.03
Airline transport of radioactive materials	Passengers	$7 \times 10^{6\,d}$	Whole body	~0.3	0.01
Airline transport of radioactive materials	Crew members and flight attendants	40,000	Whole body	~3	<0.001

[a] Based on personnel dosimeter readings; because of the relatively low energy of medical x rays, actual whole-body doses are probably less.

[b] Average dose rate to the approximately 40,000 workers who received measurable exposures was 600–800 mrem/yr.

[c] Total number of revenue passengers per year is 210×10^6, however, many of these are repeat travelers.

[d] About 1 in every 30 airline flights includes the transportation of radioactive materials; assuming 210×10^6 passengers per year (total), approximately 7×10^6 would be on flights carrying radioactive materials.

From *The Effects on Population of Exposure to Low Level Ionizing Radiations: 1980* by Committee on Biological Effects of Ionizing Radiation (BEIR), National Research Council, National Academy Press, Washington, D.C., 1980. Copyright 1980 by National Academy Press. Reprinted by permission.

function of dose. Stochastic effects are commonly regarded as having no threshold; that is, any dose, however small, has some effect, provided a large enough population is exposed. Genetic effects and somatic effects, such as cancer, are considered to be stochastic. The term *nonstochastic* is used to describe effects whose severity *is* a function of dose. For such effects, there may be a threshold below which there is no effect. Examples of nonstochastic somatic effects are cataracts, nonmalignant skin damage, hematologic deficiencies, and impairment of fertility.

2.4.1 Physical Aspects of the Biological Effects of Ionizing Radiation

All ionizing radiation affects cells by the action of charged subatomic particles, which dislodge electrons from atoms in the irradiated material, thereby producing ions and transferring energy. Radiation may be directly or indirectly ionizing depending on whether or not it carries an electric charge. A fundamental characteristic of charged particles produced directly or indirectly is their linear energy transfer (LET), which is energy loss per unit of distance traveled (kiloelectron volts per micrometer). Radiation is classified in terms of this characteristic as follows.

Low-LET radiation (less than a few keV/μm). In general, x rays and gamma rays, which are electrically neutral, are characterized by a low LET. They generate ions sparsely along their tracks and penetrate deeply into tissues. Low-LET radiation is responsible for most of the absorbed doses received by the general population and by radiation workers.

High-LET radiation (on the order of hundreds of keV/μm). Alpha particles emitted by internally deposited radionuclides constitute the most important directly ionizing high-LET radiation; neutron radiation is the principal indirectly ionizing high-LET radiation.

Intermediate-LET radiation. This radiation is from low-energy electrons produced by both directly and indirectly ionizing radiation.

The damage mechanisms can be any of the following:

1. Disruption of the cell, leading to failure of the cell to continue to perform its function.
2. Destruction of the cell's reproductive facility, leading to termination of that cell's line and future loss of its function.
3. Modification of the cell's reproductive facility, leading to the production of incompetent or destructive progeny (e.g., cancer).
4. Disruption of the individual's reproductive facility.
5. Modification of the individual's genetic code as a result of modification of the chromosome structure in the reproductive cells, leading to the generation of usually undesirable mutants among the individual's progeny.

Some radiation effects are apparently due to damage to individual autonomous cells[*] such as the mature gametes in the gonads. Other effects, such as cataractogenesis,

---

[*]Autonomous cells are cells whose response to radiation is unaffected by the irradiation of other cells or by any other entities.

are due to injury of several cells. For the most important somatic radiation hazard, car-cinogenesis, it is often assumed, because the number of cells at risk is very large, that transformation of an individual cell does not necessarily result in cancer. Various inhibitory mechanisms have been considered, including the requirement that several con-tiguous cells be transformed or the action of immunologic or other host defenses be im-paired.[10] In these cases, the dose–effect curve could have various forms at low absorbed doses. For both high-LET and low-LET radiation in the dose range where the single-cell response is linear, a multicellular mechanism of cancer induction would theoretically produce a dose–effect relation with upward curvature (slope increasing with dose).[10]

Cells are injured by the molecular changes caused by radiation-induced ions and free radicals. Among the many types of molecules that are affected by ionizing radiation, the most critical is DNA because of the limited redundancy of the genetic information en-coded in it. The total amount of energy deposited by an acutely lethal dose of x rays af-fecting the whole body (about 3–5 Sv) can cause hundreds of breaks in DNA molecules in every cell of the body. Figure 2.3 illustrates the low-level radiation damage to DNA by an electron from an ionized hydrogen atom, which can have direct or indirect effects.[19] The schematic representation of DNA at the right shows the types of damage that can oc-cur. Since the simple types of lesion can to a considerable extent be repaired by enzymes in the cells, the damage to DNA may be amplified many times as the DNA is transcribed and translated, ultimately to countless daughter cells. Damage to chromosomes and genes appears to figure more prominently than any other type of damage in the injuries. Chromosomal abnormalities include changes in the number and structure of chromo-somes. The frequency of such chromosomal aberrations increases as a linear, nonthresh-old function of the radiation dose in the low to intermediate range, with the slope of the line being steeper for high-LET than for low-LET radiation.[19]

Information about the mutagenic action of radiation on genes is provided by experi-ments with other species that can be used to predict such effects in humans. The dose–effect relation for the induction of mutations in spermatogonia and oocytes (maturing

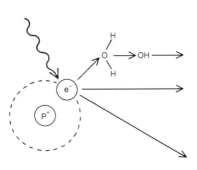

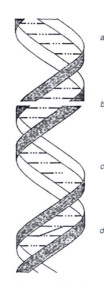

Figure 2.3 Schematic represen-tation of possible damage to DNA. (a) Normal segment of the molecule; (b) double-strand break in DNA double helix; (c) deletion of a base; (d) chemical cross-linking of the DNA strands. From *The biological effects of low level ionizing radiation* by A. C. Upton, *Sci. Am.*, vol. 246, no. 2, p. 41, 1982. Copyright 1982 by *Scien-tific American*. Reprinted by per-mission.

sperm and eggs, respectively) is similar to that for chromosomal aberrations. With high-LET radiation the frequency of mutations increases steeply in proportion to the dose but is relatively independent of the dose rate. With low-LET radiation the frequency increases less steeply as a function of the dose but is highly dependent on the dose rate. However, no heritable effects of radiation have yet been demonstrated in a human population; for example, no detectable increase in genetic abnormalities has appeared among the children of people who survived the two atomic bombings. This failure to detect an increase is not incompatible with the induction rate observed in mice, considering the smallness of the sample (78,000 children) and the low average dose (0.5 Sv) to the gonads of the parents.

2.4.2 Somatic Effects

Cancer induction. Cancer induction is usually considered the most important somatic effect of low-dose ionizing radiation. As mentioned in Section 2.3.5, a cancer in an individual cannot be attributed with certainty to radiation as opposed to other causes; the induction of cancer by radiation is therefore detectable only by statistical means.[20] There are good observational data related to cancer induction in humans over a range of high doses, but little direct evidence is available for doses of a few rads. Consequently, estimation of the excess risks at these low doses involves extrapolation from observations at higher doses based on assumed dose–response relations that cannot be specified with any certainty. The following conclusions are extracted from the 1980 BEIR Committee report.[10]

Induction in all tissues Cancer may be induced by radiation in nearly all the tissues of the human body, although tissues and organs vary considerably in their sensitivity to this induction.

Dependent factors The natural incidence of cancer varies over several orders of magnitude, depending on the type and site of origin of the neoplasm, age and sex of the individual, and other factors.

Solid tumor versus leukemia Solid tumors are now known to be of greater numerical significance than leukemia in considering the excess risk of cancer from whole-body exposure to radiation. Characteristically, solid tumors have long latent periods (they seldom appear before 10 years after radiation exposure and may continue to appear for 30 years or more after exposure). The excess risk of leukemia, in contrast, appears within a few years after exposure and largely disappears within 30 years.

Risks in the two sexes The incidence of radiation-induced breast and thyroid cancer is such that the total cancer risk is greater for women than for men. With respect to other cancers (e.g., of the lung and some digestive organs), the radiation risks in the two sexes are approximately equal.

Age factor There is considerable evidence from studies of humans that age is a major factor. If risks are given in absolute form, that is, number of cancers induced per unit of population per unit of radiation exposure, then a single value independent of age may be inappropriate.

Other host or environmental factors Host or environmental factors that interact with radiation to affect cancer incidence in different tissues include hormonal influences, immunologic status, exposure to oncogenic agents, and nonspecific stimuli to cell proliferation in tissues.

Dose–response relationship The variety of possible biological mechanisms for human cancer suggests that the dose–response relationship may not be the same for all types of radiation-induced cancer. However, epidemiologic studies of widely differing human populations exposed to radiation have given reasonably concordant results for some cancer sites and for a broad range of radiation dose and give considerable support to the dose–response information currently available. For low-LET radiation, linear extrapolation from the known effects in humans of larger doses delivered at high dose rates in the range of rising dose–incidence curve may overestimate the risks at low dose rates and may therefore be regarded as upper limits of risk for this case. For high-LET radiation, such as from internally deposited alpha-emitting radionuclides, application of the linear hypothesis is less likely to lead to overestimates of risk and may, in fact, lead to underestimate.

Figure 2.4 shows the lifetime risk of induction of various types of cancer by low-level radiation based on estimates from a number of investigators.[19] The risk is per 10,000 person-Sv, the equivalent of 1 Sv of radiation to each of 10,000 people over a lifetime.

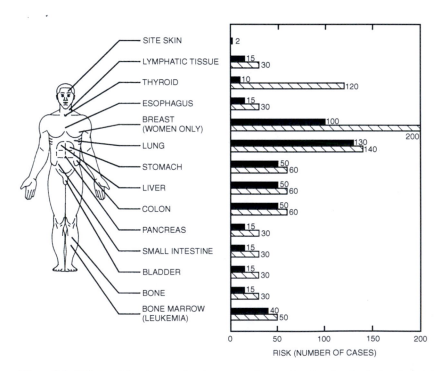

Figure 2.4 Lifetime risk estimates of various types of cancer from low-level radiation. From The biological effects of low level ionizing radiation by A. C. Upton, *Sci. Am.*, vol. 246, no. 2, p. 41, 1982. Copyright 1982 by *Scientific American*. Reprinted by permission.

The figures given are the maximum estimates for fatal cancers (filled bars) and number of cancer cases (shaded bars). In each case the estimate covers a range in which the minimum figures are much lower. Estimates of cancer incidence are less firm than mortality estimates. Estimates of excess risk for individual organs and tissues depend in large part on partial-body irradiation and are based on a wider variety of data sources.

Other somatic effects of radiation. Among the somatic effects of radiation other than cancer, developmental effects on the unborn child are of greatest concern. Exposure of an embryo or fetus to relatively high doses of radiation can cause death, malformation, growth retardation, or functional impairment. Measurable damage can be produced by doses of 1–9 rads.[10] Such effects are related to the developmental stage at which exposure occurs in humans and other mammals. Where developmental effects of radiation can be measured at the cellular level, as in the case of oocyte killing during fetal or early postnatal stages, thresholds may not be demonstrable. Threshold doses for some effects have already been demonstrated but vary for different abnormalities. Most of the perceived abnormalities produced by radiation probably result from damage to more than a single cell, so it is unlikely that such effects are linearly related to dose.

For induction of cataract of the lens, there is radiobiological and chemical evidence for a nonlinear relationship between effect and dose for low-LET radiation. There appear to be no nonspecific effects of radiation at low doses that lead to shortening of the life span, although the existence of specific effects in addition to cancer cannot yet be excluded.[10]

2.4.3 Genetic Effects

Genetic effects are health consequences of genetic damage that result when human populations are exposed to low levels of ionizing radiation in addition to natural background radiation. Estimates of genetic risk to humans are based largely on studies of animals, as the few data available for humans were derived from limited observations and estimated rather than precise dosimetry. The genetic effects of radiation are gene mutations and chromosome aberrations, and the effect on the well-being of the future population is a consequence of these changes. The following conclusions are again extracted from the 1980 BEIR report.[10]

Results of genetic change. Some results of genetic change are conspicuous, others invisible; some are tragic, others so mild as to be trivial; some occur in the first generation after the gene or chromosome change, others tens or hundreds of generations in the future. Furthermore, most of the effects that are produced by mutation are mimicked by other effects of nongenetic origin.

Sources of population gonadal exposure. Table 2.9 summarizes the radiation sources. For estimation of genetic effects, additional physical and demographic factors must be considered. Two particular sources are of concern: the transuranic actinide radionuclides resulting from nuclear power and weapons activities and the radionuclides that can be directly incorporated into DNA, principally tritium and carbon-14.

Table 2.9 Estimated annual average genetically significant dose equivalents

Source	Dose equivalent rate (mrem/yr)
Natural radiation	
Cosmic radiation	28
Radionuclides in the body	28
External gamma radiation from terrestrial sources	26
Subtotal	82
Radiation from human activities	
Medical and dental x rays	
Patients	20
Occupational	<0.4
Radiopharmaceuticals	
Patients	2–4
Occupational	<0.15
Commercial nuclear power	
Environmental	<1
Occupational	<0.15
National laboratories and contractors, occupational	<0.2
Industrial applications, occupational	<0.01
Military applications, occupational	<0.04
Weapons testing fallout	4–5
Consumer products	4–5
Air travel	<0.5
Subtotal	30–40

From *The Effects on Populations of Exposure to Low Level Ionizing Radiation: 1980* by Committee on Biological Effects of Ionizing Radiation (BEIR), National Research Council, National Academy Press, Washington, D.C., 1980. Copyright 1980 by National Academy Press. Reprinted by permission.

Age distribution of the exposed population and mean length of a human generation. Since genetic effects are seen only in the descendants of people whose germ cells have been affected by radiation, these effects depend quantitatively on the portion of the dose that is received by the gonads of future parents. The BEIR report recommended 30 years as the mean length of a human generation. Thus the estimate should be the average of 30-year individual doses accumulated by all the parents of the new generation. When exposures are not delivered uniformly or randomly to the entire population, the age distribution of the exposed population and the probability of having children for each age and sex must be taken into account.

Genetic hazards from ^{238}Pu and other transuranic nuclides. Because of the high LET of the emitted alpha particles, there is concern over genetic hazards from these sources. Fuel reprocessing for the mixed-oxide reactor fuel cycle and the breeder reactor cycle will result in exposure that is primarily occupational. Very little of the plutonium to which the general population is exposed is deposited in gonads, and this reduces the dose that can have genetic consequences. The fraction of the plutonium to which people are exposed occupationally that *is* deposited in their gonads may be larger, but the size of the work force is

Table 2.10 Genetic effects of an average population exposure of 1 rem per 30-year generation

Type of genetic disorder	Current incidence per million liveborn offspring	Effect per million liveborn offspring (rem per generation)	
		First generation	Equilibrium
Autosomal dominant and X-linked	10,000	5–65	40–200
Irregularly inherited	90,000	—	20–900
Recessive	1,100	Very few	Very slow increase
Chromosomal aberrations	6,000	Fewer than 10	Increases only slightly

From *The Effects on Populations of Exposure to Low Level Ionizing Radiation: 1980* by Committee on Biological Effects of Ionizing Radiation (BEIR), National Research Council, National Academy Press, Washington, D.C., 1980. Copyright 1980 by National Academy Press. Reprinted by permission.

small, again minimizing the genetic consequences in the population. Special consideration of the genetic hazards of plutonium and other transuranics thus appears unnecessary.

Genetic effects from transmutation of radionuclides. Such radionuclides include the nuclides hydrogen-3, carbon-14, and phosphorus-32. There are a number of positions in DNA bases at which ^{3}H transmutation leads to appreciable mutation in microorganisms, and some small effects in fruit flies can be ascribed to transmutation of ^{32}P. The yields from such transmutations are small and the risk is far smaller than that from the radiation emitted by the decay of the same nuclides.

Estimates of relative mutation risk. Table 2.10 shows genetic effects of an average population exposure of 1 rem per 30-year generation.

2.5 HISTORICAL RADIATION PROTECTION

2.5.1 Radiation Standards and Limits

Considering the risks from ionizing radiation, it is natural to seek ways to prevent such damage as much as possible and to use radiation as cautiously as possible. Setting standards for exposure to radiation is, of course, a relative matter involving the evaluation of risk versus reward, or a cost–benefit assessment. It is mandatory to protect more-susceptible groups within the general population.[17] As early as 1921 the X-Ray and Radium Protection Committee was established in Britain, and in the following year the American Roentgen Ray Society's recommendations for x-ray protection were published. Since erythema was the promptest and most visible sign of radiation overexposure, it seemed reasonable to set allowable levels of exposure in terms of a fraction of the erythemal dose. Depending on the voltage of the x rays, this dose ranged from 200 to 800 rem.[17] The task of radiation experts in the early 1920s was far from easy as they sought to define "a maximum tolerance dose in terms of a specifiable and reproducible biological standard and if possible to express this in physical limits." Many people believe that this task has not yet been completely fulfilled.

In 1925 an International Commission on Radiological Protection (ICRP) was formed. Experts in the commission agreed that occupational exposure could be limited to 0.2 roentgen per working day (1 roentgen per 5-day week) without danger of radiation aftereffects. This recommendation was finally adopted in 1934 and was followed by most nations until 1950. In the United States, however, the National Committee on Radiation Protection (NCRP) recommended 0.1 roentgen/day (0.5 roentgen/week) in 1934 and this limit was used until 1949.[17] Setting limits for radium intake into the human body was even more difficult. A maximum permissible body burden (MPBB) of 0.1 μg (or 0.1 μCi) of radium was established by an advisory committee of the U.S. National Bureau of Standards in 1941. A new dimension in radiation risk assessment and control was introduced by the Manhattan Project, as much larger occupational groups would be concerned and the radiation risks extended into a multiplicity of new areas involving exposure to neutrons, active alpha emitters, and other penetrating radiation. In 1964 the term "tolerance dose" became unsatisfactory and "permissible dose" was used. This term was defined in ICRP Publication No. 6 as "that dose, accumulated over a long period of time, or resulting from a single exposure which, in the light of present knowledge, carries a negligible probability of severe somatic or genetic injuries. Furthermore, it is such a dose that any effects that ensue more frequently are limited to those of a minor nature that would be considered unacceptable by the exposed individual and by competent medical authorities."[17]

In the United States today, the Nuclear Regulatory Commission (NRC) is responsible for regulating the use of radioactive materials. The NRC issued advisory 10CFR20 in and revised it in 1995 to regulate exposures to radiation for both the public and workers at nuclear facilities. In general, workers at nuclear facilities may receive a total effective dose equivalent of no more than 5 rem/year. To provide a higher level of protection for the general population, an annual limit of 0.5 rem for any individual in the general population is stipulated. Thus, there is a factor of 10 difference between the occupational limit and the public limit.

2.5.2 ALARA Concept

In 1975 Regulatory Guide 8.8 of the Nuclear Regulatory Commission, entitled Information Relevant to Assuring That Occupational Radiation Exposures at Nuclear Power Plants Will Be as Low as Reasonably Achievable (ALARA), was released. The ALARA concept was based on the assumption that a proportional (linear) relationship exists between dose and biological effects and that the effect is not dependent on the dose rate. Regulatory Guide 8.8, as revised in 1977, spelled out the ALARA philosophy:

1. Merely controlling the maximum dose to the individual is not sufficient; the collective dose to the group (measured in person-rems) also must be kept as low as is reasonably achievable.
2. "Reasonably achievable" is judged by considering the state of technology and the economics of improvement in relation to all of the benefits from these improvements.

3. Under the linear, nonthreshold concept, restricting the doses to individuals at a fraction of the applicable limit would be inappropriate if such action would result in the exposure of more persons to radiation and would increase the total person-rem dose.

2.5.3 Regulation of Radiopharmaceuticals and X Rays

The Nuclear Regulatory Commission can enter into an agreement with a state allowing the state to regulate all facilities typically regulated by the NRC except nuclear power plants. Thirty states have entered into such an agreement. These states also regulate the use of radiopharmaceuticals and machines that produce radiation such as x-ray machines.

2.5.4 Protection Against Harm

Protecting people against harm from radioactive material requires that the amount of radioactive material entering the environment be minimized. Releasing the material into very large volumes of air or water can dilute the activity to below the maximum permissible concentration (MPC).[1] Specifications of the MPC for radioisotopes in air and water are given in 10CFR20. Table 2.11 lists typical values in microcuries per cubic centimeter of air or water for materials to which the general public might be exposed.[1] Other figures are given in 10CFR20 for soluble and insoluble materials, for exposure of people in restricted areas, and for many more isotopes. When there are several isotopes in a mixture in air or water, the fractions of an MPC of all materials present must total less than unity.

In protecting against radiation external to the body, three factors should be considered: distance, time, and shielding. That is, it is safer to keep farther from the source of radiation, minimize the time of exposure, and use thicker or heavier shielding.[2] The radiation warning symbol and rope barriers are used to remind workers of a potential hazard.

Table 2.11 Maximum permissible concentrations (μCi/ml) above natural background for selected radioisotopes

Isotope	Air	Water
Cesium-137	2×10^{-9}	2×10^{-5}
Cesium-144	3×10^{-10}	1×10^{-5}
Hydrogen-3	2×10^{-7}	3×10^{-3}
Iodine-129	2×10^{-11}	6×10^{-8}
Iodine-131	1×10^{-10}	3×10^{-7}
Krypton-85	3×10^{-7}	—
Neptunium-237	1×10^{-13}	3×10^{-6}
Plutonium-238	3×10^{-8}	1×10^{-4}
Plutonium-239	6×10^{-14}	5×10^{-6}
Radium-226	3×10^{-12}	3×10^{-8}
Radon-222	3×10^{-9}	—
Strontium-90	3×10^{-11}	3×10^{-7}
Uranium-235	2×10^{-11}	3×10^{-5}
Uranium-238	3×10^{-12}	4×10^{-5}

From *Understanding Radioactive Waste* by R. L. Murray, Battelle Press, Columbus, Ohio, 1982. Copyright 1982 by Battelle Press. Reprinted by permission.

Protection against radiation from radwastes is, of course, the major concern of this book. A brief discussion of the multiple-barrier approach follows. More detailed treatments will be given in various chapters for different classes of radwastes.

2.5.5 The Multiple-Barrier Concept

The goal of safe waste disposal is to isolate the radioactive waste from humans and the environment. In the design and construction of a repository for radwaste, an approach will be used in which multiple barriers are placed between the waste and habitations. This concept was emphasized by the Interagency Review Group established by the Carter Administration in 1977, and endorsed by both the EPA and NRC. It includes the waste form, the containers, the packing around them, and the surrounding rock—the geologic medium. Figure 2.5 shows the use of multiple barriers to prevent the escape of radioactive materials from a waste repository.[1] The first barrier is the waste form, which is designed to immobilize the radmaterials. The waste form can be either spent fuel bundles or a mixture of waste and a solid glass. Usually the waste form should be able to contain a reasonable amount of the waste and still maintain its strength and uniformity. It should not be damaged by heat or radiation or be readily attacked chemically by groundwater solutions. The mixture is placed in a steel canister, the second barrier, whose wall should also be resistant to leaching by water or water solutions. The third barrier is packing inserted around the container to prevent radioactive material from escaping. One type of packing might be a buffer such as bentonite, a clay that swells when it becomes wet, that can serve as a migration retardant. Another packing is the backfill,

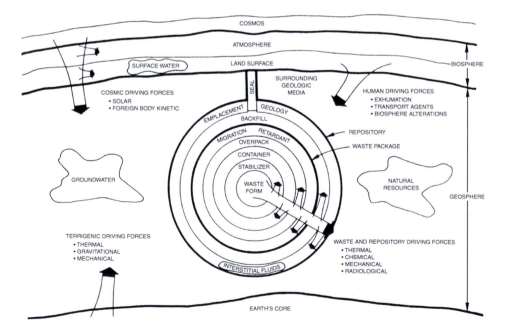

Figure 2.5 The multiple-barrier concept.[19] From *Understanding Radioactive Waste* by R. L. Murray, Battelle Press, Columbus, Ohio, 1982. Copyright 1982 by Battelle Press. Used by permission.

an absorptive or resistant substance that fills the hole from the repository to the surface of the earth. The fourth barrier is the host medium, the geologic material that separates the repository from populated regions. Figure 2.5 shows both the barriers and other forces which can affect the amount of radioactive material leaving the repository.

2.5.6 Shielding

The three important factors to consider when reducing the exposure to external radiation are time, distance, and shielding. The precautions necessary for the first two character-istics are obvious. The amount of time spent near the radioactive material should be minimized and the distance from the material maximized. The amount of shielding needed to reduce the risk to a certain level is more complicated. This section discusses shielding and provides information required to make some basic shielding calculations.

Shielding is material placed around radioactive material to absorb the radiation, thus protecting people and the environment. Many materials can be used as shielding. The effectiveness of the shield depends on the material's properties, the thickness of the shield, and the type of radiation being emitted. As discussed earlier in this chapter, alpha radiation requires very little shielding. A piece of paper or even the outer layer of skin is enough. Beta particles are also relatively easy to shield. A thin piece of aluminum is usually enough. Gamma rays, however, are more complex. They are never completely absorbed by the shielding material. Instead, the intensity of the beam is reduced by a factor that depends on the properties of the shielding material. The intensity reduction can be calculated by

$$I = I_0 e^{-\mu_l t} \tag{3}$$

where I is the intensity of the beam after passing through the shielding material, I_0 is the original intensity of the beam (before passing through the shielding material), t is the thickness of the material, and μ_l is the linear attenuation coefficient. The linear attenua-tion coefficient represents the shielding properties of the material. Mathematically, it is the slope of the line produced by plotting the fraction of gammas transmitted versus the absorber thickness.

Example 1. What is the thickness of Pb that will absorb 50% of an incident beam of 1-MeV gamma rays? The linear attenuation coefficient for the element Pb is 0.771 cm^{-1} (Table 2.12).

We have $I = I_0 e^{-\mu_l t}$, or $I/I_0 = e^{-\mu_l t}$. Thus

$$\ln \left| \frac{I}{I_0} \right| = -\mu_l t \qquad \text{or} \qquad \frac{\ln \left| \frac{I}{I_0} \right|}{-\mu_l} = t$$

Therefore

$$-\frac{\ln|0.5|}{0.771 \text{ cm}^{-1}} = t = 0.9 \text{ cm}$$

It is sometimes convenient to use mass attenuation coefficients. The mass attenua-tion coefficient μ_m is μ_l divided by the density ρ of the shielding material, and is

Table 2.12 Linear attenuation coefficient μ_l

	$\rho(\text{g/cm}^3)$	$\mu_l\,(\text{cm}^{-1})$ at given quantum energy (MeV) of γ rays										
		0.1	0.2	0.3	0.5	0.8	1.0	2.0	3.0	5.0	8.0	10.0
C	2.25	0.335	0.274	0.238	0.196	0.159	0.143	0.100	0.080	0.061	0.048	0.044
Al	2.7	0.435	0.324	0.278	0.227	0.185	0.166	0.117	0.096	0.076	0.065	0.062
Fe	7.9	2.72	1.090	0.838	0.655	0.525	0.470	0.335	0.285	0.247	0.233	0.232
Cu	8.9	3.80	1.309	0.960	0.730	0.581	0.520	0.372	0.318	0.281	0.270	0.271
Pb	11.3	59.7	10.15	4.02	1.64	0.945	0.771	0.516	0.476	0.482	0.518	0.552
Air	1.29×10^{-3}	1.95×10^{-4}	1.59×10^{-4}	1.37×10^{-4}	1.12×10^{-4}	9.12×10^{-5}	8.45×10^{-5}	5.75×10^{-5}	4.6×10^{-5}	3.54×10^{-5}	2.84×10^{-5}	2.61×10^{-5}
H_2O	1	0.167	0.136	0.118	0.097	0.079	0.071	0.049	0.040	0.030	0.024	0.022
Concrete	2.35	0.397	0.291	0.251	0.204	0.166	0.149	0.105	0.085	0.067	0.057	0.054

measured in units of cm^2/g. The equation for calculating beam intensity using the mass attenuation coefficient is

$$I = I_0 e^{-\mu_m \rho t} \tag{4}$$

Example 2. What percentage of 2-MeV gamma rays will be absorbed by 10 cm of water? The mass attenuation coefficient for 2-MeV gamma rays in water is 0.0493 cm^2/g (Table 2.13).

We have $I = I_0 e^{-\mu_m \rho t}$, or $I/I_0 = e^{-\mu_m \rho t}$.
Thus

$$\frac{I}{I_0} = e^{-(0.0493 \text{ cm}^2/\text{g})(1 \text{ g/cm}^3)(10 \text{ cm})} = 0.611$$

Therefore 61.1% of the 2-MeV gamma rays will be transmitted, and $1 - .611 = .389$ or 38.9% of the gamma rays will be absorbed.

Note that both linear and mass attenuation coefficients are somewhat dependent on the energy of the incident gamma rays.

When gamma rays travel through thick shields, some of them are scattered as they interact with the shielding material and can emerge from the shield, after several "collisions," at a reduced energy. These gamma rays are accounted for by using an empirically determined multiplier commonly referred to as the buildup factor B. The buildup factor can be incorporated into the calculation of the intensity reduction using

$$I = BI_0 e^{\mu_l t} \tag{5}$$

where $B = B(t, E, R, G)$ is a function of the thickness t of the absorber, the energy E of the original radiation, the response R of the dose detector, and the geometry G of the materials surrounding the detector, and μ_l is the linear attenuation coefficient of the absorber.

The buildup factor B is defined as $B =$ (observed dose)/(primary dose) and can be estimated using techniques beyond the scope of this book.

In theory, almost any material can be used for radiation shielding if the thickness is sufficient to attenuate the radiation to regulation limits. However, lead and concrete are among the most widely used shielding materials. Some of the factors used in determining the appropriate shield material are as follows:

1. The final desired attenuated radiation levels
2. The ease of heat dissipation and resistance to radiation damage
3. The required thickness and weight
4. Multiple use considerations (e.g., shield and/or structural)
5. The uniformity of shielding capability
6. The permanence of shielding
7. Availability

Table 2.13 Mass attenuation coefficient μ_m

ρ (g/cm³)	μ_m (cm²/g) at given quantum energy (MeV) of γ rays										
	0.1	0.2	0.3	0.5	0.8	1.0	2.0	3.0	5.0	8.0	10.0
C	0.149	0.122	0.106	0.0870	0.0707	0.0636	0.04444	0.0356	0.0270	0.0213	0.0194
Al	0.161	0.120	0.103	0.0840	0.0683	0.0614	0.0432	0.0353	0.0282	0.0241	0.0229
Fe	0.344	0.138	0.106	0.0828	0.0664	0.0595	0.0424	0.0361	0.0313	0.0295	0.0294
Cu	0.427	0.147	0.108	0.0820	0.0654	0.0585	0.0418	0.0357	0.0316	0.0303	0.0305
Pb	5.29	0.896	0.356	0.145	0.0836	0.0684	0.0457	0.0421	0.0426	0.0459	0.0489
Air	0.151	0.123	0.106	0.0868	0.0706	0.0636	0.0445	0.0357	0.0274	0.0220	0.0202
H₂O	0.167	0.136	0.118	0.0966	0.786	0.0706	0.0493	0.0396	0.0301	0.0240	0.0219
Concrete	0.169	0.124	0.107	0.0870	0.0706	0.0635	0.0445	0.0363	0.0287	0.0243	0.0229

2.6 COMPUTER CODES

Many decay and shielding calculations, especially for complex systems, are made using widely available computer codes. In addition, there are standard codes for calculating doses to workers and the public. Some of these codes are briefly described in this section. This information is taken from the website www.umich.edu/~radinfo.

MCNP. This "is a general-purpose Monte Carlo N-Particle code that can be used for neutron, photon, electron, or coupled neutron/photon/electron transport, including the capability to calculate eigenvalues for critical systems. The code treats an arbitrary three-dimensional configuration of materials in geometric cells bounded by first- and second-degree surfaces and fourth-degree elliptical tori." (www-xdiv.lanl.gov/XCI/PROJECTS/MCNP)

PHOTOCOEF. "For many years now PHOTCOEF, the full-featured version of our nuclear physics applications programs, has been used worldwide for accurate calculations of dose deposition in multi-layered absorbers, shielding, detector response and interaction coefficients, by use of Work Stations or Personal Computers based on Intel X86 microprocessors. Researchers in the fields of radiation effects, detector development, nuclear medicine and health physics, routinely perform investigations with PHOTCOEF that they would not have conducted at all if this program had not been available to them, even when they have access to mainframe computers and Monte Carlo programs. They use PHOTCOEF because of its easy interactive data inputs, speed of calculation, and publication-ready graphs." (www.photcoef.com/211.html)

GENII. "The GENII computer code was developed at Pacific Northwest National Laboratory (PNNL) to incorporate the internal dosimetry models recommended by the International Commission on Radiological Protection (ICRP) into updated versions of existing environmental pathway analysis models. The resulting second generation of environmental dosimetry computer codes is compiled in the Hanford Environmental Dosimetry System (Generation II or GENII). The GENII system was developed to provide a state-of-the-art, technically peer-reviewed, documented set of programs for calculating radiation doses from radionuclides released to the environment. Although the codes were developed for use at Hanford, they were designed with the flexibility to accommodate input parameters for a wide variety of generic sites." (www.pnl.gov/eshs/software/genii.html)

CAP88-PC. This "is a personal computer software system used for calculating both dose and risk from radionuclide emissions to air. CAP88-PC is an approved system for demonstrating compliance with 40 CFR 61 Subpart H, the Clean Air Act standard which applies to U.S. Department of Energy (DOE) facilities that emit radionuclides to air. The CAP88-PC software package allows users to perform full-featured dose and risk assessments in a personal computer environment. CAP88-PC can be used for assessments of both collective populations and maximally-exposed individuals, and allow full editing of many environmental transport variables." (www.er.doe.gov/production/er-80/cap88)

RESRAD. "The only code designated by DOE in Order 5400.5 (Acrobat pdf format) for the evaluation of radioactively contaminated sites. NRC has approved the use of RESRAD for dose evaluation by licensees involved in decommissioning, NRC staff evaluation of waste disposal requests and dose evaluation of sites being reviewed by NRC staff. EPA Science Advisory Board reviewed the RESRAD model. EPA used RESRAD in their rulemaking on radiation site cleanup regulations." (web.ead.anl.gov/resrad/resrad.html)

Space Radiation. "Space Radiation is a widely-used Space Environment and Effects modeling tool for Windows 95/98/NT on the PC. Space Radiation models the ionizing radiation environment in space and the atmosphere including trapped protons and electrons, solar protons, galactic cosmic radiation, and neutrons. The environments may be integrated along any orbit or trajectory. Radiation effects include single-event upsets, total ionizing dose, solar cell damage, and single-event latchup." (www.spacerad.com)

MIRDOSE. "The Radiation Internal Dose Information Center (RIDIC) at the Oak Ridge Institute for Science and Education (ORISE) has been distributing various versions of the MIRDOSE software since 1987. MIRDOSE performs internal dose calculations according to the MIRD technique for many radionuclides commonly used in nuclear medicine (NOTE: the software is NOT in any way associated with the MIRD Committee of the Society of Nuclear Medicine). Its main purpose is to perform the calculations that are needed to obtain dose estimates for the various organs of the body once the kinetics of an agent are established and the residence times or areas under the time–activity curves for the various source organs are established. The code's other purpose is to help the user apply standardized, recognized models and techniques for dosimetry into the calculations" (www.orau.gov/ehsd/mirdose.htm)

2.7 DISCUSSION QUESTIONS AND PROBLEMS

1. Determine the activity of an 80-gal barrel of water if 0.5% of the water molecules contain one ^3H atom.

2. Compute the number of disintegrations per second in a 300-mg sample of ^{40}K.

3. To how many curies does 500 MBq correspond? 100 mBq?

4. A researcher is ordering 3.4 mCi of ^{131}I for an experiment; how many grams of ^{131}I should the researcher order?

5. How long would it take for 99.9% of ^{241}Am to decay if its half-life is 432 years?

6. How long will it take for each of the following radioisotopes to decrease to 0.5% of its initial activity? (a) ^{35}S. (b) ^{137}Cs. (c) ^{210}Pb.

7. What thickness of shielding material is needed to reduce a 5-MeV photon beam by 50% if the material is aluminum? Concrete? Ignore buildup.

8. A beam of photons includes two energy groups. One group, of 1.5-MeV photons, includes 90% of the total intensity. The remaining 10% of the photons have an energy of 5 MeV. (a) What will be the relative proportion of the two groups after passing through 25 cm of water? (b) What would be the relative proportion of the two groups after passing through a slab of lead of the same thickness?

9. If one has 500 mCi of a radionuclide on May 10 and by June 15 the material has decayed to 350 mCi, how much of the radionuclide will be present on July 4 of the same year?

10. If a radionuclide undergoes eight half-lives, what percentage of the original activity remains? 0.4% or 1/256.

11. In the equation $^{210}_{84}\text{Po} \rightarrow {}^{4}_{2}\text{He} + {}^{Y}_{X}\text{Pb}$, what are the correct values for Y and X?

12. In the equation $^{Y}_{X}\text{Z} \rightarrow {}^{32}_{16}\text{S} + {}^{0}_{-1}\text{e} + 1.71$ MeV, what are the correct values for Y and X and atomic symbol for Z?

13. Do you think that the general public has an accurate understanding of radiation? If so, how was the information provided? If not, what improvements are needed in the education of the general public about radiation?

14. Does the average American receive a higher radiation dose from natural or man-made radiation? What is the largest contributor to the population's dose?

15. What are the four principal components of natural background radiation?

16. What factors contribute to the range in background radiation received by individuals around the world?

REFERENCES

1. Murray, R. L., *Understanding Radioactive Waste*, Battelle Press, Columbus, Ohio, 1982.
2. Lederer, C. M., and V. S. Shirley, *Table of Isotopes*, 7th ed., Wiley, New York, 1976.
3. Cember, H., *Introduction to Health Physics*, 3rd ed., McGraw-Hill, New York, 1996.
4. General Electric Company, *Nuclides and Isotopes*, 14th ed., General Electric Company, San Jose, Calif. 1989.
5. Knoll, G. F., *Radiation Detection and Measurement*, 2nd ed., Wiley, New York, 1989.
6. Committee on Biological Effects of Ionizing Radiation, National Research Council, Health Effects of Exposure to Low Levels of Ionizing Radiation: 1990, National Academy Press, Washington, D.C., 1990.
7. U.N. Scientific Committee on the Effects of Atomic Radiation, Sources and Effects of Ionizing Radiation, Report to the General Assembly, United Nations, New York, 1977.
8. Yallow, R. S., Radioactivity in the service of man, *J. Chem. Educ.*, vol. 59, p. 735, 1982.
9. American Nuclear Society, World list of nuclear power plants, *Nucl. News*, p. 67, February 1986.
10. Committee on Biological Effects of Ionizing Radiation, National Research Council, *The Effects on Populations of Exposure to Low Level Ionizing Radiation: 1980*, National Academy Press, Washington, D.C., 1980.
11. Nuclear Energy Policy Study Group, Nuclear Power Issues and Choices, Ballinger, Cambridge, Mass., 1977.
12. Schleien, B., T. T. Tacker, and D. W. Johnson, The Mean Active Bone Marrow Dose to the Adult Population of the U.S. from Diagnostic Radiology, Publ. FDA 77-8013, Bureau of Radiological Health, Rockville, Md., 1976.
13. U.S. Office of Radiation Programs, Radiological Quality of the Environment in the U.S. 1977, Report 520/1-77-009, Environmental Protection Agency, Washington, D.C., 1977.
14. Klement, A. W., Jr., C. R. Miller, R. P. Minx, and B. Schleien, Estimates of Ionizing Radiation Doses in the U.S. 1960–2000, Report ORP/CSD 72-1, Environmental Protection Agency, Washington, D.C., 1972.
15. Cohen, B. L., Analysis , critique, and re-evaluation of HLW water intrusion scenario studies, *Nucl. Technol.*, vol. 48, p. 63, 1980.
16. National Council on Radiation Protection and Measurement, Radiation Exposure from Consumer Products and Miscellaneous Sources, Report No. 56, National Council on Radiation Protection, Washington, D.C., 1977.
17. Lapp, R. E., *The Radiation Controversy*, Reddy Communications, Greenwich, Conn., 1979.
18. U.S. General Accounting Office, Report CED-78-27, General Accounting Office, Washington, D.C., 1978.
19. Upton, A. G., The biological effects of low level ionizing radiation, *Sci. Am.*, vol. 246, no. 2, p. 41, 1982.
20. Bond, V. P., The need for probabilities in cancer litigation, *Nucl. News*, p. 62, August 1986.

ADDITIONAL READINGS

Ahn, T. N., R. Daya, and R. J. Wilke, Evaluation of backfill as a barrier to radionuclide migration in a high-level waste repository, Appendix A in Nuclear Waste Management Technical Support in Development of Nuclear Waste Form Criteria, NUREG/CR-2163, Brookhaven National Laboratory, 1982.

AIC Software, Inc., PHOTCOEF Introduction, 3 July 1999. Available at http://www.photcoef.com/211.html [19 November 1999].

American Board of Health Physics, American Board of Health Physics Examination Preparation Guide, October 1999. Available at http://www.hps1.org/aahp/abhp/prepman/sec7-9.htm#Sec-7 [22 November 1999].

Argonne National Laboratory, Department of Energy, The RESRAD Family of Computer Codes, 15 July 1999. Available at http://web.ead.anl.gov/resrad/resrad.html [19 November 1999].

Baity, N. A., and J. E. Steigerwalt, Fluctuations in energy loss and their implications for dosimetry and radiobiology, in *Proc. National Symposium on Natural and Manmade Radiation in Space*, NASA tM X-2440, p. 157, 1972.

Brodzinski, R. L., The measurement of radiation exposure of astronauts by radiochemical techniques, in Proc. National Symposium on Natural and Manmade Radiation in Space, NASA TM X-2440, p. 162, 1972.

Evergreen Curriculum, Option Unit VIII: Atomic Physics, 24 August 1999. Available at http://www. sasked. gov.sk.ca/docs/physics/u8b3phy.html [23 November 1999].

Gesell, T. F., and H. M. Pichard, The contribution of radon in tap water to indoor radon concentrations, Presented at the 3rd International Symposium on the Natural Radiation Environment, Houston, Texas, April 1978.

Gloyna, E. F., and J. O. Ledbetter, *Principles of Radiological Health*, Marcel Dekker, New York, 1969.

Goldstein, H., *The Attenuation of Gamma Rays and Neutrons in Reactor Shields*, Nuclear Development Corporation of America, White Plains, New York, 1957.

Lead Industries Association, Radiation Shielding, 22 October 1998. Available at http://www.leadinfo.com/ARCH/rad.html [22 October 1998].

Los Alamos National Laboratory, Department of Energy, MCNP Directory 30 June 1990. Available at http://www.xdiv.lanl.gov/XCI/PROJECTS/MCNP/ [19 November 1999].

McIntyre, A. B., D. R. Hamilton, and R. C. Grant, A Pilot Study of Nuclear Medicine Reporting through the Medically Oriented Data System, Publ. FDA 76-8045, Department of Health, Education and Welfare, Rockville, Md., 1976.

Napier, B. A., GENII—The Hanford Environmental Radiation Dosimetry Software System, Pacific Northwest National Laboratory, Department of Energy, 10 November 1999. Available at http://www.pnl.gov/eshs/software/genii.html [19 November 1999].

National Council on Radiation Protection and Measurements, Scientific Committee "Natural Background Radiation in the United States," Report No. 45, NCRP, Washington, D.C., 1975.

Oak Ridge National Laboratory, Department of Energy, Radiation Internal Dose Information Center's MIRDOSE, 28 October 1999. Available at http://www.orau.gov/ehsd/mirdose.htm [19 November 1999].

Schleien, B., *The Health Physics and Radiological Health Handbook*, Scinta, Silver Spring, Md., 1992.

Space Radiation Associates, 2 November 1999. Available at http://www.spacerad.com [19 November 1999].

Transport Methods Group, Los Alamos National Laboratory, Department of Energy, MCNP—A General Monte Carlo N-Particle Transport Code, Version 4B, 19 November 1999. Available at http://www. xdiv.lanl.gov/XTM/xtm1/world1/docs/mcnp-manual/pdf/abstract.pdf.

U.S. Bureau of Radiological Health, Gonad Doses and Genetically Significant Dose from Diagnostic Radiology U.S. 1964 and 1970, Publ. FDA 76-8034, Bureau of Radiological Health, Rockville, Md., 1976.

U.S. Department of Energy, CAP88-PC, 10 November 1999. Available at http://www.er.doe.gov/production/er-80/cap88 [19 November 1999].

U.S. Nuclear Regulatory Commission, Ninth Annual Occupational Radiation Exposure Report 1976, NUREG-0322, Nuclear Regulatory Commission, Washington, D.C., 1977.

THREE

SPENT FUEL MANAGEMENT

3.1 INTRODUCTION

When commercial nuclear power plants were built in the United States, the spent fuel pools were usually designed to hold two to three cores because at that time the spent fuel was expected to be reprocessed within a few years of discharge. Reprocessing of spent fuel did not occur in this country, however, and reactors are having problems providing for storage of the spent fuel. Repositories and/or interim storage facilities for spent fuel are needed, and according to the Nuclear Waste Policy Act[1] such facilities are the responsibility of the federal government. However, such facilities have not been made available and utilities have had to make provisions for on-site spent fuel storage for the lifetime of the reactors. This means that utilities must provide storage capacity that is many times what the plant was originally designed to store. This need has resulted in the development of several options for increasing the capacity of existing pools and the development of systems for storing the fuel outside the spent fuel storage pool. Currently the Nuclear Regulatory Commission has approved seven metal or concrete cask designs for this purpose. In other countries, such as France and Japan, fuel is being reprocessed. Several other countries are also committed to reprocessing spent nuclear fuels. The waste generated from the reprocessing plant is the major source of high-level waste (HLW) and will be treated and eventually disposed of as such. Thus there are two different approaches to spent fuel management, depending on whether or not the fuel is reprocessed. Despite the nonexistence of commercial reprocessing plants in the United States, research and development activities related to the treatment of waste generated in fuel reprocessing are continuing; examples are the programs to vitrify commercial HLW liquids at the West Valley Plant (WVP) and defense HLW at the Savannah River Plant (SRP). Efforts are also underway to initiate the vitrification of the defense HLW at Hanford and at the Idaho Engineering Labs.

This chapter describes the fission process, nuclear fuel assemblies, and refueling cycle in Section 3.1; spent fuel storage requirements and storage options in Section 3.2; dry cask storage of spent fuel at reactors in Section 3.3; legislative and regulatory requirements in Section 3.4; federal interim storage and monitored retrievable storage in Section 3.5; spent fuel packaging for disposal in Section 3.6; transportation of spent fuel in Section 3.7; cooperative demonstration programs for dry storage in Section 3.8; experimental programs for storage systems in Section 3.9; and economic evaluation of spent fuel management systems in Section 3.10; applicable computer codes are given in Section 3.11 and discussion questions and problems in Section 3.12. The fuel reprocessing methodology and the immobilization of waste generated in reprocessing are discussed in Chapter 4. Disposal of the un-reprocessed spent fuel as well as the waste from fuel reprocessing operations, which will take place in geologic repositories, is discussed in Chapter 5.

3.1.1 Fission Process

A typical core of a 1000-MWe pressurized-water reactor (PWR) nuclear power plant will contain about 180 fuel assemblies arranged in a configuration such as that shown in Figure 3.1. As the control rods are withdrawn from the core, a nuclear chain reaction occurs. The reaction is sustained as long as the control rods remain at the withdrawn position or until other poisons absorb enough neutrons to cause the chain reaction to stop. The chain reaction is sustained by the fissioning of a fissile material such as ^{235}U in the fuel when it absorbs a neutron and becomes unstable. The ^{235}U atom splits into two fragments of different masses that have a total mass almost (but not) equal to the mass of the ^{235}U atom. These two fragments travel in opposite directions at very high velocities. As these fragments are being slowed down, they give up their energy in the form of heat, which is used for power generation. In addition to the two fission fragments, two to three (on the average 2.5) neutrons are emitted when the ^{235}U atom is fissioned. For the chain reaction to be sustained, one of these neutrons must be absorbed by another ^{235}U atom, which will in turn cause it to fission and so on. The neutrons emitted by the fission process are also initially traveling at very high velocities, so the probability of their being absorbed by ^{235}U atoms is very low. A moderator is therefore required, and the water surrounding the core acts as such to slow down the neutrons in addition to acting as a coolant to remove the heat from the core. The fission fragments, on the other hand, are very large and are all retained in the fuel. Some of these fission products, which are atoms of barium, cesium, zirconium, niobium, and so on, also capture some of the fission neutrons (but without fissioning) and thus compete with the ^{235}U atoms for the neutrons. Eventually, as these fission products build up in the fuel and the number of ^{235}U atoms is decreased by the fission process, the chain reaction can no longer be sustained and the reactor will be shut down for refueling. Historically, fuel has been removed when it has reached a

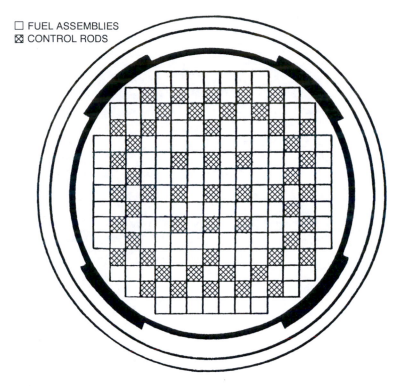

□ FUEL ASSEMBLIES
⊠ CONTROL RODS

Figure 3.1 Typical PWR reactor core. Open squares, fuel assemblies: filled squares, control rods.

burnup of about 33,000 MWD/MT [MW(t)-day per metric ton of total uranium in the fuel] and becomes a "spent" fuel.

3.1.2 Fuel Assemblies

Nuclear fuel material is contained in fuel assemblies, which consist of a large number of fuel elements. Each element consists of fuel pellets, cladding for the fuel, and structural support material to maintain the fuel in the desired configuration. The cladding material is generally a zirconium alloy and the structural members are generally stainless steel. Pellets having 98% of theoretical density (10.96 g/cm^3), which are about 0.89 cm (0.35 in.) in diameter by about 0.89 cm long for PWR elements (fuel elements for boiling-water reactors are slightly larger), are stacked in 300-cm-long stacks, which are loaded into about 457-cm (15-ft)-long zircaloy seamless tubes of slightly larger inside diameter than the outside diameter of the pellet. For a 17 × 17 fuel assembly for a PWR about 270 fuel elements are placed in a structural framework consisting of four corner support rods, 15 control rod thimble tubes, and stainless steel spacer grids spaced about 28 in. apart along the length. The cross section of a PWR fuel assembly is about 21.3 cm (8.4 in.) square and that of a BWR fuel assembly is 14 cm (5.5 in.) square; both are shown in Figure 3.2. The

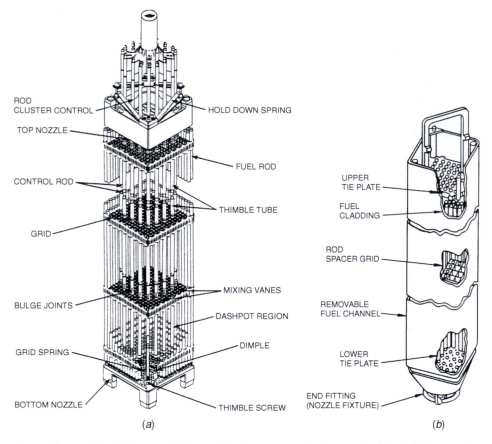

Figure 3.2 (a) Typical light water PWR core. The 17 × 17 assembly has a total weight of 658 kg, contains 461 kg U, has a fuel element outer diameter of 0.95 cm, and has 264 fuel elements per assembly. (b) Schematic drawing of a typical BWR fuel assembly. The 8 × 8 assembly has a total weight of 320 kg, contains 189 kg U, has a fuel element outer diameter of 1.25 cm, and has 63 fuel elements per assembly.

fresh fuel assemblies are stored in containers and sufficiently separated that a nuclear chain reaction could not be supported, even under the most adverse conditions. The same separation is provided during shipment of the fuel assemblies and storage at the power plant until they are placed in the core of the reactor. (Most plants have storage space in their water pools for core reloads.) This new fuel is stored and shipped on about 53.3-cm (21-in.) centers instead of the 21.6-cm (8.5-in.) center in an operating PWR reactor core. The spacing required for storing and shipping BWR fuel assemblies is smaller.

3.1.3 Refueling Cycle

After the fuel is removed from the reactor core, it is stored in a spent fuel storage pool. The refueling cycle occurs every 12–18 months, depending on plant operations and the fuel usage goals of the operating utility. The general procedure in core refueling consists of removing fuel assemblies from the center area of the core, followed by shuffling of the fuel from the two designated zones surrounding the central zone. The new fuel is then placed in the zone farthest from the center (zone 3). Some power plants have very complex shuffling schemes to provide for more uniform burnup of the fuel and a more even temperature distribution across the core, both of which improve the overall operating efficiency of the plant. In a typical 1000-MW PWR, about 60 fuel assemblies (one-third of the core) are replaced at each refueling.

Before bringing new fuel into the core, the depleted fuel assemblies must be removed using the fuel handling system, as shown in Figure 3.3. The manipulator crane is positioned over the fuel assembly, and the assembly is lifted by engaging the grippers from the crane's hoist in the upper nozzle of the fuel assembly and moved from the core area to the area of the fuel transfer system (Figure 3.4) and reactor core control assemblies (RCCA) change fixture. If the fuel assembly contains a thimble plug or spider rod assembly, the insert may be removed in the RCCA change fixture before the fuel assembly is placed in the carriage of the conveyor. The carriage is then lowered from the vertical position to the horizontal one by an upender and is moved through the tunnel. On the other end (spent fuel pit side) of the tunnel, the carriage is raised again by another upender. The fuel assembly is then removed from the carriage by a long-handled spent fuel tool and placed in the storage rack of the spent fuel pool. The spent fuel pool is full

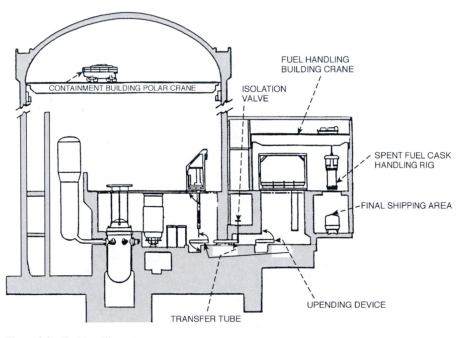

Figure 3.3 Fuel handling system.

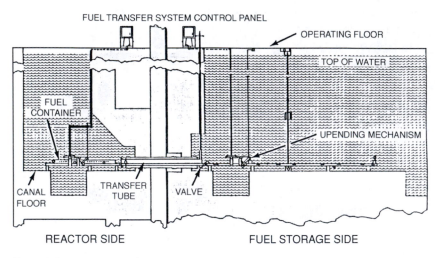

FUEL TRANSFER SYSTEM CONTROL PANEL

Figure 3.4 PWR fuel transfer system.

of water, which serves as radiation shielding and coolant for spent assemblies. Burnable poison rod assemblies are also transferred to the spent fuel pool.

3.2 SPENT NUCLEAR FUEL STORAGE

3.2.1 Spent Fuel Storage Requirement and Generation Rates

Spent fuel storage can be classified into two broad groups: water pool (wet) and dry storage. Based on its location, the storage may be at the reactor site (AR) or away from the reactor site (AFR). Almost all spent fuel in the United States has been in AR storage. The existing and projected amounts of such storage are shown in Table 3.1.[2] Figure 3.5 shows the results of a survey[3] conducted by the National Association of Regulatory Utility Commissioners which indicates the number of nuclear generating units that will

Table 3.1 Spent fuel storage requirements

Year	Metric tons of uranium
1995	31,952
2000	43,300
2005	54,200
2010	63,400
2015	72,400
2020	78,500
2025	84,200
2030	86,700

From Integrated Data Base Report-1996, U.S. Spent Fuel and Radioactive Waste Inventories, Projections and Characteristics, DOE/RW-0006, Rev. 13, U.S. Department of Energy, Washington, D.C., December 1997.

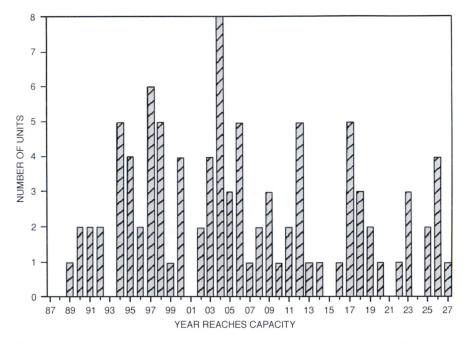

Figure 3.5 Year that spent fuel pools reach capacity (with fuel consolidation and reracking). From Spent-fuel storage. NARUC: Pool capacity dwindling in U.S. by American Nuclear Society, *Nucl. News*, May 1988, p. 64. Copyright 1988 by American Nuclear Society. Reprinted by permission.

reach their storage capacity each year until the year 2027. We now know that by using available storage technologies, there is no need for any plant to run out of storage capacity until the end of the plant life.

Before discussing ways to provide additional storage, two methods for minimizing the need for such additional storage should be mentioned:

1. Increasing fuel utilization. This means extending the burnup of the fuel from the current value of 33,000 MWD/MT for PWRs (27,500 MWD/MT for BWRs) to 45,000 MWD/MT, as demonstrated by the Rochester Gas and Electric Company's PWR systems. Other programs are attempting to extend the burnup to 55,000 or 60,000 MWD/MT, which would almost double the utilization and thus halve the amount of fuel being discharged. Such high burnup, however, would increase problems of transportation and disposal of such fuels.
2. Special shuffling schemes. Shuffling results in more uniform burnup and temperature distribution across the entire core. This helps in two ways: by generating more power because of the even distribution of the core and by reducing the amount of spent fuel to be discharged.

With due consideration of the foregoing, a utility can look for maximum use of existing facilities. The spent fuel storage facility is described in the next section.

3.2.2 Spent Fuel Storage Facility

Most of the nuclear industry's spent fuel is now being stored in water pools at the reactor sites. Water is a convenient storage medium because it is inexpensive and available, can cool by natural circulation, provides shielding from radiation, and provides visibility for handling.

The spent fuel storage facilities typically consist of a spent fuel cooling and cleanup system and equipment to handle the spent fuel being removed from the reactor or being transferred to an alternative storage location. These facilities are designed to limit radiation exposure to less than 2.5 mrem/hr during normal operations and less than 10 mrem/hr during fuel handling operations. Gaseous activity above the spent fuel pool is maintained below the limit set by 10CFR20.[4]

The spent fuel pool contains a nominal water volume of approximately 55,200 ft^3 and holds 413,000 gal within the steel-lined reinforced concrete structure. Borated reactor makeup water with a boron concentration of 2000 parts per million (ppm) is used to fill the spent fuel pool, which has an initial installed capacity of approximately 4.4 cores (850 spent fuel assemblies). The assemblies are stored on racks (Figure 3.6) in a lattice array that was originally designed with a center-to-center distance of 53 cm (21 in.) in both directions to ensure a k_{eff} of 0.95 even if the pool is filled with unborated demineralized water. The storage racks are also designed to withstand the impact of a falling fuel assembly under normal loading and unloading conditions. For storage of failed fuel assemblies, specially designed stainless steel containers are used in the pool, which is designed so that inadvertent draining of the pool is impossible. As shown in Figure 3.6, half of the boxes have lead-in guides attached at their tops to prevent the insertion of fuel assemblies into them and to guide the assemblies into the next boxes without lead-in guides. Adjacent to the spent fuel pool are the fuel transfer canal, loading pool for putting spent fuel in shipping casks, and washdown pit (Figure 3.7a). The fuel transfer canal is connected to the refueling canal by the fuel transfer tube, and the new fuel assemblies are transferred from the new fuel storage facility to the fuel transfer canal by the fuel handling machine. The washdown pit is used for decontamination of shipping casks after cask loading.

3.2.3 Spent Fuel Pool Cooling and Cleanup Systems

Cooling system. The fuel pool cooling system (Figure 3.7a) is designed to remove an amount of decay heat in excess of that produced by the spent fuel assemblies stored in the pool. The system is capable of removing the heat generated from a full core 196 hr after reactor shutdown and the fuel assemblies from two previous refuelings while maintaining the pool temperature below 160°F, corresponding to a heat load of 10.4 MW (35 × 10^6 Btu/hr). For redundancy, the system also contains two cooling trains with 100% capacity with redundant emergency makeup water supplies provided by the refueling water storage tank. Each train consists of a pump, a heat exchanger, and valves, piping, and instrumentation.

Cleanup system. The fuel pool cleanup system (Figure 3.7b) is connected to the cooling system upstream of the makeup water connections. The cleanup system is designed to

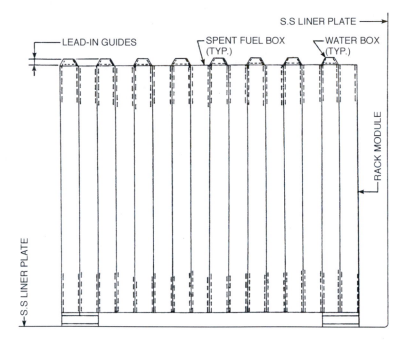

TYPICAL MODULE ELEVATION

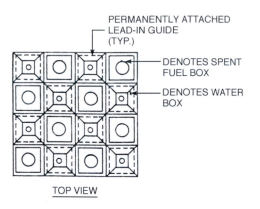

TOP VIEW

Figure 3.6 Early spent fuel storage racks.

maintain optical clarity and to limit the activity of the pool water so that the dose rate at the surface with stored fuel is less than that allowed for continuous occupational exposure (i.e., 2.5 mrem/hr). It is made up of two centrifugal pumps, filters, a mixed-bed demineralizer, and a wye-type strainer. The suction for the cleanup system is taken through six surface skimmers in the spent fuel pool, one in the refueling canal, and five in the refueling pool.

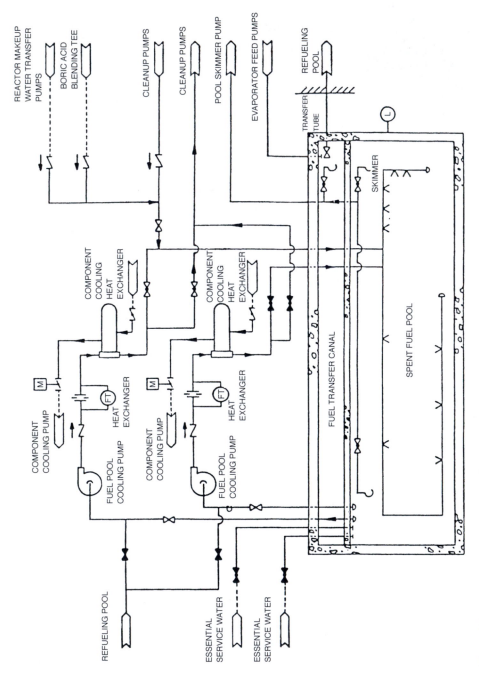

Figure 3.7 (a) Fuel pool cooling and transfer system.

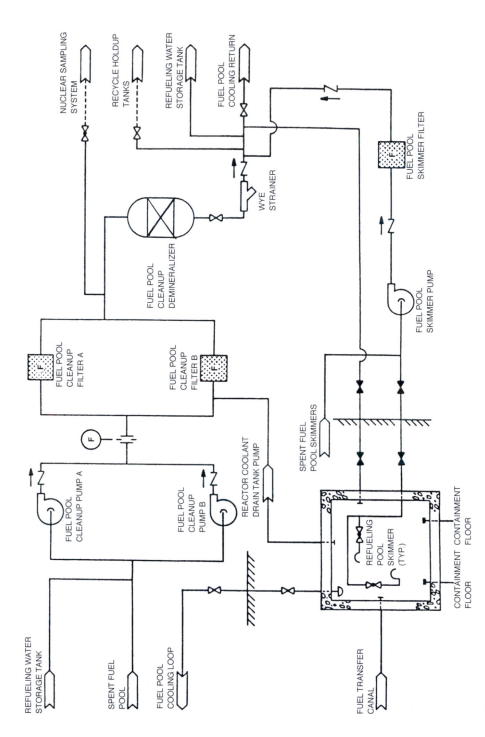

NUCLEAR SAMPLING SYSTEM

RECYCLE HOLDUP TANKS

REFUELING WATER STORAGE TANK

FUEL POOL COOLING RETURN

WYE STRAINER

FUEL POOL CLEANUP DEMINERALIZER

FUEL POOL SKIMMER FILTER

FUEL POOL CLEANUP FILTER A

FUEL POOL CLEANUP FILTER B

FUEL POOL SKIMMER PUMP

SPENT FUEL POOL SKIMMERS

REACTOR COOLANT DRAIN TANK PUMP

FUEL POOL CLEANUP PUMP A

FUEL POOL CLEANUP PUMP B

REFUELING POOL SKIMMER (TYP.)

CONTAINMENT FLOOR   CONTAINMENT FLOOR

CONTAINMENT FLOOR

REFUELING WATER STORAGE TANK

SPENT FUEL POOL

FUEL POOL COOLING LOOP

FUEL TRANSFER CANAL

Figure 3.7 (continued) (b) Fuel pool cooling and cleanup system.

3.2.4 Options for Increasing Storage Capacity

The maintenance of reserve capacity to accommodate the full reactor core in the spent fuel storage pool at a nuclear plant is not a safety matter. Many plant owners consider the full-core reserve capacity to be desirable for operational flexibility, as the ability to fully unload a reactor is useful in making modifications and repairs to reactor structural components and for periodic reactor vessel inspections. Such reserve capacity is nevertheless unused space in the spent fuel storage pool and has the effect of reducing the available fuel storage capacity. Several options to increase storage capacity are being considered.[5] These are to expand pool storage, rerack the pool, consolidate fuel, transship to other pools, use dry storage, and use chemical dissolution, separation, and solidification.

Pool expansion. This means building another pool or expanding an existing one. This approach requires the least development, as water pool storage is a well-proven technology. However, it is very expensive and is not usually recommended.

Rerack. Partly because a need for large storage capacity was not foreseen and partly because the NRC was conservative in its storage requirements, fuel was stored on 53-cm (21-in.) centers in the pools. Experience in spent fuel storage and improved calculation techniques have shown that many plants can increase their storage capacity significantly by changing to storage racks with closer spacings, such as 23-cm (9-in.) centers, and still meet the seismic requirements and the additional floor loadings. Such a reduction in space between fuel assemblies makes it possible to store five times as much fuel in the same storage pool. The power plant is able to do this safely by taking credit for the burnup of fuel, which not only reduces the amount of fissile material present, but also introduces fission products that compete with the remaining fissile material for the available neutrons. Figure 3.8 shows the advantage of reracking and taking the burnup credits, as well as the effects of fuel consolidation, which is discussed in the next section. The high-density racks thus divide the pool into two separate regions, the first designated for storing fresh, unirradiated fuel and the second for storing fuel that has sustained 85% of design burnup. In this way the number of storage locations can be increased. In the meantime, a neutron absorber can be added to the rack by sandwiching boron containing B_4C between stainless steel plates to make up the walls of each storage cell. The B_4C has a very high neutron absorption cross section, and it absorbs a high percentage of the available neutrons that could otherwise cause fissioning.

Fuel consolidation. In fuel consolidation the fuel assembly is dismantled and the spent fuel rods are rearranged into a close-packed geometry in a storage canister. This process has the potential to increase the existing capacity of a spent fuel storage pool by a factor of two, and it may be a relatively inexpensive alternative for storage pools that are strong enough to support the added weight. Fuel consolidation has not yet been licensed by the NRC.[6]

Fuel consolidation in conjunction with maximum use of high-density racks allows a pool storage density of up to 1.04 metric tons of uranium (MTU) per square foot for PWR fuel and up to 1.25 MTU per square foot for BWR fuel. Fuel rods occupy about 40% of the space in typical LWR fuel assemblies. An increase of 1.7–2.0 in the volumetric efficiency of fuel storage can be achieved by consolidating fuel rods according to the process outlined in Figure 3.9. Several companies have designed and some have built

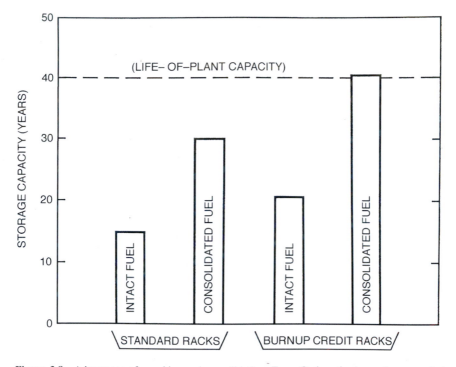

Figure 3.8 Advantages of reracking and consolidation. From Options for increasing spent fuel storage capacity by Y. S. Tang and J. H. Saling, in *Proc. American Nuclear Society First Regional Conference*, Pittsburgh, Penn., 1986. Copyright 1986 by American Nuclear Society. Reprinted by permission.

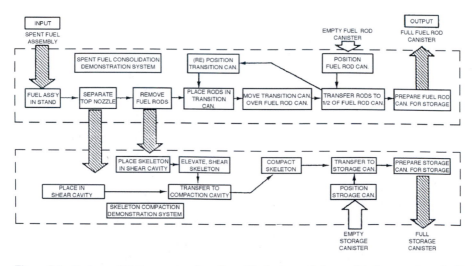

Figure 3.9 Fuel consolidation process flow. From Westinghouse fuel consolidation experience by E. A. Bassler, Presented at the Spent Fuel Storage Seminar, Institute of Nuclear Materials Management, Washington, D.C., January 1984. Reprinted by permission.

equipment to consolidate fuel. In 1982 a successful demonstration with four PWR assemblies in two storage canisters was performed by Duke Power at its Oconee Station.[6] This demonstration proved that a two-to-one reduction in storage space for fuel rods can be made safely. Figure 3.10 shows how the fuel rods were changed over from the square pitch to the hexagonal close-packed array and the skeletons from the fuel bundles were compacted into one-sixth of the original volume and packaged remotely for off-site disposal. Other demonstrations were planned under a cooperative agreement with DOE initially at the Tennessee Valley Authority (TVA) Browns Ferry Plant (not implemented because of delay in NRC approval); at West Valley, New York (1986); and the National Engineering Laboratories, Idaho Falls, Idaho.[7,8] The demonstrations included single-rod versus multiple-rod pulls and different methods of packaging the fuel rods. Consolidation of spent fuel is not widely used.

Transshipment. Some utilities plan to ship fuel between reactor pools at different sites to delay the need for additional storage within the utility system. However, in other cases, transshipments have been barred by state laws or local ordinances. These legal impediments are being challenged in court but the results are not yet final. Transshipment also requires NRC licensing, which can take considerable time. Only a few transshipments are being made. For example, General Electric has a storage facility in Morris, Illinois, that was built for reprocessing (Chapter 4) but was closed down; consequently, the storage pool at this facility became available for SNF storage. General Electric is shipping spent fuel from the plants it built and in which it owned the fuel to the Morris facility. Although transshipment delays the need for additional storage, it does not significantly change the long-term storage requirements.

3.2.5 Dry Storage

In dry storage, the spent fuel is stored in a shielded container outside the reactor containment building.[3,8–10] Studies indicate that dry storage may be competitive in cost with other options such as fuel consolidation and wet (in-pool) storage.[11] Dry storage technology has the advantages of offering flexibility in system types, so that it can be tailored to the needs of a specific site, and of providing long-term storage with low maintenance and ready expandability.[12] It would be even more attractive to utilities if the storage cask could also be used as a shipping container,[13–15] which would eliminate the need to put the fuel back into the reactor fuel transfer pool to transfer it to a shipping container. In general, utilities are concerned about operations that could contaminate the fuel pool; the plant owner is always mindful that any mishap in handling spent fuel could cause a "reportable" (to the NRC) incident that would affect the record of the entire nuclear industry. Thus the plant owner is concerned about having any operations conducted at the plant site, and for this reason it is possible that fuel consolidation at the reactor site will be viewed with concern.

3.2.6 Chemical Dissolution, Separation, and Solidification

The last available option is chemical dissolution of the fuel with subsequent treatment of the resulting liquid and solidification of the end products. This option has not been

Figure 3.10 Consolidated rod bundle and the compacted fuel assembly skeleton. From Westinghouse fuel consolidation experience by E. A. Bassler, presented at the Spent Fuel Storage Seminar, Institute of Nuclear Materials Management, Washington, D.C., 1986. Reprinted by permission.

used in the United States but is being used in other countries. Without separating or removing any constituents of the dissolved fuel, this method provides the ultimate in fuel consolidation. It may be possible to store six or more times as much fuel in the same volume after the treatment. This process can provide the additional advantage to the total waste management system of separating the short-half-life or heat-producing materials from the long-half-life materials, which can significantly affect the packaging, transportation, and disposal parts of the waste management system.

3.3 DRY CASK STORAGE OF SPENT FUEL AT REACTORS

Over the next 30 years every operating nuclear reactor in the United States is scheduled to be shut down. It is quite probable that every one of these reactors will have to provide additional spent fuel storage space. It is also quite probable that this additional storage space will be provided by dry storage technologies. At the present time 15 utilities are using dry storage technologies to provide that additional storage capability and 20 other utilities are on the verge of having to do so. Figures 3.11 and 3.12[16] illustrate the locations of these utilities and the types of dry storage each utility is or will be using.

The problem of having to store spent fuel for very long periods of time will have to be faced all over the world, since Sweden is the only country that has done anything to provide a facility that provides off-site storage, and which also has the possibility of being a disposal site at some future date. The rest of the world, including Sweden, is clearly waiting for the United States to establish a system for permanent disposal of spent fuel and HLW. Table 3.2[17] provides a chronology of initiatives on the storage of spent nuclear fuel.

The systems used to provide additional dry storage capacity at utility sites have not changed. Although there are several additional manufacturers, the containers are still of the types that are described elsewhere in this chapter. There has been little or no additional fuel consolidation to help provide additional storage space. Although the law still requires the federal government to provide spent fuel storage for utilities that cannot reasonably provide it, there has been little progress by the government in this area. In view of the fact that utilities have found ways to provide the needed additional storage, it does not seem likely that the government will ever provide that kind of assistance. It is also quite likely that the rest of the world will use the technology being used here to provide any additional storage capacity that they might need.

The latest potential storage requirements for utilities in the United States are provided in Tables 3.3 and 3.4.[2]

3.4 LEGISLATIVE AND REGULATORY REQUIREMENTS CONCERNING SPENT NUCLEAR FUEL

A number of legislative and regulatory requirements have been promulgated by Congress and the government agencies that have jurisdiction over different parts of the system. Sometimes different groups have overlapping responsibility for regulation

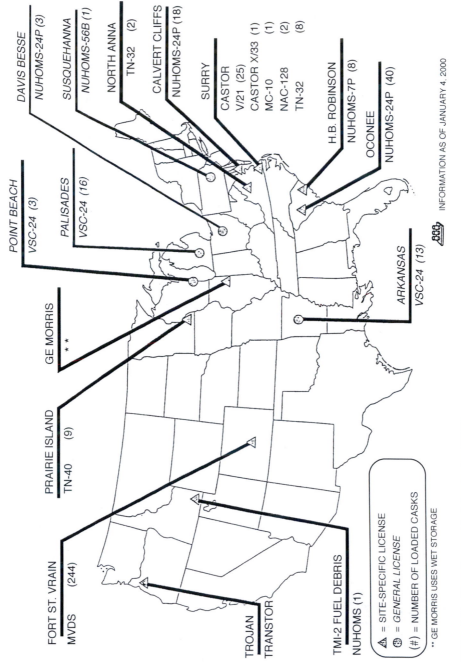

Figure 3.11 Operating spent fuel storage sites (ISFSI).
Source: Reference 16.

FORT ST. VRAIN
MVDS   (244)

TROJAN
TRANSTOR

TMI-2 FUEL DEBRIS
NUHOMS (1)

PRAIRIE ISLAND
TN-40   (9)

GE MORRIS
* *

POINT BEACH
VSC-24 (3)

PALISADES
VSC-24 (16)

DAVIS BESSE
NUHOMS-24P (3)

SUSQUEHANNA
NUHOMS-56B (1)

NORTH ANNA
TN-32   (2)

CALVERT CLIFFS
NUHOMS-24P (18)

SURRY
CASTOR
V/21 (25)
CASTOR X/33 (1)
MC-10   (1)
NAC-128   (2)
TN-32   (8)

H.B. ROBINSON
NUHOMS-7P (8)

OCONEE
NUHOMS-24P (40)

ARKANSAS
VSC-24 (13)

INFORMATION AS OF JANUARY 4, 2000

⬛ = SITE-SPECIFIC LICENSE
⊕ = GENERAL LICENSE
(#) = NUMBER OF LOADED CASKS

** GE MORRIS USES WET STORAGE

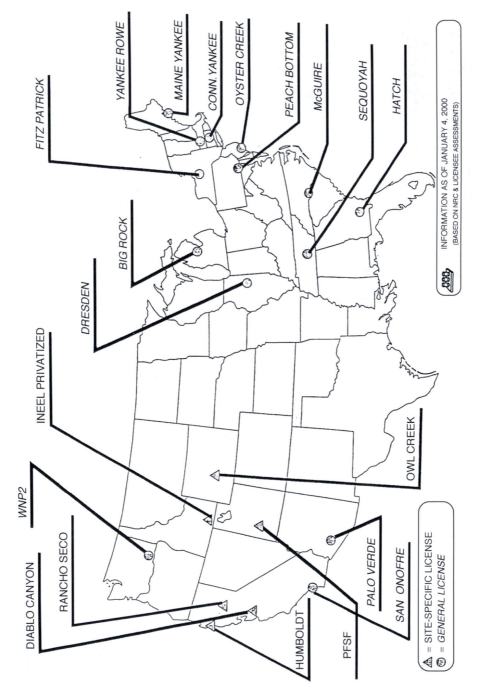

FITZ PATRICK

YANKEE ROWE

MAINE YANKEE

CONN. YANKEE

OYSTER CREEK

PEACH BOTTOM

McGUIRE

SEQUOYAH

HATCH

BIG ROCK

DRESDEN

INEEL PRIVATIZED

WNP2

DIABLO CANYON

RANCHO SECO

HUMBOLDT

PFSF

PALO VERDE

SAN ONOFRE

OWL CREEK

INFORMATION AS OF JANUARY 4, 2000
(BASED ON NRC & LICENSEE ASSESSMENTS)

△ = SITE-SPECIFIC LICENSE
⊕ = GENERAL LICENSE

Figure 3.12 Potential near-term new ISFI sites.
Source: Reference 16.

Table 3.2 Chronology of initiatives taken on storage of spent nuclear fuel

1972	AEC announces retrievable surface storage
1975	Energy Research and Development Administration (ERDA) gives up plan for early construction of a retrievable surface storage facility
1976	Department of Energy (DOE) announces plan for the government to accept custody of commercial spent fuel and to store it in away-from-reactor (AFR) facilities
1986	DOE submits a proposal on monitored retrievable storage (MRS) for spent nuclear fuel (SNF) and high-level waste (HLW) (Congressional authorization for the construction is required)
1987	Congress passes the NWPAA of 1987 establishing an MRS Review Commission

Source (up to 1980): Fred C. Shapiro, *Radwaste: A Reporter's Investigation of a Growing Nuclear Menace*, Random House, New York.

Table 3.3 Historical mass (MTIHM) of permanently discharged commercial SNF by reactor type[a]

	BWR		PWR		Total LWR	
End of CY	Annual	Cumulative	Annual	Cumulative	Annual	Cumulative
1968–1970	—	16	—	39	—	55
1971	65	81	44	83	109	164
1972	146	226	100	183	246	410
1973	94	320	67	250	161	570
1974	242	562	208	458	449	1,020
1975	226	787	322	780	548	1,567
1976	298	1,085	401	1,181	699	2,266
1977	383	1,469	467	1,648	850	3,116
1978	384	1,852	699	2,346	1,082	4,199
1979	400	2,252	721	3,068	1,121	5,320
1980	620	2,872	618	3,686	1,238	6,558
1981	459	3,331	676	4,362	1,135	7,692
1982	357	3,688	640	5,002	998	8,690
1983	491	4,179	771	5,773	1,263	9,952
1984	498	4,677	841	6,614	1,339	11,291
1985	532	5,209	861	7,475	1,393	12,684
1986	458	5,667	996	8,472	1,454	14,139
1987	597	6,264	1,109	9,581	1,706	15,844
1988	536	6,799	1,117	10,697	1,652	17,497
1989	698	7,497	1,215	11,913	1,913	19,410
1990	633	8,130	1,504	13,417	2,137	21,547
1991	588	8,718	1,271	14,688	1,859	23,406
1992	695	9,413	1,596	16,284	2,291	25,697
1993	700	10,113	1,532	17,816	2,232	27,929
1994	675	10,788	1,207	19,024	1,882	29,811
1995	627	11,415	1,514	20,538	2,141	31,952[b]
1996[c]	690	12,105	1,610	22,148	2,300	34,252

[a] Based on refs. 21 and 22. CY, calendar year; BWR, boiling-water reactor; PWR, pressurized-water reactor; LWR, light-water reactor.

[b] Excludes 70 MTIHM of discharged fuel assemblies that are expected to be reinserted.

[c] Data reported are based on projection for CY1996.

Table 3.4 Current and projected mass (MTIHM) of permanently discharged commercial LWR SNF for the DOE/EIA reference case[a]

End of CY	Annual	Cumulative
1996[b]	2,300	34,252
1997	2,100	36,300
1998	2,300	38,600
1999	2,400	41,000
2000	2,300	43,300
2001	2,100	45,500
2002	2,200	47,600
2003	2,200	49,800
2004	1,900	51,700
2005	2,500	54,200
2006	1,600	55,800
2007	2,000	57,800
2008	1,800	59,600
2009	1,800	61,400
2010	2,000	63,400
2011	1,300	64,700
2012	2,100	66,800
2013	2,200	69,000
2014	2,400	71,400
2015	1,000	72,400
2016	1,800	74,100
2017	1,100	75,200
2018	1,200	76,400
2019	900	77,300
2020	1,100	78,500
2021	1,100	79,500
2022	1,400	81,000
2023	900	81,800
2024	1,500	83,300
2025	1,000	84,200
2026	1,300	85,500
2027	600	86,100
2028	200	86,300
2029	300	86,600
2030	100	86,700

[a] Assumes no future fuel reprocessing. Note that cumulative levels reported may not equal the sum of annual additions because of independent rounding.
[b] Data reported as based on projection for CY1996.

of certain functions. In general, the regulatory system works in the following manner. Congress enacts all laws and legislative requirements that govern the implementation of the system; the Environmental Protection Agency (EPA) is responsible for establishing the requirements that must be met to protect the health and safety of the public. For example, the EPA establishes the maximum radiation dose to which the public may be exposed at the boundary of a spent fuel storage facility or in transporting spent fuel from one point to another. The EPA also establishes the emission limits, under both normal operation and accident conditions, for any facility in operation involving radioactive materials. The regulatory organizations such as the Nuclear Regulatory Commission (NRC) and the Department of Transportation (DOT) then promulgate requirements, rules, and guidelines that must be satisfied to meet the EPA requirements. How well equipment or a facility meets the NRC or DOT requirements is evaluated during the licensing process and reviewed periodically, in many cases even continuously, to ensure that the equipment is maintained or the operations are performed in a manner that meets the regulatory requirements and thus protects the health and safety of the public. The Department of Energy (DOE) is responsible for the design, construction, and operation of all the equipment and facilities that are needed to handle, package, store, transport, and dispose of the spent nuclear fuel or high-level waste (HLW) generated by commercial nuclear programs after DOE takes the title of such wastes. The DOE is also responsible for the transportation and disposal of the HLW generated by Department of Defense (DOD) programs. Other government agencies have less significant roles in the management of nuclear materials, such as the Department of Commerce, the Department of the Interior, and the State Department.

3.4.1 Legislative Requirements—Nuclear Waste Policy Act

In December 1982 Congress passed the Nuclear Waste Policy Act (NWPA), which required that the DOE implement a system that would meet the regulatory requirements to protect the health and safety of the public and provide for the initiation of deep geologic disposal of commercial spent nuclear fuel and HLW by January 31, 1998.[1] It required that all equipment and facilities be designed, constructed, and operated to meet the licensing requirements established by the NRC. The Act also provided for funding the programs required for the disposal of SNF and HLW by assessing each nuclear power plant a fee of 1 mill per kilowatt-hour of electricity generated by the plant after April 7, 1983. A charge equivalent to 1 mill per kilowatt-hour of electricity generated by the plant would also be assessed against each plant owner for spent fuel discharged before that date. The latter charge could be paid as a lump sum or in installments but had to be paid in full within 5 years from April 7, 1983.

The NWPA also required the DOE to evaluate and make a recommendation to Congress by June 1986 on the need for and feasibility of a monitored retrievable storage (MRS) facility. The recommendation was to be accompanied by design drawings and cost estimates in sufficient detail for Congress to evaluate and authorize construction of the facility should it be deemed advisable; the DOE complied with this requirement. The NWPA further stated that the DOE may provide federal interim storage (FIS) for spent fuel up to a total of 1900 metric tons of heavy metal for nuclear power plants that have done everything possible to provide for storage of their spent fuel and through no fault of their own cannot do so and would have to shut down if additional storage were not provided. The responsibility for deciding whether the power plant owner had met the obligation was given to the NRC. Other provisions of the NWPA addressed research and development programs to support technology development and provided a time table for implementation of the total program.

In 1987 the NWPA was amended by the Nuclear Waste Policy Amendment Act (NWPAA),[18] which redirected the program on the first geologic repository with the result that the schedule was moved 5 years later than the initial NWPA milestones. The NWPAA also nullified the proposal to site the MRS at Oak Ridge, as submitted by DOE, and established an MRS Review Commission to evaluate the need for such a facility as part of the nation's nuclear waste management system. Some of the major milestones for the design, construction, and operation of the MRS and the first geologic repository, based on this Amendment Act, are shown in Figure 3.13.

3.4.2 Environmental Protection Agency Standards

The EPA has issued 40CFR191, Environmental Standards for the Management and Disposal of Spent Fuel, HLW and Transuranic (TRU) Radwastes. These standards define limits on radiation exposure, radiation levels, and concentrations or quantities of radioactive materials in the general environment outside the boundaries of locations under the control of persons possessing or using radioactive materials. A key provision of the standards is a limit on the amount of radioactivity that may enter the environment now and 10,000 years into the future. The standard establishes limits on annual doses to members of the public of 25 mrem to the whole body, 75 mrem to the thyroid, and 25 mrem to any other organ from planned exposures associated with management, storage, and preparation for disposal of any of these materials.

3.4.3 Nuclear Regulatory Commission Regulations

The NRC promulgates rules based on the EPA standards which ensure that all equipment and facilities are designed, constructed, and operated in a manner that complies with the EPA standards. Although there are many NRC regulations that govern the handling, storage, transportation, and disposal of SNF and HLW, the following represent the key regulations dealing with this subject.

10CFR, Part 60, consists of rules that establish procedures for the licensing of geologic disposal of SNF and HLW and provides technical criteria for the evaluation of license applications under those procedural rules. The objective of the criteria is to provide reasonable assurance that geologic repositories will isolate the radioactive materials for at least 10,000 years without undue risk to public health and safety. The technical criteria include the following:

1. The waste package is to contain the waste for 300–1000 years.
2. The rate of radionuclide release from the engineered system is not to exceed 1 part in 100,000 per year after the containment period for each significant radionuclide.
3. The pre-waste-emplacement groundwater travel times from the repository (more precisely, from the disturbed zone around the repository) to the accessible environment are to exceed 1000 years.

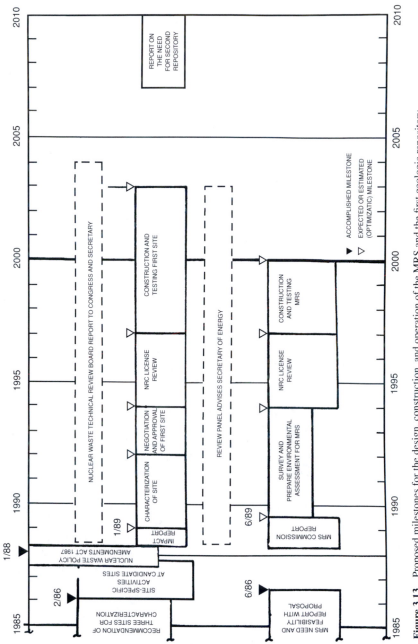

Figure 3.13 Proposed milestones for the design, construction, and operation of the MRS and the first geologic repository.

10CFR, Part 71, governs transportation of SNF and HLW and has the following key provisions:

1. The radiation level at any point on the external surface of the shipping container may not exceed 200 mrem/hr and also may not exceed 10 mrem/hr at 2 m (6.6 ft) from that surface.
2. The shipping container must survive a 30-ft drop at $-40°C$ ($-40°F$) onto an unyielding surface without loss of integrity.
3. The shipping container must survive a completely enveloping fire at a temperature of 800°C (1475°F) for 30 min without loss of integrity.
4. The shipping container must survive immersion in water immediately after the fire test for 8 hr at a depth of 0.9 m (3 ft) without loss of integrity.

10CFR, Part 72, governs the storage of SNF. These regulations are similar to the requirements for a shipping container except that they are much less stringent. For example, the drop test is from 1.8 m (6 ft) instead of 9 m (30 ft). There is one additional requirement for storage of SNF; that is, the maximum fuel rod cladding temperature must be maintained below 385°C (725°F) while stored under an inert atmosphere.

10CFR, Part 51, concerns licensing and regulatory policy that includes handling and storage of the spent fuel in the water pools at the reactor site.

10CFR, Part 20, contains standards for protection against radiation.

3.5 FEDERAL INTERIM STORAGE AND MONITORED RETRIEVABLE STORAGE

The NWPA[1] states that the federal government is responsible for providing not more than 1900 metric tons of capacity for interim storage of SNF for civilian nuclear power reactors that cannot reasonably provide adequate storage capacity at their sites (see Section 3.4.1). Thus Congress has directed the DOE to provide for federal interim storage of SNF. At the same time, Congress has stated that the owners and operators of nuclear power plants have the primary responsibility for providing interim storage of SNF from their reactors by maximizing, to the extent possible, the effective use of existing storage facilities and by adding new on-site storage capacity in a timely manner where practical.

To provide better assurance that the DOE could meet its goal of accepting spent fuel from reactors by January 31, 1998, the NWPA also required that the DOE provide Congress with a plan for one or more MRS facilities for long-term storage of spent fuel, which was completed in 1986. However, as stated previously, the NWPAA of 1987[18] nullified the proposal made by DOE and instead established an MRS Review Committee to evaluate the need for such a facility.

3.5.1 Federal Interim Storage

The DOE expects the increased efficiency of on-site fuel storage resulting from fuel rod consolidation and dry storage to be sufficient to preclude the need for FIS.[19] However, in compliance with the NWPA and as a backup in case unexpected problems arise, the DOE has developed a plan to provide FIS if any utility requests it and the NRC determines that the utility is eligible under NRC regulation 10CFR, Part 53. Because of the time and financial constraints imposed by the NWPA, only existing federally owned sites with storage and handling facilities are being considered. These will minimize the impact on the public health and safety and on the environment as well as minimize the cost of storage to the utility owner. Unless the storage capacity was already provided through the use of available capacity at one or more such facilities on the date of enactment of the NWPA (January 7, 1983), the storage facility and its operation must be licensed by the NRC. If 300 or more metric tons of storage capacity is provided at any one federal site, an environmental impact statement (EIS) must be prepared in accordance with requirements of the National Environmental Policy Act of 1969. Provision of less than 300 metric tons of storage at any one federal site requires only an environmental assessment (EA).

The NWPA also required that if FIS is established, it must be handled as a stand-alone program with full cost recovery, separate from the permanent disposal program established by the Act. The NWPA established a separate fund (the Interim Storage Fund) that will be activated to receive the fee for FIS services and from which the costs of establishing and operating the FIS program will be paid. The funding plan for the service will distribute the costs of the service equitably among all users on a pro rata basis.

3.5.2 Monitored Retrievable Storage (MRS)

Monitored retrievable storage is the long-term isolation of SNF and HLW in facilities that permit continuous monitoring, ready retrieval, and periodic maintenance as necessary to ensure containment of the radioactive materials. The initial NWPA[1] provided the impetus for DOE's work on MRS facilities, and DOE submitted to Congress a detailed study of the need for and feasibility of MRS, along with a proposal for the construction of one MRS facility for civilian HLW and SNF. The facility is to permit continuous monitoring, management, and maintenance of these wastes; provide for their ready retrieval for further processing or disposal; and safely store them as long as necessary by maintaining the MRS facility.[19,20] DOE concluded that an MRS facility could be an integral and important part of the overall waste management system.[19] The performance of the waste management system could be enhanced by having such an MRS facility that is centrally located for most of the commercial nuclear reactors, and has the following principal functions[21]:

1. Prepare spent fuel for emplacement in a repository (may include rod consolidation as well as canister loading).
2. Serve as the central receiving station for the waste management system.
3. Provide limited temporary storage for SNF (with a capacity of up to 15,000 MT of uranium).

Studies indicated that technologies for passive dry storage, which do not require external power to provide cooling, are preferable for long-term MRS.[21] The MRS concepts considered by the DOE were the eight concepts included in the MRS

Research and Development Report to Congress[22]: field drywell, concrete cask, open cycle vault, metal cask (stationary and transportable), concrete cask-in-trench, closed cycle vault, tunnel drywell, and tunnel rack vault. A description of each of these concepts follows.

Field drywell. A field drywell MRS facility uses stationary, in-ground, dry sealed containers for storage of spent fuel or reprocessing wastes. The facility consists of the following components:

1. An array of near-surface drywells in a field into which the canisters of radioactive material are placed for storage and from which the canisters can be retrieved for final disposition.
2. An on-site transporter containing a shielded transfer cask to carry the canisters from the receiving and handling facility and insert them into the proper drywell location or vice versa.
3. A system for detecting any releases of radioactive material from the stored canisters.

A typical drywell is 0.4–0.7 m in diameter and extends to 6–8 m into the ground (Figure 3.14). The surrounding soil serves to attenuate any emitted radiation and to transfer the decay heat from the stored material to the atmosphere. Drywells may be inserted into the soil at an existing site, or an engineered berm may be built to ensure consistent soil characteristics and alleviate water intrusion problems. The arrays of drywells are enclosed within a second fence area to minimize accidental or intentional public intrusion.

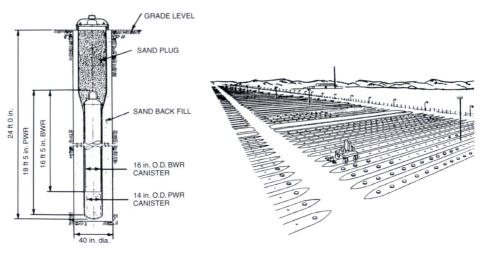

Figure 3.14 Open-field drywell concept. From Selection of concepts for MRS by W. F. Ashton et al., in *Proceedings, Civilian Radwaste Management Information Meeting*, 1983, p. 9, DOE OCRWM. Reprinted by permission.

Concrete cask. A concrete-cask MRS facility uses large cylinders of reinforced concrete for the storage of canisters of spent fuel or reprocessing wastes. The facility consists of the following elements:

1. Large reinforced concrete casks for storage of the wastes with sufficient shielding to reduce the radiation dose at the cask exterior to an acceptable level, with heat removal capability adequate to keep the stored wastes at an acceptable temperature level, and with proper seals on the cask cavity to ensure containment of the radioactive material.
2. An on-site transporter to carry the loaded storage casks from the receiving and handling facility to the storage area and vice versa.
3. A mobile crane to transfer the loaded casks from the transporter to a storage pad and vice versa.
4. A suitable foundation (storage pad) for the storage of casks.
5. A system for detecting releases of radioactive material.

A typical concrete cask is about 3 m in diameter and 7 m high and weighs about 200 tons (Figure 3.15). Heat from radioactive decay of the stored material is conducted through the concrete and transferred to the atmosphere through surface connection and thermal radiation. The reinforced concrete structure and the sealed waste canister are designed to withstand credible man-caused and natural events. The storage arrays are enclosed within a second fence area to minimize accidental or intentional public intrusion.

Open cycle vault. An open-cycle-vault MRS facility uses a large, shielded warehouse for storage of canisters of spent fuel or reprocessing wastes. The facility consists of the following components:

1. A large building with thick concrete shielding to house the canisters, with large-volume ventilation stacks extending 6–15 m above the building.
2. A crane or other mechanical transporter to move the canisters into the storage location and place them vertically in storage tubes that keep them rigidly positioned 0.3–1 m apart.
3. A system of air ducts that directs and distributes outside air past the storage tubes for cooling and discharge to the atmosphere by means of the natural draft.
4. A system for monitoring the air in the interior of the storage tubes and the airflow through the vault to detect any leakage of radioactive material.

In the open vault concept (Figure 3.16), multiple barriers are designed to prevent radioactive material releases to the atmosphere. For spent fuel storage, these include the fuel cladding, the steel canister, and the storage tube; and for HLW storage, the barriers are the high-integrity waste form (e.g., glass), the steel canister, and the storage tube. Additional overpack canisters can be used. The facility is contained within a secured, fenced area to minimize accidental or intentional public intrusion.

Metal cask (stationary and transportable). A metal cask MRS facility uses large metal casks for the storage and/or transportation of canisters of spent fuel or reprocessing

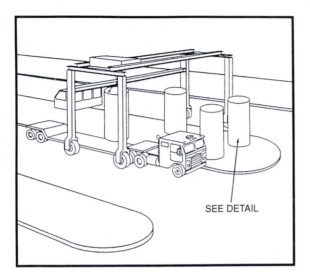

SEE DETAIL

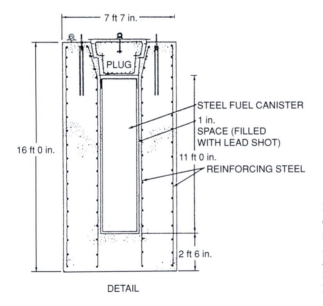

7 ft 7 in.

PLUG

STEEL FUEL CANISTER

1 in.
SPACE (FILLED
WITH LEAD SHOT)

16 ft 0 in.

11 ft 0 in.

REINFORCING STEEL

2 ft 6 in.

DETAIL

Figure 3.15 Surface cask concept. From selection of concepts for MRS by W. F. Ashton et al., in *Proceedings, Civilian Radwaste Management Information Meeting*, 1983, p. 93, DOE OCRWM. Reprinted by permission.

wastes. The facility consists of the following elements:

1. Large metal casks into which the spent fuel or reprocessing wastes are placed for storage. Being sufficiently shielded and with adequate cooling capability, a loaded cask will keep both the radiation dose to its exterior and the temperature in the stored wastes at acceptable levels.

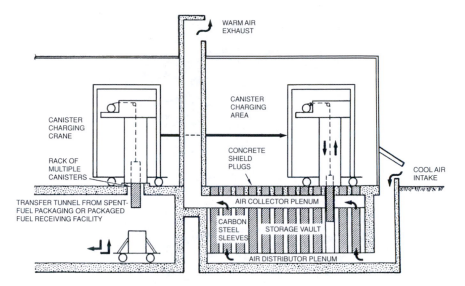

Figure 3.16 Air-cooled surface vault concept. From Spent fuel dry storage: A look at the past, present, and future by J. B. Wright, Presented at Fuel Cycle Conference, Los Angeles, 1981. Also reproduced in Expected Performance of Spent LWR Fuel Under Dry Storage Conditions, EPRI NP-2735, Electric Power Research Institute, Palo Alto, Calif., 1982. Reprinted by permission.

2. An on-site transporter to carry the loaded casks from the receiving and handling facility to the storage area or vice versa.
3. A mobile straddle crane to remove the loaded cask from the transporter and place it on the storage pad and vice versa.
4. A suitable foundation (e.g., a reinforced concrete pad) for storage of arrays of casks.
5. A system for monitoring the integrity of the cask seals and for detecting releases of radioactive materials.

A typical metal cask is about 2.5 m in diameter and 5 m high and weighs about 100 tons (Figure 3.17). Heat from the radioactive decay of the stored material is conducted through the metal cask wall and transferred to the atmosphere by surface convection and thermal radiation. The transportable metal cask can be envisioned as simply the stationary metal cask with the addition of appropriate overpacks and impact limiters as required to license a loaded cask for transport. These casks again are stored in a fenced, secured area to minimize accidental or intentional public intrusion; the storage area could be open or enclosed within simple structures.

Concrete cask-in-trench. A variant of the concrete cask concept is the cask-in-trench, as illustrated in Figure 3.18. A cask similar in configuration to the concrete cask is placed in a burial trench, or berm, that is subsequently backfilled to level with the top of the cask.

Closed cycle vault. Closed cycle vaults (Figure 3.19) are similar in concept to open cycle vaults in that both provide relatively large, shielded enclosures for storage and

(a)

(b)

Figure 3.17 (a) Castor cast iron cask. (b) TN-1300 cast iron cask. From Spent fuel dry storage: A look at the past, present, and future by J. B. Wright, Presented at Fuel Cycle Conference. Los Angeles, 1981. Also reproduced in Expected Performance of Spent LWR Fuel Under Dry Storage Conditions, EPRI NP-2735, Electric Power Research Institute, Palo Alto, Calif., 1982. Reprinted by permission.

both rely on natural circulation of air to remove the decay heat passively from the stored radioactive material. In the closed cycle vaults, however, decay heat is transferred from the waste packages to the air by an intermediate fluid and there is no direct contact of the air with the waste packages. A number of different closed cycle vault designs have been developed. The facility consists of the following components:

1. Multiple concrete storage modules, each containing nine vertical silos for storage of emplaced materials. Heat pipes are used in combination with cooling passages in the concrete to dissipate the heat passively to the air by natural convection.
2. Large, sealed containment canisters that can accommodate multiple spent fuel or waste canisters.
3. A canister transfer cask for transporting the loaded storage canisters from the receiving and handling facility to the silo loading machine.
4. A silo loading machine that transports the storage canister to the storage location, inserts it into the storage module, and retrieves it when needed.

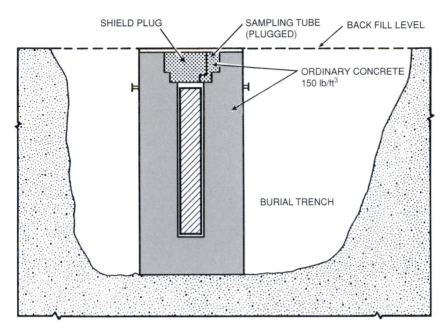

Figure 3.18   Concrete cask-in-trench concept. From Selection of concepts for MRS by W. F. Ashton et al., in *Proceedings, Civilian Radwaste Management Information Meeting*, 1983, p. 98, DOE OCRWM. Reprinted by permission.

Tunnel drywell. A tunnel drywell MRS facility (Figure 3.20) uses underground storage of spent fuel or reprocessing wastes in dry, sealed containers located within a mined tunnel. It is assumed that the storage facilities will be located in a near-surface tunnel in hard rock and above the underground water table. The facility has the same major components as the field drywell.

Tunnel rack vault. The tunnel rack vault (Figure 3.21) uses the same natural draft cooling principle as the open cycle vault. The facility would be built with conventional tunneling equipment and techniques. The facility consists of the following elements:

1. Waste canister storage racks that are unshielded and transportable.
2. Remotely operated transfer machines for moving the loaded storage racks from the hot cell to the storage tunnels.
3. The storage tunnels, which are accessed from a main transfer tunnel. The tunnels must be in a highly stable soil or rock mass. (Preferred orientation is with the receiving and handling facility on the surface and the storage facility at the same elevation under a nearby hill.)
4. Ventilation tunnels to provide air passages for natural convective cooling of the stored material.
5. A system for continuous monitoring of air to detect leakage of radioactive material from the stored canisters and a visual monitoring system using remotely controlled cameras.

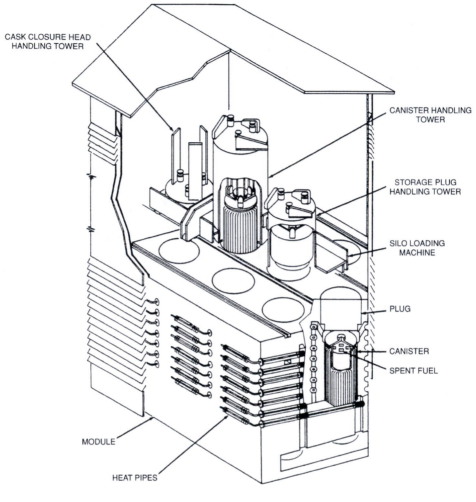

CASK CLOSURE HEAD
HANDLING TOWER

CANISTER HANDLING
TOWER

STORAGE PLUG
HANDLING TOWER

SILO LOADING
MACHINE

PLUG

CANISTER

SPENT FUEL

MODULE

HEAT PIPES

SILO LOADING

Figure 3.19 Closed-cycle vault concept. From Selection of concepts for MRS by W. F. Ashton et al., in *Proceedings, Civilian Radwaste Management Information Meeting*, 1983, p. 98, DOE OCRWM. Reprinted by permission.

Selection of concepts. After a conceptual design analysis in which these concepts were compared on the basis of relative performance, the concrete storage casks and field dry-well concepts were selected as the primary and alternate concepts, respectively, for further design by DOE and were included in the DOE proposal to Congress.[21] However, the proposal also selected a site on the Clinch River in the Roane County part of Oak Ridge, Tennessee, with alternative sites on the Oak Ridge reservation of DOE and on the former site of a proposed nuclear power plant in Hartsville, Tennessee. This proposal was annulled and revoked in the NWPAA of 1987.[18]

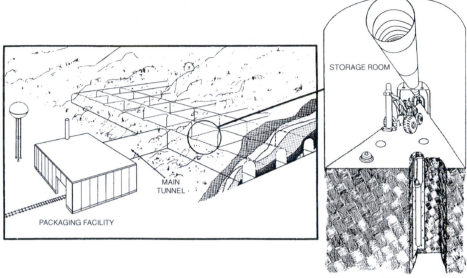

Figure 3.20 Tunnel drywell concept.

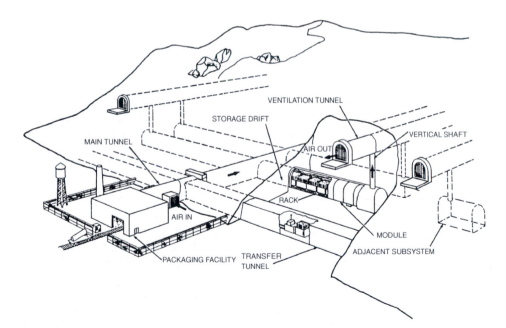

Figure 3.21 Tunnel rack concept.

3.6 SPENT FUEL PACKAGING FOR DISPOSAL

Two basic design concepts were initially proposed for disposal packages for spent fuel or vitrified HLW: the reference concept and the alternate concept. Three different geologic media (i.e., salt, basalt, and tuff) were considered for a nuclear waste repository. The conditions that must be considered to ensure that the waste package meets EPA and NRC requirements include lithostatic and hydrostatic pressures, corrosive environments, groundwater travel times, radionuclide retardation factors, and thermal limits. Each geologic medium has different properties, and the waste package must be designed accordingly of suitable material and dimensions.[23] The reference concept for each geologic medium initially under consideration was the borehole concept, where the waste package is placed in a hole bored in either the floor or the wall of the tunnel and the geologic medium provides the radiation shielding. The alternate concept involved a self-shielded package, where the walls of the package are thick enough to provide the necessary radiation shielding. The following sections describe the SNF waste packages that were considered for the salt, basalt, and tuff geologic media. Since the NWPAA dictated that only the Nevada site, where tuff is the geologic medium, be characterized, most of the recent work has been focused on packages for tuff.

3.6.1 Salt Repository Spent Fuel Packages[24]

The reference waste form for spent fuel is an array of individual fuel rods with no designated geometry. Thus the package size and waste loading could be adjusted depending on thermal, radiation, and economic factors. For the borehole (reference) concept, two configurations of spent fuel rods were used. One contained the spent fuel from six PWR or 17.5 BWR assemblies, tightly bound in a circular array that was placed in the overpack reinforcement. (This was designated as the reference package.[24]) In the second configuration the fuel rods were compacted in such a way that the rods from two assemblies were packed in a box that had the dimensions of a single intact assembly. This design was similar to the circular or reference design, except that the diameter was increased so that the package would hold three boxes containing the equivalent of six PWR assemblies or nine boxes containing the equivalent of 18 BWR assemblies. In the self-shielded (alternate) concept, the spent fuel rods were stored in the boxed form, as the self-shielded package (SSP) design was configured to be a "system package" with the internal cavity and wall thickness compatible with the size and radioactivity of compacted, boxed fuel rods. Thus the SSP contained eight compacted PWR and BWR fuel boxes. The reference container sizes were defined for the most abundant size of PWR and BWR rods and can contain 1584 rods from PWR assemblies or 1100 rods from BWR assemblies with a thermal equivalent of 3.3-kW limits. This first container placed around the waste form, commonly called the canister, was to serve as a structural reinforcement for the corrosion-resistant overpack or subsequent containers. Thus, for the reference design the term overpack reinforcement was used rather than canister. A titanium (Ti Code-12) shell, or overpack, encapsulated the spent fuel rods completely and was surrounded by the crushed salt backfill. The generic waste packages for disposal of spent fuel in a salt repository are discussed in Chapter 4 (Section 4.4.3).

3.6.2 Basalt Repository Spent Fuel Packages[25]

For disposal in basalt medium, only borehole packages were considered. The spent fuel rods for packages containing SF2* were placed in sealed canisters sized to hold 792 rods from three PWR assemblies or 441 rods from seven BWR assemblies. This container size could accommodate the most abundant size of PWR and BWR rods, with minor modifications of dimensions to accommodate rod size variations. Spent fuel rods from domestic commercial nuclear power plants vary in length, diameter, weight, and number per fuel assembly. For instance, the most common rods from a PWR fuel assembly are about 3.86 m (152 in.) long and 0.95 cm (0.37 in.) in diameter and weigh 2.5 kg (5.5 lb); there are 264 rods per assembly with each rod containing 1.74 kg (3.18 lb) of uranium. Those from a BWR fuel assembly are about 4.1 m (161 in.) long and 1.2 cm (0.47 in.) in diameter and weigh 4.4 kg (9.8 lb); there are 63 rods per assembly, each containing 3 kg (6.6 lb) of uranium. The Basalt Waste Isolation Project (BWIP) specified that the spent fuel rods received at the repository would be sealed in carbon steel canisters with a wall thickness of 0.95 cm (0.375 in.), an inside diameter of 30.5 cm (12 in.), and a usable length of 389 cm (153 in.) for housing rods from three PWR assemblies or a length of 411 cm (162 in.) for seven BWR assemblies.[25] The conceptual design for these packages is discussed in Chapter 4.

3.6.3 Tuff Repository Spent Fuel Packages[26]

The spent fuel form considered for disposal in tuff is the SF2 with no containment credit taken for the zircaloy cladding on the fuel rod, as there is no guarantee that the cladding will still be intact after the consolidation operation. For disposal of SF2 in tuff, the waste form configuration used with the reference design was one in which the rods from six PWR (or 18 BWR) assemblies were consolidated in a canister that is divided into compartments, with rods from one PWR (or three BWR) in each compartment. The dividing webs for compartments also served as conduction paths to enhance heat removal. An alternate design considered removing the titanium alloy overpack and using instead a thicker carbon steel overpack reinforcement as the corrosion barrier. Another alternate design for SF2 used the consolidated square box form. Concern about the anticipated performance of the waste form SF2 after the 1000-year containment period led to still another alternate design that provided an additional component to help limit radionuclide releases to 1 part in 10^5 per year. This component was highly compacted sodium bentonite, which, when hydrated, swelled to form a barrier with low hydraulic conductivity, which would limit the movement of water sufficiently for acceptable radionuclide release rates.

3.7 TRANSPORTATION OF SPENT FUEL

The mere fact that radmaterial is transported often makes people apprehensive; this is discussed in Chapter 11. This section only covers the specific features in transporting spent fuels. According to the NWPA of 1982, the DOE is responsible for transportation of commercially generated SNF (and HLW). The DOE will take title to the spent fuel at

---

*The DOE designation of a special spent fuel bundle, Form 2, containing consolidated fuel rods.

the commercial power reactor sites and transport it to federally owned and operated storage or disposal facilities. The DOE has issued a transportation business plan that provides information and schedules for the development of such a transportation system, which was supposed to start accepting fuel from nuclear power plants in January 1998, or 2003 as amended.[27] Delays in repository construction have been accompanied by delays and changes in the transportation plan.

3.7.1 Transport Casks

Studies of the capability of nuclear power plants to package and handle spent fuels indicate that 70% of the reactor plants will be able to handle the large and more efficient transport casks designed for rail shipment and 30% will be forced to use the smaller truck casks.[28] DOE has invited private industries to provide conceptual designs and cost estimates for transport casks to meet the system requirements. Two transportation systems are being considered by DOE and are referred to as the authorized and improved systems. The first system, as shown in Figure 3.22, provides transportation of the SNF from the reactor to the repository or to FIS if it is required. The second system (Figure 3.23) includes an integrated MRS facility that will store and package the SNF to meet the package requirements for repository disposal and to be ready for shipping to the repository for emplacement and permanent disposal. Studies have been made of the total number of shipments that would be required, the number of accidents that would probably occur based on normal shipping statistics, and the costs that would be associated with such systems.[14,29] Typical examples of transportation casks for truck and rail are shown in Figures 3.24 and 3.25. The truck casks (legal weight and overweight) can be designed to transport one to seven PWR assemblies and a little more than twice as many BWR assemblies. The rail casks can be designed to transport 12–30 PWR assemblies or over twice as many BWR assemblies.

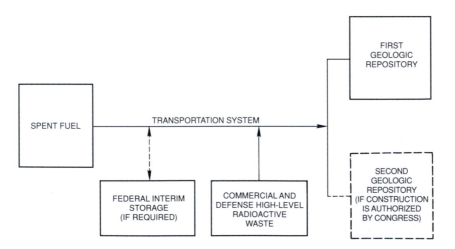

Figure 3.22 Waste disposal system currently authorized by Congress. From Mission Plan for the Civilian Radwaste Management Program by U.S. DOE, DOE/RE-0005, vol. 1, June 1985, Fig. 2-1, p. 16. Reprinted by permission.

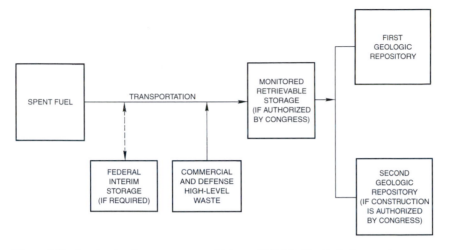

Figure 3.23 Improved performance system with integral MRS facility. Dashed arrow shows transport option to be determined. From Mission Plan for the Civilian Radwaste Management Program by U.S. DOE. DOE/RE-0005, vol. 1. June 1985, Fig. 2-2, p. 18. Reprinted by permission.

3.7.2 Risk Analysis

The risk associated with transporting spent fuel can be estimated from the statistics accumulated over the years on commercial shipments by truck and rail. The most pertinent statistics are provided in Table 3.5.[29] As shown, one could expect an accident severe enough to deform the container and potentially release any radmaterial to occur for every 400 million miles by truck, or 660 million miles by rail. It should be noted that all shipments of spent fuel are made in type B containers. When translated into the number of years of operation of the proposed transportation systems (for both the authorized and improved systems), the results are as shown in Table 3.6. These results show that the risks associated with the transport of SNF are predictably small. However, the federal government has taken additional steps to ensure the health and safety of the public by establishing eight emergency response centers with emergency support teams that will be available quickly to provide expert advice and assistance in the event of any accident involving radmaterials. Analysis of the risk of shipping radwaste is discussed further in Chapter 11.

3.8 COOPERATIVE DEMONSTRATION PROGRAMS FOR DRY STORAGE

In addition to DOE's dry storage demonstrations for single-bundle and multiassembly storage (Section 3.8.1), there have been cooperative agreements between DOE and utilities to demonstrate dry storage technologies and spent fuel consolidation in the reactor storage pool. The programs include the Virginia Power Company (VPCO) Cooperative, Carolina Power & Light (CP&L) Cooperative, and NorthEast Utilities Spent Fuel Consolidation Demonstration Programs.

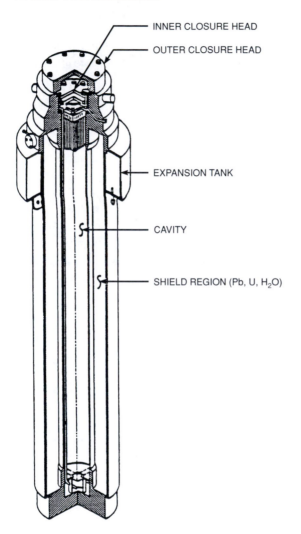

INNER CLOSURE HEAD

OUTER CLOSURE HEAD

EXPANSION TANK

CAVITY

SHIELD REGION (Pb, U, H$_2$O)

Figure 3.24 Legal weight truck cask, shown with impact limiters removed. From Transportation of radioactive waste by R. M. Burgoyne, in *Radioactive Waste Technology*, edited by A. A. Moghissi et al., Fig. 11-1. Copyright 1986 by American Society of Mechanical Engineers. Reprinted by permission.

3.8.1 VPCO Cooperative Demonstration Program

Virginia Power Company was one of the first utilities to experience spent fuel storage problems, and it elected to provide additional storage by purchasing metal storage casks that could be licensed for transportation, which would eliminate the need to handle the fuel again for future shipment. In this program VPCO provided nine large metal storage casks for a storage demonstration. Four were used to demonstrate storage of both intact and consolidated fuel at the Idaho National Engineering Laboratories (INEL) and the other five to demonstrate storage at the utility's site.[10] The demonstration provided about two more years of storage for VPCO's Surry Plants 1 and 2 and also yielded data that allowed VPCO to obtain an NRC license to provide the additional storage needed for the lifetime of these two plants. Figure 3.26 illustrates one of the storage casks used in this program.

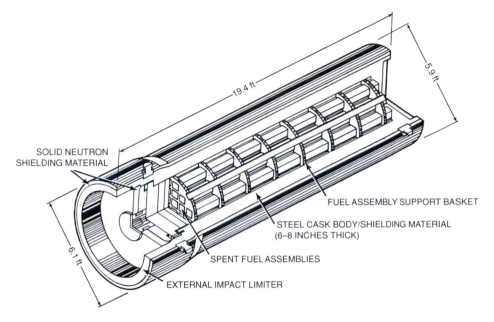

Figure 3.25 Conceptual drawing of a 75-ton rail cask. From DOE's program for transportation of civilian HLW—An overview by W. W. Bixly, in *Proceedings, 27th Annual Meeting of the Institute of Nuclear Materials Management*, June 1986; also U.S. DOE Transporting Spent Nuclear Fuel: An Overview, DOE/RW-0065. Office of Civilian Radwaste Management, Washington, D.C., 1986. Reprinted by permission.

3.8.2 Carolina Power & Light Demonstration Program

The CP&L–DOE demonstration program involved a horizontal silo concept for dry storage. As shown in Figure 3.27, the NRC-licensed shipping cask IF-300 is used as a transfer cask to transport the fuel from the reactor storage pool to the modular storage container. Each storage container is loaded with seven PWR fuel assemblies.[30] The NRC has granted CP&L a license for this demonstration, and at least one other utility is considering this concept to increase its spent fuel storage capacity.

Table 3.5 Transportation risk data

400,000 truck miles per accident resulting in more than $250 damage
660,000 rail car miles per accident resulting in more than $1500 damage
1 in 100 of these are severe enough to approach design conditions
1 in 10 of these severe accidents could deform the cask so that there was a potential for some release
All analyses show that this release would be small
To date, there has never been an accident that caused release from a type B package

From The Probability of Spent Fuel Transportation Accidents by J. D. McClure, Report SAND 80-1721, Sandia National Laboratories, 1981. Reprinted by permission.

Table 3.6 Spent fuel transportation risks (in terms of serious-accident-free years)[a]

	Number of years per serious accident	
System	Trucks	Rail
System without MRS	22	181
System with MRS	43	280

[a]MRS, monitored retrievable storage.

3.8.3 NorthEast Utilities Spent Fuel Consolidation Demonstration

NorthEast Utilities has a cooperative agreement with DOE to demonstrate spent fuel consolidation at its Millstone Unit 2.[12] The consolidation process was demonstrated on simulated fuel and the consolidation of eight spent fuel assemblies at Millstone 2

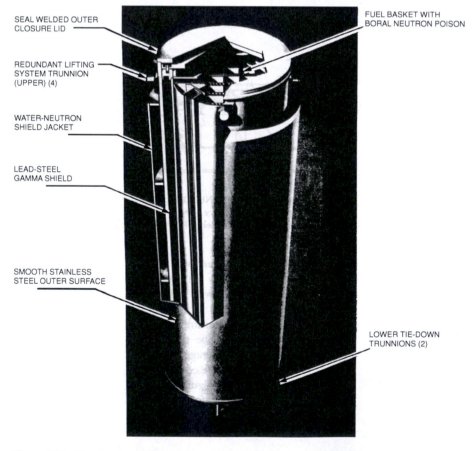

SEAL WELDED OUTER CLOSURE LID

FUEL BASKET WITH BORAL NEUTRON POISON

REDUNDANT LIFTING SYSTEM TRUNNION (UPPER) (4)

WATER-NEUTRON SHIELD JACKET

LEAD-STEEL GAMMA SHIELD

SMOOTH STAINLESS STEEL OUTER SURFACE

LOWER TIE-DOWN TRUNNIONS (2)

Figure 3.26 Dry storage casks for on-site storage of spent nuclear fuel. From Status of development of the REA 2023 dry storage cask for LWR spent fuel by R. E. Best, Presented at the Spent Fuel Storage Seminar II, Washington, D.C., January 1985. Institute of Nuclear Materials Management. Reprinted by permission.

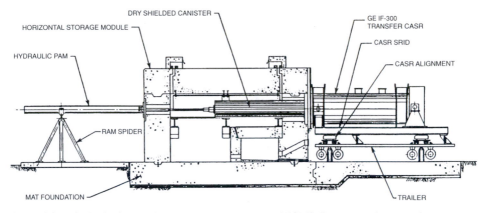

Figure 3.27 CP&L horizontal silo concept. From CP&L–NUHOMS: Dry storage demonstration program, in *Proceedings, Joint ASME–ANS Nuclear Power Conference*, Myrtle Beach, S.C., p. 143, April 1988. Copyright 1988 by American Society of Mechanical Engineers. Reprinted by permission.

during the third quarter of 1987. This demonstration illustrated the technology that will be used to consolidate all of the spent fuel in the NorthEast Utilities systems.

3.9 EXPERIMENTAL PROGRAMS FOR STORAGE SYSTEMS

3.9.1 Experimental Programs for Spent Fuel Storage

Most of the experiments were performed by the Spent Fuel Handling and Packaging Program (SFHPP), which was sponsored by DOE to develop the capability to encapsulate LWR spent fuel assemblies and to establish the suitability of surface and near-surface interim dry storage.[31] Technology development testing on dry storage of spent fuel was carried out at the Engine-Maintenance Assembly and Disassembly (E-MAD) facility at the Nevada Test Site, using three system types (isolated drywell, concrete cask, and air-cooled vault) and stainless steel canisters. In addition to the maximum temperatures of fuel clad, canister, and liner (except for the system with an air-cooled vault, where no liner was used), the axial and radial fuel assembly temperatures under varying gas coolant conditions inside the canister (via air, helium, and vacuum) were measured with external canister thermal impedance conditions representing a variety of dry system types. The maximum fuel cladding temperature was well below the 380°C (715°F) limit imposed on these tests with helium as a medium in the canister. This value was calculated conservatively as a temperature to which the fuel could be exposed for an infinite time without clad rupture due to creep. An accurate thermal power measurement was provided by a boiling-water calorimeter. The clad integrity was indicated by the gas samples taken from canisters representing storage periods of 5–18 months with helium and air environments, at estimated clad temperatures ranging from 180°C to 270°C (356–518°F). Figure 3.28 shows the typical fuel temperature test results for imposed drywell storage conditions, performed in the hot cell with spent fuel in the canister. For a specified lifetime (time before rupture) of cladding under dry storage conditions with air or inert gas in the canister, the isothermal storage temperature corresponding to room temperature and internal rod pressure have

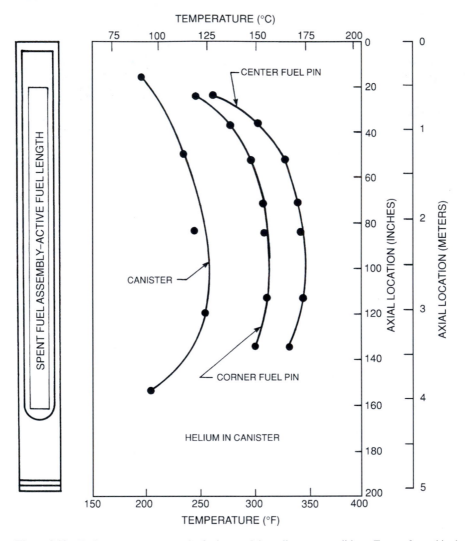

Figure 3.28 Fuel temperature test results for imposed drywell storage conditions. Test performed in the hot cell with spent fuel. From Heat transfer associated with dry storage of spent LWP fuel by G. E. Driesen et al., in *Heat Transfer in Nuclear Waste Disposal*, ASME HTD vol. 11, p. 16, ASME Winter Annual Meeting, 1980. Copyright 1980 by American Society of Mechanical Engineers. Reprinted by permission.

been calculated (Figure 3.29).[32] Thermal performance tests of spent fuel in casks have also been performed. Such performance test data for single-canister, consolidated spent fuel rods[33] and multiassembly REA casks[34] were obtained by Ridihalgh, Eggers, and Associates (REA). Performance tests on the CASTOR-1C storage cask were done at the Wurgassen Nuclear Power Plant by the German Association for the Reprocessing of Nuclear Fuels in conjunction with the Preussen Elektra Utility.[35] The CASTOR-V/21 PWR spent fuel storage cask,[36] a fully loaded (i.e., 21-kW) TN-24P cask,[37] and the MC-10 PWR SF storage cask[38] were tested in the United States.

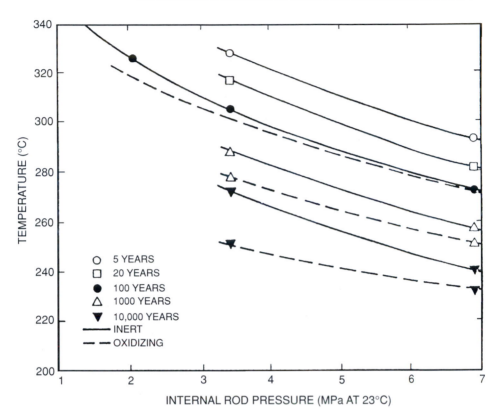

Figure 3.29 Isothermal storage temperature for designed lifetime of cladding. From Low-temperature stress rupture behavior of PWR spent fuel rods under dry storage conditions by R. E. Einziger and R. Hohli, Presented at the Commercial Spent Fuel Management Program Annual Technical Review Meeting, Richland, Wash., October 1983. Reprinted by permission.

3.9.2 Spent Fuel Storage Performance Analysis

Computer codes have been developed for thermohydraulic analysis of a spent fuel assembly contained within a canister, considering the coupled heat transfer modes of conduction, convection, and radiation and various spatial boundary conditions, thermophysical properties, and power generation rates. Examples of such codes are HYDRA-1,[40] TRUMP,[41,42] HEATING5,[43] COBRA-SFS,[44] SPECTROM-41,[39] and RTEDD for consolidated rod temperature calculations.[45]

3.10 ECONOMIC EVALUATION OF SPENT FUEL MANAGEMENT SYSTEMS

The preceding sections have discussed the options utilities may have for increasing their spent fuel storage capabilities. The only option that has been used commercially to increase spent fuel storage pool capacity is reracking of the spent fuel storage pools. Depending on the utility and the amount of additional storage space that can be obtained, this system typically costs about $4–8 per kilogram of additional heavy metal stored. Comparative cost analyses of dry storage options[42] to provide additional storage have consistently shown that dry wells will be most economical. However, the fuel thus stored will eventually have to be loaded into a shipping container by transferring it back into the reactor storage pool. Costs for storage of spent fuel in different types of sealed metal casks are estimated to be $65–100 per kilogram of heavy metal stored.[46,47] As mentioned in Section 3.2.5, storage in such casks has the potential to eliminate additional handling of the fuel if the casks can be licensed for shipping and thereby to reduce the cost.

Several systems for spent fuel consolidation have been designed and the costs estimated.[14,47] The estimates vary from $10 to $35 per kilogram of heavy metal stored and it is not certain whether they include the cost of disposal of the hardware from the consolidation process, which is quite variable. It may be some time before commercialization of dry storage and fuel consolidation settles the issue of which option will best serve the storage needs of the utilities.

3.11 COMPUTER CODES

There are many computer codes that are used in conjunction with the packaging, handling, and storage of spent nuclear fuel. The listing provided below is a sampling of these codes and is not meant to be complete. These codes were selected from a software catalog maintained by the U.S. Department of Energy (DOE), Office of Scientific and Technical Information, Energy Science and Technology Software Center, P.O. Box 1020, Oak Ridge, TN 37831. The catalog can be accessed on the Internet at URL *www.doe.gov/waisgate/estsc.html*. Descriptions of several other computer codes are provided elsewhere in this book.

CAN. Canister model, system analysis. This package provides a computer simulation of a system model for packaging spent fuel in canisters.

CINCAS. Nuclear fuel cycle cost and economics. This is a nuclear fuel cycle cost code, which may be used for either engineering economy or predictions of fuel cycle costs or for accounting forecasting of such costs.

COBRA-SFS. Thermal analysis, spent fuel storage. This code is used for steady state and transient thermal hydraulic analysis of spent fuel storage systems.

COBRA-SFS CYCLE 3. Thermal hydraulic analysis of spent fuel casks. This code is used for thermal hydraulic analysis of multiassembly spent fuel storage and transportation systems.

COMRADEX 4. Accidental released radiological dose. This code is used to evaluate potential radiological doses in the near environment of radioactive releases.

DCHAIN V13. Radioactive decay and reaction chain calculations. This code calculates the time-dependent daughter population in radioactive decay and nuclear reaction chains.

FRA. Fuzzy risk analyzer. This is a general-purpose code for risk analysis using fuzzy, not numeric, attributes.

KEFFMGBSTGAN. Nuclear criticality safety. This code is used for criticality safety calculations.

ORMONTE. Monte Carlo sensitivity analysis code. This code produces a probability histogram for each output variable of interest so that the risk associated with the attainment of a given deterministic value can be assessed.

PATH. Gamma dose calculations and shielding analysis. This is a highly flexible shielding code utilizing the common point-kernel integration technique.

PRODCOST. Utility generating cost simulation. This code simulates the operation of an electric utility generation system. Through a probabilistic simulation, the expected energy production, fuel consumption, and cost of operation for each unit in the system are determined.

RSAC-5.1. Radiological safety analysis computer code. This code calculates the consequences of the release of radionuclides to the atmosphere.

TOAD. Processing of analyzer gamma-ray spectra. This code is used to process and analyze gamma-ray spectra.

3.12 DISCUSSION QUESTIONS AND PROBLEMS

1. Calculate the annual volume of wastes from a 1000-MW nuclear power plant, assuming that the spent fuel is not reprocessed. In Chapter 4 a problem is given for calculating the waste volume if the spent fuel is reprocessed, for comparison with the volumes obtained here.

2. Discuss the disposition of spent fuel that is being stored at the reactor site at the point in time that the reactor terminates operation. Consider the following options;

(a) The utility should be required to continue storing the fuel at the reactor site until a storage site is provided by the government.

(b) The government should be required to take over and maintain the at-reactor site until it can be moved to a government site.

(c) The government should pay a private company to maintain the at-reactor site until a government site is available.

REFERENCES

1. U.S. Congress, 97th, Nuclear Waste Policy Act of 1982, Public Law 97-425. January 1983.
2. U.S. Department of Energy, Integrated Data Base Report-1996, U.S. Spent Fuel and Radioactive Waste Inventories, Projections, and Characteristics, DOE/RW-0006, Rev. 13, Washington, D.C., December 1997.
3. American Nuclear Society, Spent-fuel storage, NARUC: Pool capacity dwindling in U.S., *Nucl. News*, p. 64, May 1988.
4. U.S. Code of Federal Regulations, Title 10, Section 20, Standards for Protection against Radiation.
5. Tang, Y. S., and J. H. Saling, Options for increasing spent fuel storage capacity, in *Proc. Am. Nucl. Soc. First Regional Conference*, Pittsburgh, 1986.
6. Bassler, E. A., Westinghouse fuel consolidation experience, Presented at the Spent Fuel Storage Seminar, Institute of Nuclear Materials Management, Washington, D.C., January 1984.
7. Johnson, C. R., TVA rod consolidation equipment design, Presented at the Commercial Spent Fuel Management Program Review Meeting, Richland, Wash., October 1983.
8. Fischer, M. W., Dry rod consolidation advancements in the OCRWM program, Presented at the Institute of Nuclear Materials Management Meeting, Washington, D.C., January 1988.

9. Saling, J. H., and R. Unterzuber, At-reactor dry well storage, Presented at the Spent Fuel Storage Seminar, Institute of Nuclear Materials Management, Washington, D.C., January 1984.

10. Smith, M. L., Dry cask storage: A VEPCO/DOE/EPRI cooperative demonstration program, in *Proc. 1983 Civilian Radwaste Management Information Meeting*, February 1984; also B. H. Wakeman, Status of metal cask dry storage at the Surry power station, in *Proc. Joint ASME/ANS Nuclear Power Conf.*, Myrtle Beach, S.C., April 1988, p. 177.

11. Johnson, E. R., Economics of monitored retrievable storage, Presented at the Spent Fuel Storage Seminar, Institute of Nuclear Materials Management, Washington, D.C., January 1984.

12. Isakson, R., The NUSCO/Fuel Consolidation Program—A progress report, Presented at the 1985 Spent Fuel Storage Seminar, Institute of Nuclear Materials Management, Washington, D.C., 1985.

13. Pasupathi, V., and D. Stahl, Expected Performance of Spent LWR Fuel Under Dry Storage Conditions, EPRI NP-2735 Final Report, Electric Power Research Institute, Palo Alto, Calif., 1982.

14. Saling, J. H., Evaluation of metal cask systems to provide packaging, handling, storage, transportation and disposal of spent fuel and high-level waste, Presented at the Spent Fuel Storage Seminar, Institute of Nuclear Materials Management, Washington, D.C., January 1986.

15. Saling, J. H., and Y. S. Tang, Spent nuclear fuel casks development, Presented at the Southeastern Symposium on In Situ Immobilization of Hazardous and Radioactive Waste, University of Tennessee, Knoxville, Tenn., 1986.

16. Hodges, M. W., Presentation at the INMM Technical Conference, Washington, D.C., January 13, 2000.

17. Shapiro, F. C., *A Reporter's Investigation of a Growing Nuclear Menace*, Random House, New York, 1981.

18. U.S. Congress, 100th, *Nuclear Waste Policy Amendments Act of 1987*, Omnibus Budget Reconciliation Act for FY 1988, Public Law 100-203, December 21, 1987.

19. U.S. DOE, Mission Plan for the Civilian Radwaste Management Program, DOE/RE-0005, vols. 1 and 2, U.S. Department of Energy, Washington, D.C., 1985.

20. Woods, W. D., and D. S. Jackson, Functional design criteria for monitored retrievable storage, in *Proc. 1983 Civilian Radwaste Management Information Meeting*, 1984.

21. U.S. DOE, Monitored Retrievable Storage Submission to Congress, DOE/RW-0035, vols. 1–3, U.S. Department of Energy, Washington, D.C., 1986.

22. Ashton, W. B., W. S. Kelly, R. I. Smith, and M. B. Triplett, Selection of concepts for monitored retrievable storage, in *Proc. 1983 Civilian Radwaste Management Information Meeting*, 1984.

23. Kircher, J. F., and D. J. Bradley, NWTS waste package design and materials testing, in *Proc. Materials Research Symposium, Scientific Basis for Nuclear Waste Management*, vol. 15, p. 383, 1983.

24. Westinghouse Advanced Systems Division, Waste Package Reference Concept Designs for a Repository in Salt, WTSD-TME-001. Rev. A, Pittsburgh, Penn., 1984.

25. Westinghouse Advanced Systems Division, Waste Package Concepts for Use in the Conceptual Design of the Nuclear Waste Repository in Basalt, AESD-TME-3142, Pittsburgh, Penn., 1982.

26. Westinghouse Advanced Systems Division, Conceptual Waste Package Design for Disposal of Nuclear Waste in Tuff, AESD-TME-3138, Pittsburgh, Penn., 1982.

27. Bixly, W. W., DOE's program for transportation of civilian HLW—An overview, in *Proc. 27th Annual Meeting of the Institute Nuclear Materials Management*, New Orleans, La., June 1986; also U.S. DOE, Transporting Spent Nuclear Fuel: An Overview, DOE/RW-0065, Office of Civilian Radwaste Management, Washington, D.C., 1986.

28. Viebrock, J. M., and W. J. Lee, Facility Interface Capability Assessment Project Draft Interim Assessment Report on Test Visits, ORNL/SUB/86-97393/3, Oak Ridge National Laboratory, 1987.

29. McClure, J. D., The Probability of Spent Fuel Transportation Accidents, Report SAND 80-1721, Sandia National Laboratories, 1981.

30. Koss, D. M., and J. C. McLean, CP&L—NUHOMS: Dry storage demonstration program, in *Proc. Joint ASME–ANS Nuclear Power Conf.*, Myrtle Beach, S.C., April 1988, p. 143.

31. Driesen, G. E., D. F. Moran, P. S. Sherba, and R. J. Steffen, Heat transfer associated with dry storage of spent LWR fuel, in *Heat Transfer in Nuclear Waste Disposal*, ASME HTD vol. 11, p. 9, ASME Winter Annual Meeting, 1980.

32. Einziger, R. E., and R. Hohli, Low-temperature stress rupture behavior of PWR spent fuel rods under dry storage conditions, Presented at the Commercial Spent Fuel Management Program Annual Technical Review Meeting, Richland, Wash., October 1983.

33. Eggers, P. E., Thermal Test Results for Simulated BWR Unconsolidated and Consolidated Fuel, Final Report, Eggers, Ridihalgh Partners, Inc., Columbus, Ohio, 1985.
34. McKinnon, M. A., J. W. Doman, J. E. Tanner, R. J. Guenther, J. W. Creer, and C. E. King, BWR Spent Fuel Cask Performance Test: Vol. I, Cask Handling Experience and Decay Heat, Heat Transfer and Shielding Data, PNL-5777, Pacific Northwest Laboratory, Richland, Wash., 1986.
35. GNS, Topical Safety Analysis Report for the CASTOR-1C Cask Independent Spent Fuel Storage Installation (Dry Storage), Rev. 3, U.S. NRC Project Docket M-34, Gesellschaft für Nuklear Service mbH, West Germany; see also COBRA-SFS Thermohydraulic Analysis of the CASTOR-1C and REA-2023 BWR casks containing CSF, U.S. DOE Report PNL-5802, 1986.
36. EPRI, The Castor-V/21 PWR Spent Fuel Storage Cask: Testing and Analyses, EPRI NP-4887, Interim Report by VPCO, PNL and EG&G Idaho for Electric Power Research Institute, Palo Alto, Calif., 1986; see also Wakeman, B. H., Status of metal cask storage program at the Surry power station, Presented at the INMM Spent Fuel Storage Seminar, January 1987.
37. Creer, J. M., et al., TN-24P PWR Spent Fuel Storage Cask Performance Testing and Analysis, PNL-6054, Pacific Northwest Laboratory, Richland, Wash., 1986; see also EPRI NP-5128, by PNL, VPCO, and EG&G INEL, Electric Power and Research Institute, Palo Alto, Calif., April 1987.
38. McKinnon, M. A., et al., The MC-10 PWR Spent Fuel Storage Cask: Testing and Analysis, PNL-6139, Pacific Northwest Laboratory, Richland, Wash., 1987; also EPRI NP-5268 Interim Report, Electric Power and Research Institute, Palo Alto, Calif., July, 1987.
39. Krause, W. B., L. L. Van Sambreek, and R. G. Stickney, In-situ brine migration experiments at the Avery Island salt mine, in *Heat Transfer in Nuclear Waste Disposal*, ASME HTD vol. 11, p. 27, ASME Winter Annual Meeting, 1980.
40. McCann, R. A., Thermohydraulic analysis of a spent fuel assembly contained within a canister, in *Heat Transfer in Nuclear Waste Disposal*, ASME HTD vol. 11, p. 9, ASME Winter Annual Meeting, 1980.
41. Altenbach, T. J., and W. E. Lowry, Three-D thermal analysis of a baseline spent fuel repository, in *Heat Transfer in Nuclear Waste Disposal*, ASME HTD vol. 11, p. 43, 1980.
42. Edwards, A. L., TRUMP: A Computer Program for Transient and Steady State Temperature Distributions in Multidimensional Systems, UCRL-14754, Rev. 3, Lawrence Livermore National Laboratory, 1972.
43. Beyerlein, S. W., and H. C. Claiborne, The possibility of multiple temperature maxima in geologic repositories for spent fuel from nuclear reactors, in *Heat Transfer in Nuclear Waste Disposal*, ASME HTD vol. 11, p. 49, ASME Winter Annual Meeting, 1980.
44. Rector, D. R., C. L. Wheeler, and N. J. Lombardo, COBRA-SFS: A Thermal Hydraulic Analysis Computer Code. Vol. 1: Mathematical Models and Solution Method, PNL-6049, Pacific Northwest Laboratory, Richland, Wash., 1986.
45. Lee, Y. T., Theoretical evaluation of consolidated rod temperatures in spent fuel storage canister, in *Heat Transfer Problems in Nuclear Waste Management*, ASME HTD vol. 67, p. 45, National Heat Transfer Conference, Pittsburgh, Penn., 1987.
46. Johnson, E. R., A Preliminary Assessment of Alternative Dry Storage Methods for the Storage of Commercial Spent Nuclear Fuel, DOE/ET/47929-1, U.S. Department of Energy, Washington, D.C., 1981.
47. Moscardini, R. L., Status of combustion engineering fuel consolidation program, Presented at the Institute of Nuclear Materials Management Meeting, Washington, D.C., 1985.

FOUR

HIGH-LEVEL WASTE MANAGEMENT

4.1 INTRODUCTION

High-level waste (HLW), by definition, includes liquid wastes from the first-cycle solvent extraction of fuel reprocessing and solids into which such liquid wastes have been converted. Also included are concentrated liquid wastes from subsequent extraction cycles. When the spent nuclear fuel (SNF) is not processed, as discussed in Chapter 3, it is also considered as HLW. Because of the high radioactivity of such waste, its management has become a matter of great public concern.

Many scientists and engineers believe that nuclear waste disposal could be done safely.[1] The National Research Council, through the Waste Isolation Systems Panel, indicated that the waste technology identified in their study of the isolation system for geologic disposal of radwaste should be more than adequate for isolating radwastes from the biosphere and for protecting public health and safety.[2] Such a view has been supported by most of the organized technical societies.[3]

As indicated in Chapter 3, Congress, by passing the NWPA in 1982 and the NWPAA in 1987, directed the Department of Energy to accomplish specific goals at specified times. Public acceptance and cooperation are indispensable for the success of such an important program. Thus the management technologies, design philosophy, and performance evaluation related to processing, packaging, storing, transporting, and disposing of radwaste should be examined thoroughly. The risk assessment and safety analysis involved should be performed with demonstrable accuracy. Furthermore, improved confidence must be sought in the prediction of long-term performance of radmaterial disposal. These topics are covered in this chapter. In addition, the generation of HLW (the fuel reprocessing methodology), the cumulative heavy metal in the HLW with or without fuel reprocessing, and the solidification processes are discussed. Major development programs abroad and in the United States are described at the end of the chapter.

4.1.1 Government Policy

The government policy regarding the disposal of HLW (including NSF) was stated in 1980 in President Carter's statement to Congress. For disposal of HLW, the administration adopted an interim planning strategy focused on the use of mined geologic repositories capable of accepting waste from both reprocessed and unreprocessed commercial spent fuel. DOE has sponsored an expanded and diversified program of geologic investigations. Attention is focused on research and development and on locating and characterizing potential repository sites.

Table 4.1 shows the chronology of initiatives on the treatment of HLW for storage and ultimate disposal.[4]

The first serious goal-setting legislation was the NWPA of 1982. Of significance are the conclusions of the National Research Council's study[2] of the isolation system, in which conceptual repositories in basalt, granite, salt, and tuff were considered. The study was completed in 1983. These conclusions are outlined below.

1. Borosilicate glass and unreprocessed spent fuel are the waste forms appropriate for further testing and for repository designs.
2. A simulated repository environment is necessary to develop an adequate prediction of the long-term performance of waste packages in a geologic repository.
3. Site-specific data on geology, hydrology, and geochemical properties were evaluated and used to define parameters for estimating long-term releases.
4. The overall criterion to be used by the DOE in designing a geologic waste isolation system and evaluating its performance had not been specified. As a guideline, the panel selected an average annual dose of 10^{-4} Sv (10^{-2} rem) to a maximally exposed individual at any future time.
5. The following are favorable contributors to a geologic isolation system: (a) slow dissolution of key radioelements as limited by solubility and by diffusion/convection

Table 4.1 Chronology of initiatives taken for the disposal of HLW

1963	Calcining demonstrated
1964	Atomic Energy Commission (AEC) begins to convert liquid wastes to salt cake
1968	Committee on Radioactive Waste Management (CRWM) review programs for solidification and disposal
1969	AEC declares that all commercially generated HLW must be solidified within 5 years and delivered to a federal repository within 10 years
1962	Congress passes the NWPA of 1982, which becomes Public Law 97-425
1987	Congress passes the NWPAA of 1987, redirecting the nuclear waste program to phase out site specific activities at all candidate sites other than Yuca Mountain site and establishing an MRS Review Commission
1992	DOE decides to phase out the domestic reprocessing of irradiated nuclear fuel for the recovery of enriched uranium or plutonium in support of defense activities

in ground water, (b) long water travel times from the waste to the environment, and (c) sorption retardation in the media surrounding the repository.

Current DOE plans are to immobilize and package HLW for disposal in NRC-licensed underground geologic repositories. These high-level wastes are currently being temporarily stored at four different DOE-controlled sites,[5] where the HLW will be immobilized to a form acceptable to the DOE Office of Civilian Radioactive Waste Management (DOE/RW), which has responsibility for accepting the waste for ultimate emplacement in a repository. Table 4.2[6-9] provides the historical and projected cumulative volumes of HLW stored at the above-mentioned sites.

Table 4.2 Historical and projected cumulative volume (10^3 m^3) of HLW stored in tanks, bins, and capsules, by site[a,b]

End of year[c]	Hanford	INEEL	SRS	WVDP	Total
1990	227.4	12.0	131.7	1.2	372.3
1991	230.6	10.4	127.9	1.7	370.7
1992	231.1	11.2	126.9	1.6	370.7
1993	233.6	10.5	129.3	2.0	375.4
1994	215.3	11.0	126.3	2.2	354.8
1995	209.6	11.2	126.5	2.2	349.5
1996	207.3	10.5	127.5	2.0	347.3
1997	208.9	9.8	121.9	1.1	341.7
1998	202.1	9.7	116.4	0.5	328.7
1999	198.7	8.8	110.8	0.4	318.7
2000	196.0	8.8	105.3	0.2	310.2
2001	196.1	8.9	99.8		304.7
2002	195.5	8.9	94.2		298.7
2003	194.7	8.9	88.7		292.2
2004	193.6	8.9	83.1		285.6
2005	192.5	8.8	77.6		278.9
2006	191.4	8.6	72.0		272.1
2007	190.3	8.3	66.5		265.2
2008	189.2	8.1	61.0		258.3
2009	188.1	7.6	55.4		251.0
2010	187.0	7.4	49.9		244.2
2011	185.9	7.1	44.3		237.3
2012	184.0	7.1	38.8		229.9
2013	178.0	7.1	33.3		218.3
2014	169.2	7.1	27.7		204.0
2015	156.0	7.1	22.2		185.3
2016	142.9	7.1	16.6		166.6
2017	129.7	7.1	11.1		147.9
2018	116.5	7.1	5.5		129.1
2019	103.3	6.7			110.0
2020	90.2	6.2			96.4
2021	77.0	5.7			82.7
2022	65.5	5.2			70.7

(continued)

Table 4.2 (continued)

End of year[c]	Hanford	INEEL	SRS	WVDP	Total
2023	53.9	4.7			58.6
2024	42.4	4.2			46.6
2025	30.8	3.7			34.5
2026	19.3	3.2			22.5
2027	7.8	2.8			10.6
2028	2.0	2.3			4.3
2029	2.0	1.8			3.8
2030	2.0	1.4			3.4
2031	2.0[d]	1.0			3.0
2032	2.0[d]	0.6			2.6
2033	2.0[d]	0.3			2.3
2034	2.0[d]	0.0			2.0
2035	2.0[d]	0.0			2.0

[a] Historical inventories for HLW volume are taken from the U.S. Department of Energy report DOE/RW-0006, Rev. 12, December 1996. The inventories for 1996 and the projections through 2035 are taken from ref. 6.

[b] Numbers shown as 0.0 are less than 50 m³. Values of 0.0 or blank do not imply tank cleanout will be 100%.

[c] Data for 1990–1995 are on EOCY basis; data for 1996–2035 are on an EOFY basis.

[d] These volumes (2000 m³) represent the residual amount (<1.0%) of HLW which will remain in tanks until 2035 or later, as per agreement among DOE, the Washington State Department of Ecology, and the EPA.[10]

The federal government has spent well over 15 billion dollars in an attempt to provide for the safe permanent disposal of HLW wastes. It has drilled exploratory boreholes in tuff at Yucca Mountain, a site being studied to determine whether it is suitable for a deep geologic repository. However, no HLW has been placed in the facility, and the DOE is now projecting that wastes will not be placed there before 2010.

4.1.2 HLW Generation from Reprocessing Plants

Although the United States has operated nuclear reprocessing plants at Hanford, Idaho Falls, and Savannah River as part of the defense program, no commercial reprocessing plant other than the West Valley Demonstration Project (WVDP) exists in this country, nor is one expected in the foreseeable future. This will make the cumulative amount of HLW in the form of SNF higher in the United States than in any other country with nuclear power. In countries such as France, Germany, and Japan, where HLW solidification plants are available or planned, the cumulative irradiated heavy metal will mostly be recycled, leaving a small fraction in the waste.[10] As shown, the United States will accumulate the largest amount in terms of heavy metal tonnage or in terms of curies from long-lived fission products and

Table 4.3 National plans for HLW disposal of LWR fuels

	France	West Germany	Japan	United States
Reprocess fuel	Yes	Yes	Yes	No
Waste form	Glass	Glass	Glass	Spent fuel
First waste form plant	1978	1986	1992	1988[a]
Repository	Geologic	Geologic	Geologic	Geologic
Repository startup	~2000	~2000	~2000	1998

[a] West Valley Vitrification Plant for existing reprocessing waste.

From Management of Commercially Generated Radioactive Waste by U.S. Department of Energy, Environmental Impact Statement, DOE/EIS-0046F, U.S. Department of Energy, Washington, D.C., 1980, p. 108. Reprinted by permission.

tonnage of waste actinides. In countries with reprocessing, the cumulative amount of HLW from irradiated fuel to the year 2000 is less than 0.5% of the amount available for recycling.

The plans for HLW disposal of LWR fuels are shown in Table 4.3, where the years for the first waste form (solidification) plant and for the HLW repository are indicated. Typically, the volume of HLW generated from a reprocessing plant is about 5000 liters (L) per metric ton of uranium (MTU) processed, concentrated to about 500–1000 L/MTU processed in the storage. The waste solution goes through a further concentration and precipitation process and is reduced to 300–500 L/MTU processed before being solidified. The mass of the solid waste form produced is 300–500 g solid per liter of feed. Thus the waste form generated from spent fuel reprocessing can be assumed to be about 200 kg waste (glass)/MTU processed, or 20% by weight of the heavy metal processed. A commercial plant generating 1 GW-year of electric energy would require 33 MTU of fuel [at a burnup of 33,000 MW(thermal)-day/MTU]. The volume of the irradiated fuel elements (including cladding) would be 15,000 L, which would generate a volume of HLW solution from reprocessing of 16,000–33,000 L. The resulting vitrified glass (waste form) would occupy 2000 L (2 m^3) or weigh 6600 kg for disposal (13% by volume of the spent fuel). This related volume of HLW is shown in Table 4.4.

Table 4.4 HLW volume from commercial plants

Electric energy generated	1 GW-yr
Fuel required at burnup 33,000 MWD/MTU	33 MTU
Volume of irradiated fuel (actual fuel element including cladding)	15,000 L
HLW from reprocessing for storage	16,000–33,000 L
Vitrified glass (waste form) for disposal	2,000 L (2 m^3)[a]

[a] Or 60 L of waste form/MTU processed or 13% by volume of irradiated fuel.

4.2 FUEL REPROCESSING METHODOLOGY

Reprocessing is a series of physical and chemical operations that separate uranium and plutonium from spent nuclear fuel (Figure 4.1). The uranium recovered from these operations can be in the oxide form for subsequent fuel fabrication or the hexafluoride form for gas diffusion treatment. Plutonium is normally produced as the dioxide. Various waste streams are also produced whose quantity and characteristics depend on the process. A number of separation techniques have been evaluated, and they can be divided into two groups: aqueous processes and pyrometallurgical processes. A brief review is given in the following sections; the solvent extraction process is emphasized because of its importance and widespread use.

4.2.1 Aqueous Processes

Bismuth phosphate/lanthanum fluoride carrier precipitation. The carrier precipitation process was used on a production scale in 1945 to produce plutonium. It takes advantage of the change in solubility between the +4 and +6 oxidation states of Pu to successively precipitate the Pu metal, leaving the uranium and fission products in solution. Hexavalent Pu is reduced to tetravalent Pu with sodium nitrite, and Pu is coprecipitated as a phosphate $[Pu_3(PO_4)_4]$ with bismuth phosphate $(BiPO_4)$ by the addition of bismuth nitrate and sodium phosphate. After separation and redissolution in nitric acid, the Pu is oxidized to the hexavalent state using a suitable oxidizing agent, and the fission products are removed in the same manner in subsequent repeated cycles until the last cycle, when lanthanum fluoride (LaF_3) is used as the carrier precipitate. Because of its inability to extract uranium, this process has been replaced by solvent extraction.

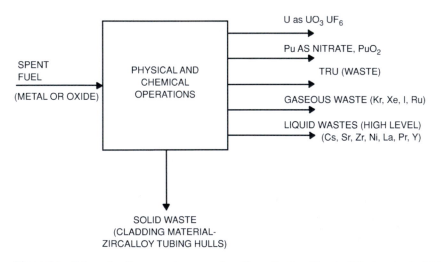

Figure 4.1 Schematic of reprocessing separations. From *Nuclear Chemical Engineering* by M. Benedict, T. H. Pigford, and W. H. Levi, 2nd ed., McGraw-Hill, New York, 1981. Copyright 1981 by McGraw-Hill Book Company. Reprinted by permission.

Solvent extraction.[12] Solvent extraction methods for separating U and Pu from the spent fuel were developed in the late 1940s, and solvent extraction remains the process of choice for fuel reprocessing today. Uranium and plutonium as tetravalent and higher valent cations are extracted into an organic phase, and the trivalent forms and fission product nitrates remain in an aqueous phase of a contact extraction column. A salting agent, such as concentrated nitric acid or aluminum nitrate, is added to the aqueous phase to prevent back extraction of the heavy metals. The lean aqueous solution contains the bulk of the fission products and is treated as HLW. A decontamination factor (the ratio of the contaminant, fission product, in the feed to that in the final product) of about 1000 is obtained across each extraction column. Figure 4.2 shows one complete solvent extraction cycle, which includes partition and stripping after U and Pu are extracted from the feed (nitrate) solution into the extraction column. The heavy-metal-rich organic solvent from the extraction column is sent to the partitioning column, where a dilute nitric acid stream containing a reductant (e.g., ferrous sulfamate or hydroxylamine nitrate) reduces the Pu to the trivalent state. This is preferentially stripped into the aqueous phase and is subsequently purified further by solvent extraction or ion exchange. The U-rich organic phase is stripped with nitrate-free water and purified by a later process. Typically, three full solvent extraction cycles are required. Overall decontamination factors are in excess of 1 million, and product recoveries exceed 99%. Lean organic solvent from the stripping column

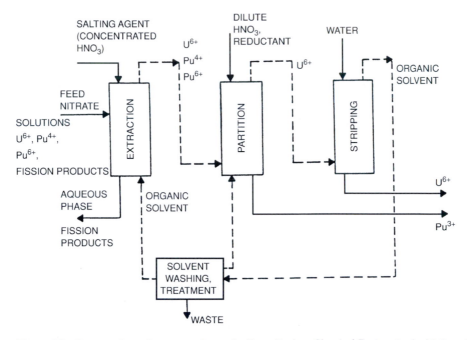

Figure 4.2   One complete solvent extraction cycle. From *Nuclear Chemical Engineering* by M. Benedict, T. H. Pigford, and W. H. Levi, 2nd ed., McGraw-Hill, New York, 1981. Copyright 1981 by McGraw-Hill Book Company. Reprinted by permission.

is treated to remove radiolysis products before being recycled. Several solvent processes are available, and are described below.

Hexone/Redox solvent processes Hexone (methyl isobutyl ketone) is the solvent for the Redox process, and sodium dichromate is the oxidizing agent added to the feed nitrate solutions. (A later process improvement uses potassium permanganate as the oxidizing agent, which coprecipitates some of the fission products.) In the extraction column, aluminum nitrate is used as the salting agent. In the partition column, ferrous sulfamate [$Fe(H_2NSO_3)_2$] is the reducing agent, and aluminum nitrate is added to prevent stripping of the U, which is recovered in the stripping column by contacting with a dilute nitric acid stream. Fission products are further separated from the heavy metals by additional solvent extraction cycles and silica gel treatment. The Redox process was developed at Argonne National Laboratory (ANL), and a pilot plant was tested at Oak Ridge in 1948–1949 and installed at the Hanford Works by General Electric in 1951. A modification of the process was used at the Idaho Falls Chemical Processing Plant (ICPP) for reprocessing highly enriched ^{235}U fuels from the Material Testing Reactor (MTR) and subsequently from the naval reactor program. Although the process worked well, it was replaced by the Purex process (described later) because of the highly volatile and flammable hexone solvent and the large amount of nonvolatile reagents added to the radioactive wastes.

Butex process[13] The Butex process, developed in the late 1940s, uses dibutyl carbitol ($C_4H_9OC_2H_4OC_2H_4OC_4H_9$) because of its lower vapor pressure and greater stability in nitric acid than the hexone solvent. Concentrated nitric acid is added as the salting agent, although in some variations ammonium nitrate is also used. The extraction and decontamination effectiveness are somewhat lower than with hexone but the processing steps are similar. In the late 1950s this process was adopted for reprocessing operations at the Windscale plant in England, where it was retained as a front-end processing step for power reactor oxide fuel until two incidents in 1974–1976 forced the redesign of the front-end facility.

TBP/decane solvents (Purex and Thorex processes) The Purex process uses a mixture of tributyl phosphate (TBP), (n-C_4H_9)$_3PO_4$, in a paraffinic hydrocarbon diluent as the extracting solvent (decane and dodecane are the preferred diluents, with a TBP concentration of about 30% by volume). Tetravalent Pu is more extractable in the Purex solvent than the hexavalent form, and the trivalent form is not extractable. For this reason, the feed is sometimes adjusted to have all of the Pu in tetravalent form by adding nitrogen tetroxide or hydroxylamine. Nitric acid is again used as a salting agent in the extraction column and ferrous sulfamate as the reducing agent in the partitioning column. The rest of the process is similar to the Redox process. The Purex process has the following significant advantages over the Redox process:

1. Waste volumes are much lower, as the nitric acid salting agent can be evaporated and recovered.
2. The TBP solvent is less volatile and flammable than hexone.

3. The TBP is more stable against attack by nitric acid.
4. Process operating costs are lower as the TBP solvent resists radiolysis and extracts heavy metals better than either the Butex or hexone solvents.

Developed at the Knolls Atomic Power Laboratory and tested at Oak Ridge, the process was adopted for the Savannah River Plant in 1954 and replaced the Redox process at Hanford in 1956. Since that time, the Purex process has become the universal choice for separation of U and Pu from the fission products in irradiated nuclear fuel.

The Thorex process is a modification of the Purex process that allows reprocessing of thorium oxide fuels for recovery of ^{233}U. Typically, such fuels contain thorium and uranium with very little plutonium, which is usually not recovered. Because ThO_2 fuels are harder to dissolve in nitric acid and Th has a lower distribution coefficient between TBP and an aqueous solution than U or Pu, and because of other unfavorable chemical properties, the Thorex process becomes increasingly complex. For instance, a low fluoride ion concentration, from weak hydrofluoric acid or potassium fluoride, added to the concentrated nitric acid will dissolve thorium fuels; the addition of aluminum nitrate will reduce the corrosion; and adding dilute phosphoric acid to the top of the extraction column reduces the extraction of Zr and Nb. The protactinium (^{233}Pa) activity in the Th fuel means that the spent fuel is aged before reprocessing (>300 days for PWR-like burnup), so that less ^{233}Pa has to be recovered from the aqueous waste stream.

Developed at Knolls Atomic Power Laboratory and Oak Ridge National Laboratory (ORNL) in the late 1950s, the Thorex process was performed at Savannah River and Hanford during the 1960s. The West Valley Nuclear Fuel Services (NFS) Plant used the Thorex process to recover U from spent U–Th fuel discharged from the Indian Point-1 plant. West German researchers tested the process on high-temperature gas reactor (HTGR)-type fuels in a hot cell facility with good results.

4.2.2 Pyrometallurgical Processes

Halide volatility processes. The various uranium, plutonium, and fission product halides have different vapor pressures and can conceivably be separated from each other on the basis of that property difference. The fluoride-based process has received the most attention. At reasonable temperatures (50–200°C), U, Np, and Pu form volatile hexafluorides, although the latter two hexafluorides are less stable. For fuels aged longer than 100 days after discharge, the only volatile fission product fluorides would be Te, Ru, and I, which can be separated by a distillation column from U and Np hexafluorides (Figure 4.3). Oxide fuel can be converted to the fluorides by using fluorine in a fluidized-bed reactor. The reactor itself or a separate vessel can be used as the evaporator, from which the off gases are sent to a distillation column, where the U and Np hexafluorides are recovered as the bottoms product. However, the Pu tends to remain in the nonvolatile tetrafluoride state and thus cannot be recovered by this process.

The uranium fuel cycles for the fluoride volatility and solvent extraction processes are compared in Figure 4.4, which shows that the fluoride process is simpler but has the

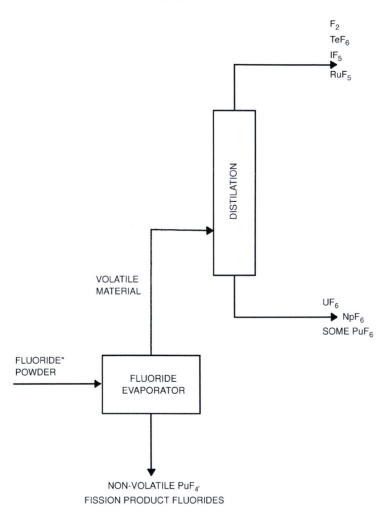

F_2
TeF_6
IF_5
RuF_5

DISTILLATION

VOLATILE
MATERIAL

UF_6
NpF_6
SOME PuF_6

FLUORIDE*
POWDER

FLUORIDE
EVAPORATOR

NON-VOLATILE PuF_4,
FISSION PRODUCT FLUORIDES

Figure 4.3   Fluoride volatility process.

drawback of low or no recovery of plutonium. General Electric developed the Aqua-fluor process for oxide fuel reprocessing, which combined the best features of both the solvent extraction and fluoride volatility technologies. As originally outlined,[14] the process would concentrate and recover U, Pu, and Np by first going through a single Purex solvent extraction step. Then two anion exchange columns in series would be used to recover Pu and Np, while the U-containing aqueous stream was concentrated and calcined to UO_3 in a fluidized-bed reactor followed by high-temperature fluorination and distillation. Unfortunately, the process never worked out at the proposed Midwest Fuel Recovery Plant (MFRP) at Morris, Illinois, before it was abandoned.

Powder process/classification schemes. The Airox process is the most significant "dry" reprocessing scheme; it is a low-decontamination approach without any aqueous

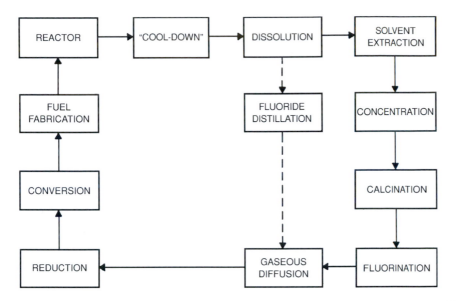

Figure 4.4 Comparison of fluoride and solvent extraction processes. From Report to the American Physical Society by the Study Group on Nuclear Fuel Cycle and Waste Management, *Rev. Mod. Phys.*, vol. 50, no. 1, p. II, 1978. Copyright 1978 by American Physical Society. Reprinted by permission.

separations or high-purity streams. The process was initially developed in the 1960s and because of its inherent proliferation-resistant nature was later reconsidered as a useful front-end operation for the conventional Purex process. In the Airox process, individual fuel elements are exposed to air (at 400°C) by being sheared into sections 2–10 cm (0.79–3.9 in.) long or having holes punched into them every 2.5–4 cm (1–1.6 in.) along their length. The UO$_2$ is thus thermally converted to U$_3$O$_8$ with a 30% volume increase. This swelling stresses the compound sufficiently to rupture the cladding and partially pulverize the fuel pellet, which exposes unreacted fuel to the oxygen and quantitatively releases the volatile fission products. Although it is good for volatile fission product removal, the Airox process suffers from lack of actual reprocessing and irradiation testing. The low fission-product decontamination factor would necessitate remote fuel fabrication and handling and the associated capital expense.

Chelox reprocessing. The Chelox reprocessing concept uses the selectivity and volatility of certain organometallic complexes to separate U and Pu from the other fission products, such as diketonates, which form unusually volatile metal complexes. Quantitative separation of the actinides from the rare earth and lanthanide series elements can be obtained by distillation, thereby recovering the U and Pu and leaving other transuranics with the fuel. Valuable lanthanide series elements and noble metals (e.g., rhodium) can also be recovered by fractional distillation. This process can have a low decontamination factor and be proliferation resistant. Figure 4.5 illustrates one such process in which the organic complexing agent (chelate) is used to dissolve the powder.[15] An evaporator quantitatively separates the more volatile cesium, strontium, and lanthanide series complexes from the less volatile

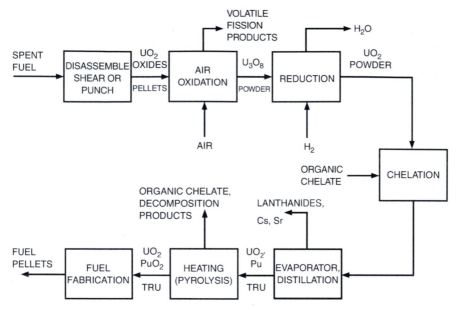

Figure 4.5 Chelox reprocessing concept.

actinide series complexes. Separate distillation of the overhead product can discriminate between individual species, while rapid heating of the bottoms product containing U, Pu, and transuranic constituents can drive off the organic chelate and any decomposition products. This is a new approach and the technology is incomplete, although favorable experimental results have been obtained on nonirradiated fuel pellets spiked with stable fission products.

4.2.3 Description of a Purex Process Plant

Because of the universal use of the Purex process, the operations and equipment involved in such a reprocessing plant with PWR-like oxide fuel are described in this section. Fuel is aged a minimum of 160 days after reactor discharge before reprocessing, allowing for ^{131}I decay and lower fuel activity. The plant can be divided into three parts: head-end operations, solvent extraction, and uranium and plutonium conversion.

Head-end operations. Figure 4.6 shows the head-end operations, including mechanical shearing for decladding and fuel dissolution in hot concentrated nitric acid in a semibatch mode. Before dissolution, voloxidation may be performed to oxidize the UO_2 to U_3O_8; this results in a swollen fuel and releases the balance of the fission product gases, tritium and ^{14}C, which allows recovery of the gases. Subsequent release of some nitrogen oxides, the ^{131}I, and any remaining fission gases takes place in the dissolution step. The leached cladding hulls are removed from the dissolver and sent to solid waste treatment, while the dissolved solution is clarified and the clear liquid sent to the accountability tank.

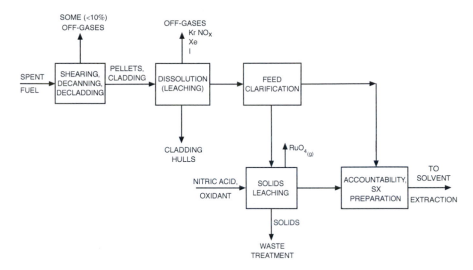

Figure 4.6 Purex process plant, head-end operations. From The Exxon Nuclear Fuel Recovery and Recycling Center Process Design by G. L. Ritter, *Nucl. Tech.*, 1979, vol. 43, p. 196. Copyright 1979 by American Nuclear Society. Reprinted by permission.

Solvent extraction. The solution in the accountability tank contains actinides—for instance, U (in the uranyl state), Pu (in the plutonyl nitrate state), and Np (in a mixture of extractable and unextractable neptunium nitrate states)—and the fission products. The rest of the lanthanides and actinides are in the unextractable (with TBP) trivalent oxidation state. The adjusted feed nitrate solutions flow to the solvent extraction cycle (as shown in Figure 4.2) by gravity or forced circulation. As an example, the Barnwell Nuclear Fuel Plant (BNFP) uses two solvent extraction cycles followed by silica gel treatment for uranium recovery and purification, while three full solvent extraction cycles are used to recover the plutonium; in both cases the recovery is expected to exceed 98.5%.

Uranium and plutonium conversion. Uranium is normally wanted in the oxide (UO_3) or hexafluoride form. Figure 4.7 shows the uranium conversion operations, where the nitrate solution is concentrated to uranyl nitrate hexahydrate, calcined to the oxide, and then converted to UF_6 by fluorination with gaseous fluorine in a fluidized-bed reactor. To meet gas diffusion plant feed specifications, purification is required.

Plutonium nitrate can be directly converted to the oxide by calcination (Figure 4.8), although this has been demonstrated only on a pilot plant scale. Currently, the plutonium is precipitated as the oxalate, filtered, and calcined to the dioxide. Gas streams are treated in adsorbent beds (mercuric nitrate, silver zeolites for iodine, ferric oxide for ruthenium) and released from a stack.

4.2.4 Reprocessing Development in the United States

Although only one commercial reprocessing plant has been operated (for 650 MTU) in the United States, the development work in reprocessing continues but is confined to

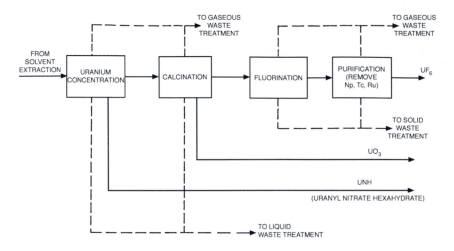

Figure 4.7 Purex process plant, uranium conversion. From The Exxon Nuclear Fuel Recovery and Recycling Center Process Design by G. L. Ritter, *Nucl. Tech.*, 1979, vol. 43, p. 198. Copyright 1979 by American Nuclear Society. Reprinted by permission.

government operations. For example, a modified Zirflex fuel dissolution process is being evaluated by ICPP to minimize the generation the HLW from fuel reprocessing.[16]

Government operations. Spent nuclear fuels for military purposes are reprocessed at three sites: Hanford area, Washington; Savannah River Plant (SRP), South Carolina; and Idaho Chemical Processing Plant (ICPP), Idaho Falls. In 40 years of operation, the Purex process has performed well at all three sites, with only 12 near-criticalities. High-level liquid wastes from the first two sites are neutralized and stored in underground tanks, while ICPP calcines all of its HLW to a free-flowing granular solid, which is stored in stainless steel tanks.

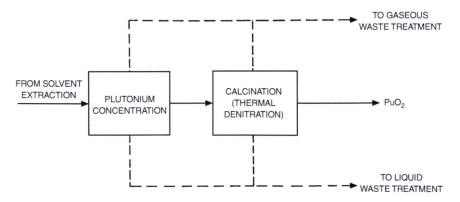

Figure 4.8 Purex process plant, plutonium conversion. From The Exxon Nuclear Fuel Recovery and Recycling Center Process Design by G. L. Ritter, *Nucl. Tech.*, 1979, vol. 43, p. 198. Copyright 1979 by American Nuclear Society. Reprinted by permission.

Commercial operations. Of the three plants designed and built in this country, only one has ever been operated. Brief descriptions are given below.

West Valley Plant (NFS) Built and licensed in 1966, the West Valley Plant had a planned capacity of 300 metric tons of heavy metal (MTHM) a year and cost 26 million dollars. A three-cycle Purex process was used on spent fuels with a minimum 150-day cooling time after discharge. The plant included a spent fuel receiving/storage area, a chop/leach fuel dissolution process, Purex extraction in pulsed columns, evaporators, analytical laboratories, two 750,000-gal mild steel storage tanks for neutralized HLW, two 15,000-gal stainless steel tanks for acid wastes, and a lagoon/LLW treatment system. Operated remotely with the chop/leach cells maintained remotely, the plant used direct maintenance for the rest. Uranyl and plutonium nitrate solutions were the end products.

Midwest Fuel Recovery Plant (MFRP) Designed as a large developmental plant of 300 MTHM/year, the MFRP plant at Morris, Illinois, is the only large-scale plant not built with the Purex process explicitly. It was designed to use the Aquafluor process (see Section 4.2.2) to reduce the plant equipment requirement and thus the capital cost from that of a conventional-size Purex plant. The Morris plant started to receive spent fuel shipments in 1972 but was declared "inoperable in its present form" in 1974 after preoperational tests with nonirradiated, simulated spent fuels indicated many technical problems in the remote operation of the high-temperature calcining and fluorination steps. General Electric notified the NRC that it would not reprocess fuel at the site and continues to store spent fuel at this facility.

Barnwell Nuclear Fuel Plant The plant was to be built at Barnwell, South Carolina, by Allied General Nuclear Services Company (AGNS),[17] but was only partially completed. The plant was to reprocess commercially LWR and mixed oxide fuels containing a maximum of 4% Pu at 1500 MTHM/year. A modified Purex process would be used, with an "electro-pulse" column to partition the Pu. No liquid effluent streams would leave the plant site, as the process water would be recycled or vented as vapor through the stack. The original plant, including spent fuel receiving and storage, continuous dissolution, solvent extraction, HLW waste storage tanks, and uranium conversion unit, was completed in 1976 and passed all cold testing requirements. It failed to obtain licenses to store or reprocess spent fuel at the site and was prevented from operating by the 1977 ban on commercial reprocessing and the national energy policy debate on mixed oxide fuel use. Despite the lifting of the ban on commercial reprocessing in 1981, AGNS indicated that it would no longer fund the plant.

4.2.5 Reprocessing Development and Demonstrations Abroad

There are large-scale commercial reprocessing in Britain and France; also large-scale experimental reprocessing has been done in Belgium,[18] Japan, and Germany.[19]

Reprocessing in the United Kingdom Windscale started to reprocess nuclear fuel for military purposes in the early 1950s and used the Butex solvent extraction process until

1964. The second-generation Windscale plant has been operating since then using the Purex process with mixer–settlers; it has both remote operation and maintenance and is designed to reprocess metallic fuels only, at an annual capacity of 2000–2500 MTHM. The first-generation plant was shut down and modified by adding a chop/leach facility and reducing the oxide fuel activity to comparable levels of metal fuel. The aqueous stream from the modified first plant is sent as feed to the second plant for Purex separations. Thus the first plant acts as a head end for the second plant with an annual oxide fuel capacity of 400 MTHM. High-level wastes from both plants are stored as liquids for future solidification.[20] Still operating, primarily on the lower-burnup metallic fuel, the Windscale plants reprocessed some oxide fuels for overseas customers. In 1974 there was a criticality incident in a scrap plutonium recovery area of the second plant, caused by trace Purex solvent accumulation. In 1976 an explosion occurred in the oxide head-end facility, caused by the Butex solvent. This facility was shut down, redesigned, and modified to handle the Purex process. Because of the complete commitment of the Windscale plants to reprocessing metallic fuel, a complete oxide fuel reprocessing plant, Thorp-1, was constructed nearby to handle British spent oxide fuels. Thorp-1 is designed to handle 600 MTHM/year using the Purex process.

Reprocessing in France France has two sites for reprocessing nuclear fuels: Marcoule and Cap La Hague, both of which use the Purex process and attain high fission-product decontamination factors.[12,13,19] All HLW is maintained in the acid form and is solidified in glass using the French AVM process (described in the next section).[21] Both plants use remote operation and maintenance for the front-end and solidification operations, with remote operation and direct maintenance for the balance of the plant. The Marcoule plant has been operating since 1958 with an annual capacity of 1000 MT of uranium metal fuels. The French plan is to have the Marcoule plant handle all metal fuels and the Cap La Hague plant handle LWR oxide fuels. The Cap La Hague (Cogema) facility is the only commercial reprocessing operation in the world. In 1966 the La Hague UP-2 plant began initial operation with a 2000-MT annual metallic fuel capacity, using mixer–settler contactors. A head-end facility to handle 400–800 MT of oxides a year was added in the mid-1970s; it uses a chop/leach dissolution step, followed by one Purex solvent extraction cycle, and uses the Robatel centrifugal contactor to minimize solvent radiolysis and damage. An additional plant at La Hague (UP-3A) with a capacity of 800 MT/year is financed entirely by contracts for reprocessing overseas LWR fuel, and a second 800-MT/year facility (UP-3B) is planned and has been designed. Specific French designs or costs for their reprocessing plants are not publicly disclosed.

4.3 TREATMENT OF HIGH-LEVEL WASTE

High-level wastes have been stored in storage tanks in the United States. The inherent limitations of tank storage, such as potential leakage and necessity of liquid-waste transfer for periods of hundreds of years, resulted in a vigorous research and development program in the United States (in the early 1960s) as well as in Europe. Conversion of these materials to a stable solid form offers several advantages for ultimate waste disposal, namely a major reduction in waste volume and a general simplification of long-range waste management problems.[22] As indicated in the preceding section, nuclear fuel reprocessing plants have been operated in the United States. All of them stored their reprocessing wastes, including the HLW, which consists of about 99% fission products, unrecovered heavy metals and fuel materials, and reprocessing chemicals. These wastes are stored as neutralized sludge, salt, and liquid at Hanford, Savannah River, and West Valley; as salt and separated cesium and strontium at Hanford; as acid liquid at Idaho and West Valley; and as calcine at Idaho.[23]

4.3.1 Solidification Processes

Calcination and fixation. The calcination of aqueous HLW to a stable dry solid form can be performed by the following processes as tested or used in this country.[22]

Fluidized-bed calcination A hot demonstration plant facility (Demonstrational Waste Calcining Facility, DWCF) with a capacity of 60 gal/hr was installed at the National Reactor Testing Station (NRTS), Idaho. Aqueous aluminum nitrate waste is calcined in a fluidized bed at 400–500°C (752–932°F); the water evaporates, the nitric acid decomposes to water and nitrogen oxide, and the metalic nitrates decompose to the corresponding oxides. Among the problems noted have been excessive solid carryover in the calciner off gas, feed-nozzle caking and erosion, dust leakage in the transport air blower system, and formation of alpha aluminum in the calciner product. The original waste calcination facility operated from 1963 to 1981. The New Waste Calcination Facility (NWCF) started up in September 1982.[23]

Pot calcination process This process involves evaporation to dryness and calcination at temperatures of 700–900°C (1292–1652°F) of solids in a pot that would serve as the final storage container. Developed at ORNL, this system has potential advantages because of its simplicity, versatility in processing several waste types, minimal aerosol problems, and elimination of further packaging of the final product. A pot calciner pilot plant was established at Hanford. One of the major development areas concerned the off-gas problems associated with this process, that is, physical entrainment of mixed waste solids and volatility of such waste components as ruthenium, sulfates, cesium, and sodium.

Radiant heat-spray calcination (Hanford Laboratory) This process consists in feeding liquid wastes through a pneumatic nozzle into the top of a tower while the walls are heated to 850°C (1562°F) by passing low-voltage current through the entire column. The process is technically successful, but it has been largely superseded by the simpler practice of feeding the glass melter with an aqueous waste slurry rather than with precalcined solids. (The slurry-fed ceramic melter process is described later in this section.)

Rotary ball-kiln calcination (Brookhaven National Laboratory) In this process, feed is introduced into a slowly rotating horizontal tube or kiln, externally heated with an electric resistance heater by means of a distribution nozzle, and falls on a hot, shallow bed of metal balls. The calcined solids overflow from the discharge end into a powder receiver and the calciner is designed for continuous operation.

Ceramic-sponge fixation (Coors Porcelain Company–Los Alamos Scientific Laboratory joint study) In this process, liquid radwaste is absorbed in ceramic sponges. The process involves the preparation of a highly porous clay body, which is fired to approximately 1100°C (2012°F). After being soaked in the liquid radwaste and dried several times, the clay body is finally fired at 1300°C to fix the radionuclides permanently in the ceramic material. Off gases from the kiln pass through a silica gel bed, a condenser, and finally absolute filters.

Calcination in molten sulfur (Savannah River Laboratory) In support of plutonium production operations, the Savannah River Laboratory developed a calcination process using molten sulfur at 150°C (302°F) so that the water and volatile acids are driven off and the chemical compounds are decomposed and/or chemically reduced. The resulting sulfur-waste slurry is then heated at 400–440°C (752–824°F) before being cooled back to 115–120°C (239–248°F) and transferred as a liquid to the final container, where it is allowed to solidify.

Slurry-fed ceramic melter (SFCM) process (West Valley Demonstration Project) Based on the experience at Savannah River Laboratory in support of the defense waste processing program, the commercial waste treatment program at Pacific Northwest Laboratory, and the vitrification work at ENICO (Idaho Falls), information from abroad, and the characteristics of the West Valley waste, an SFCM process was selected for the West Valley project. The feed consists of radioactive cesium, strontium, plutonium, other actinides, and technetium removed from the supernatant of the neutralized waste and from the washed sludge, the sludge itself, and the acidic Thorex waste, which are all blended in the feed preparation. The sizing of the SFCM depends on the liquid processing rate, which was determined with reference to testing results from SRL and PNL for the various melters in the DOE development programs.[24] As shown in Table 4.5, a range of liquid processing rates has been obtained with satisfactory performance, although the rate should be kept at the same level. The refractory design of such a process is shown in Figure 4.9. Several features of the melter design are outlined in Table 4.6, indicating the design preference for West Valley's SFCM.

Solidification processes in Europe. *Advanced vitrification method (AVM) (Marcoule, France)* Significant progress has been made in Europe, especially France, in the development of solidification processes and technologies. Since 1957, a research team of the Commissariat à l'Énergie Atomique (CEA) has worked on such development, and the resulting AVM process has been used by Cogema since 1978 at the Marcoule facility. In the AVM process, nitric acid solutions of fission products are introduced from the fission product storage units into a rotary tube heated at 600–900°C

Table 4.5 U.S. pilot plant experience with Joule-heated glass melters

	Pacific Northwest Laboratory				Savannah River Laboratory		Total (or maximum)
	LFCM	CFCM	PSCM	SCM	CFCM	LSFM	
Operational period	Feb 77–Sep 81	May 78–Dec 79	Nov 80–Jan 83	Oct 79–Feb 82	Aug 80–Sep 81	Jan 82–Jan 83	
Total time at temperature (months)	42	19	26	27	13	12	139
Process time (hr)							
Calcine fed	791	1211	—	4722	55	—	5,950
Slurry fed	830	—	1176	1110	44	1320	4,870
Total	1621	1211	1176	5832	99	1320	10,820
Glass produced (tons)							
Calcine fed	43	60	—	23	49	—	175
Slurry fed	23	—	30	5	18	60	136
Total	66	60	30	28	67	60	311
Production rate (L/hr)							
Slurry feeding							
Average	70		53.2	7.6	81.7	75.7	(81.7)
Maximum	150		83.2	—	123.8	93.7	(150.0)
Glass rates (kg/hr)							
Calcine fed							
Average	94	49.4	—	5.0	88.9	—	(94)
Maximum	163	85	—	—	200.0	—	(200)
Slurry fed							
Average	25	—	25.7	4.7	40.2	45.5	(45.5)
Maximum	58	—	41.6	—	61.9	62.7	(62.7)

(continued)

Table 4.5 (continued)

	Pacific Northwest Laboratory			Savannah River Laboratory			Total (or maximum)
	LFCM	CFCM	PSCM	SCM	CFCM	LSFM	
Specific rate (kg/L m²)							
Average	66.7	—	72.8	63.3	69.8	68.8	(72.8)
Maximum	142.8	—	113.9	—	105.8	84.4	(142.8)
Glass rates (kg/hr m²)							
Calcine fed							
Average	89.5	65.0	—	41.7	75.9	—	(89.5)
Maximum	155.2	111.8	—	—	170.9	—	(170.9)
Slurry fed							
Average	23.8	—	35.2	39.2	34.3	41.0	(41.0)
Maximum	55.2	—	57.0	—	52.9	56.5	(57.0)

From Design preferences for a slurry-fed ceramic melter suitable for vitrifying West Valley wastes by C. C. Chapman, in *Proc. International Symposium on Ceramics in Nuclear Waste Management*, vol. 8, p. 154, 1983. Copyright 1983 by American Ceramic Society. Reprinted by permission.

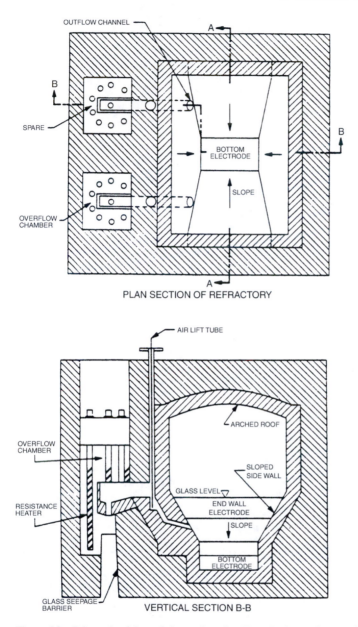

Figure 4.9 Schematic of slurry-fed ceramic melter. From Design preferences for a slurry-fed ceramic melter suitable for vitrifying West Valley wastes by C. C. Chapman, in *Proc. International Symposium on Ceramics in Nuclear Waste Management*, vol. 8, p. 149, 1983. Copyright 1983 by American Ceramic Society. Reprinted by permission.

Table 4.6 West Valley slurry-fed ceramic melter design preferences

Design feature	Basis
Sloped refractory sidewalls (30°+)	Reduces refractory corrosion Minimizes refractory loss due to cave-in Reduces glass inventory for fixed surface area
Arched refractory roof	Improves structural stability
Sloped floor	Provides space to accumulate metal and/or conductive sludge for evacuated canister removal without loss of melter Enhances two-zone Joule-heating control
Bottom electrode	Allows control over floor and bottom-glass temperature Accommodates potential problem of an electrically conductive sludge or molten metal
Direct electrode cooling	Provides operational flexibility Allows higher bulk glass temperature Minimizes electrode corrosion and prolongs electrode life
Glass airlifting	Simple, replaceable, ease of control of glass outflow
Drain outlet offset	Reduces temperature at throat and of channel materials; therefore extends life
Double overflow drains	Doubles operational life of unreplaceable and most-corrodable component
Overflow seepage flange	Ensures that glass leakage in overflow does not occur
Suction canister draining	Allows for removal of molten metals and/or sludge Eliminates problems of remote operations with bottom drain Eliminates danger of uncontrolled bottom draining Provides rapid method of completely draining melter cavity
All-refractory construction of vapor space to first quencher/scrubber refractory line off-gas piping	Avoids hot corrosion of metals by sulfur Allows use of hot gas startup/restart
Gas/plasma torch startup	Simple, controllable Needs only one connecting nozzle, which can be removed and used for another function during operation
Computer-controlled melter	With verified software will optimize and stabilize melter control Will prompt operator about potential problems in advance of difficulties

From Design preferences for a slurry-fed ceramic melter suitable for vitrifying West Valley wastes by C. C. Chapman, in *Proc. International Symposium on Ceramics in Nuclear Waste Management*, vol. 8, p. 159, 1983. Copyright 1983 by American Ceramic Society. Reprinted by permission.

(1112–1652°F), where the solutions are evaporated and the nitrates decomposed to oxides. The solid calcinate formed is mixed with borosilicate glass frit. This mixture is received, melted, and refined in an induction-heated melting pot raised to a temperature of 1100°C (2012°F). The glass formed is poured every 8 hr, effected by the melting of a cold glass plug in the drain tube, into a refractory steel container, which holds about 360 kg of glass (Figure 4.10). The method is a closely coupled calcination and vitrification technique using a continuously fed rotary calciner and thus can be considered a two-step

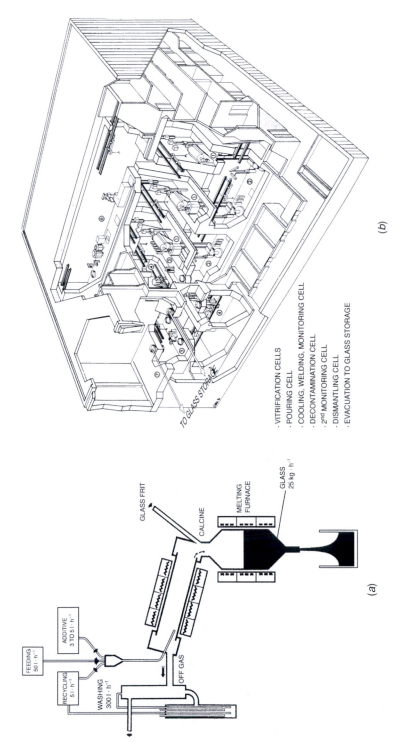

- VITRIFICATION CELLS
- POURING CELL
- COOLING, WELDING, MONITORING CELL
- DECONTAMINATION CELL
- 2nd MONITORING CELL
- DISMANTLING CELL
- EVACUATION TO GLASS STORAGE

(b)

Figure 4.10 The French continuous vitrification process. (a) AVM continuous process diagram. (b) Illustration of La Hague vitrification facility. From Present status of the French continuous fission product vitrification process by A. Jouan, C. Ladirat, and J. P. Moncouyoux, in *Nuclear Waste Management II, Advances in Ceramics*, vol. 20, p. 106. American Ceramic Society, Westerville, Ohio, 1986. Copyright 1986 by American Ceramic Society. Reprinted by permission.

process. Designed to process all kinds of solutions, AVM is flexible from an operational point of view and is relatively inexpensive from the standpoint of investment and operating costs.* Its equipment is readily operated remotely and is easy to decommission. With good knowledge of the process control, AVM can be shut down and started up with ease.[26,27] The AVH melter is a modification of the AVM melter; this design (Figure 4.11) ensures a homogeneous product by having vertical convection flux streams stir the glass during the filling of the melter.

The rising-level glass process (U.K.) This process, which underwent substantial development in the early 1960s,[28] is a batch vitrification as described in the section on the pot calcination process. The liquid waste and glass formers are fed into the canister, which serves as both the melting vessel and the disposal container, while the canister is heated to 1050°C by a cylindrical multizone furnace. The glass formers can be introduced as a slurry or as dry glass frit. The waste undergoes evaporation and calcination in a crust floating on top of the glass before reaching the melting temperature, at which point it fuses with the glass formers. The process is named for this phenomenon. The resulting off gas, which consists of steam, nitrogen oxides, and particles, exits through a cover on the canister into an off-gas cleanup system. The high particulate entrainment into the off gas is the major disadvantage. Because of the low process rate and waste glass/canister thermal stress problems, this process has not been considered, however, for production plants.

SYNROC process. The SYNROC strategy for the immobilization of HLW was developed at the Australian National University (ANU) and first described in 1978.[29,30] SYNROC is a synthetic rock that consists essentially of a small number of titanate mineral phases chosen because of their geochemical stability and collective ability to accept into their crystal structures nearly all the elements present in HLW. These characteristics yield a waste form with exceptionally high resistance to leaching by ground water, particularly at elevated temperatures.[31] A comparison of materials for HLW form will be given in Section 4.3.2. The initial publications on SYNROC stimulated widespread interest and led to rapid growth of research and development on this waste form, principally in Australia and the United States, but also on a smaller scale in Britain, Japan, and West Germany. The Australian Atomic Energy Commission (AAEC) and ANU are jointly working on a nonradioactive SYNROC fabrication demonstration plant, where ANU is developing new SYNROC formulations to meet specific objectives with the following variants[31]:

1. SYNROC-B is similar to SYNROC-C (below) but does not contain real or simulated HLW.
2. SYNROC-C is a fully dense polycrystalline titanate waste form consisting predominantly of hollandite ($BaAl_2Ti_6O_{16}$), perovskite ($CaTiO_3$), zirconolite ($CaZrTi_2O_7$), and rutile (TiO_2) and designed to immobilize up to 20 wt % HLW.

---

*The AVM facility has been operated since 1978 and has always met its assigned objectives. In January 1986 the facility had been in operation for >35,000 hr and vitrified >1000 m^3 of fission product solutions into approximately 450 MT of glass.[25]

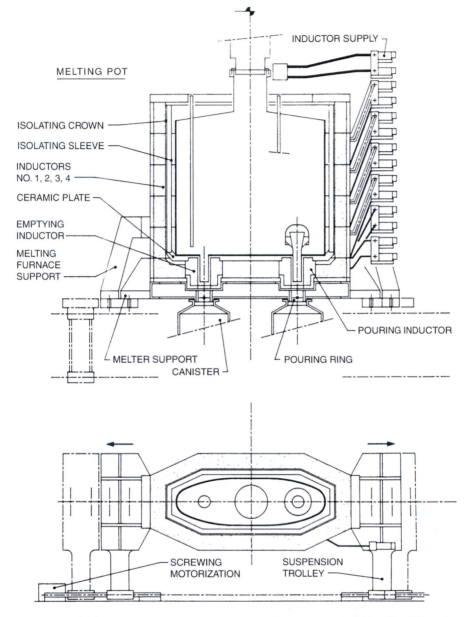

Figure 4.11 AVH-type melter. From Conceptual design for vitrification of HLW at West Valley using a rotary calciner/metallic melter by J. P. Giraud, J. P. Conord, and P. M. Savrot, in *Proc. 2nd International Symposium on Ceramics in Nuclear Waste Management, Advances in Ceramics*, vol. 8, p. 134, Chicago, 1983, American Ceramic Society, Copyright 1983 American Ceramic Society. Reprinted by permission.

3. SYNROC-D consists mainly of the mineral assemblage spinel plus zirconolite plus perovskite plus nepheline and is designed to immobilize up to 65 wt % of defense HLW.
4. SYNROC-E is a similar mineral assemblage to SYNROC-C but contains sufficient TiO_2 to form a continuous rutile matrix, thereby encapsulating hollandite, perovskite, and zirconolite. This waste form is designed to immobilize 5–10 wt % HLW and displays exceptional stability and leach resistance over a wide range of pH and temperature conditions.
5. SYNROC-F is a new SYNROC formulation. It consists of the mineral assemblage pyrochlore $[Ca(U,Zr)Ti_2O_7]$ plus minor hollandite and rutile and is designed to immobilize about 50 wt % of dissolved but otherwise unreprocessed spent commercial fuel.

The process as developed in Australia is shown in Figure 4.12.[31] A chemically reactive and homogeneous mixture of Ti, Zr, Ba, Al, and Ca oxides, intimately mixed on a submicrometer scale, is prepared outside the hot cell and then slurried with HLW. The SYNROC precursor thus formed is subsequently dried and calcined in preparation for reactive hot pressing. After calcination and baking, about 2–3 wt % fine Ti metal powder is intimately mixed with the calcined precursor to provide redox control during the subsequent hot-pressing operation.* The waste material is sealed in a stainless steel, bellows-shaped can. Densities that are 99% of theoretical values can easily be achieved in the uniaxial hot pressing as shown in Figure 4.12. The advantage of such material is its excellent aqueous leach resistance compared with borosilicate glasses. Under repository conditions, the leachability of SYNROC will be relatively insensitive to changes in groundwater composition, pH, or flow rates.[32] The major observable difference in performance between the glass and ceramic forms is for the uranium. In short-term tests, the glass forms release uranium to solution 50–1000 times faster than the ceramic forms,[33] which can be attributed to chemical binding of uranium by the ceramic's crystalline host phase (discussed in Section 4.3.2).

Treatment and conditioning of cladding waste. In the defense reprocessing plants at Hanford, Idaho Falls, and Savannah River the fuel cladding is normally dissolved along with the spent fuel and hence forms part of the HLW rather than a separate waste stream. Shear-leaching processing was used at West Valley and has been a feature of all commercial reprocessing plant designs. The latter process leaves undissolved zircaloy or stainless steel cladding hulls as a separate waste stream.

Treatment and conditioning of gaseous reprocessing wastes. The EPA regulations for the nuclear fuel cycles in the United States call for extensive treatment and recovery of gaseous reprocessing wastes in sufficiently large plants. Because of its complexity, the effluent control system is typically the largest system in the solidification facility. A decontamination factor of 10^9–10^{10} for effluents must be provided before release to

---

*The uniaxial hot pressing is one of the methods of pressing and densification in fabricating SYNROC-C and is the common means of producing small-volume samples at AAEC and ANU in Australia and Livermore and Sandia in the United States.

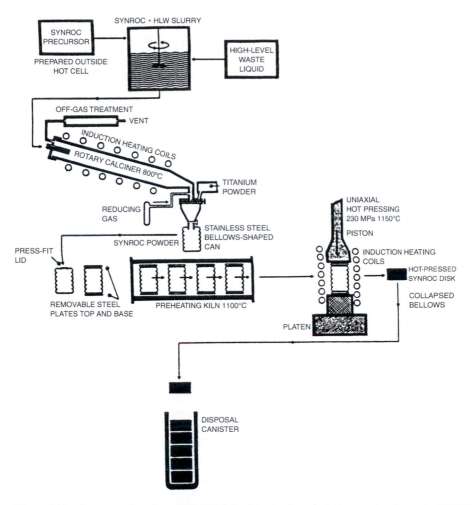

Figure 4.12 Conceptual flow sheet of SYNROC-C fabrication by uniaxial hot pressing. From The SYNROC process for immobilizing HLW by K. D. Reeve and A. E. Ringwood, in *Proc. International Conference on Radwaste Management, Seattle, May 1983*, IAEA-CN-43/127, vol. 2, p. 307, IAEA, Vienna, 1984. Copyright 1984 by International Atomic Energy Agency. Reprinted by permission.

the environment. Therefore, the off-gas system is made up of a series of different process units including condensers, scrubbers, filters, and concentrators. To date, U.S. operational facilities have used only iodine scrubbers and particulate filters in all plants and used [85]Kr recovery in the Idaho Chemical Processing Plant.

4.3.2 Material for Immobilization (Fixation) of HLW

Since it is mandatory to achieve maximum safety during the storage of HLW, the waste material must be rendered both immobile and insoluble. Thus the desirable properties of the waste form material are as follows:

1. Good capacity to accept all the elements in the waste.
2. Composition range flexible enough to accommodate variations in the waste.
3. Low melting point to facilitate production.
4. High thermal conductivity to dissipate the heat produced by radioactive decay.
5. Good resistance to leaching by waste.
6. Good mechanical integrity at elevated temperatures.
7. Good resistance to radiation damage.

Selection of Immobilization Form Materials. As early as in the 1950s the borosilicate-, phosphate-, and nepheline-syenite-based glasses and a variety of polyphase ceramic, bituminous, and concrete materials were investigated as materials for immobilization forms.[34] Based on extensive evaluations of alternative waste forms,[35-44] borosilicate glass and titanate-based polyphase ceramic (SYNROC) were selected in 1982 as the reference and alternative forms, respectively, for continued development and evaluation in the U.S. HLW program. A specific ceramic form, SYNROC-D (Section 4.3.1.3), was designated as the alternative form for SRP defense wastes. The evaluation began with 17 potential media, of which 10 were dropped from further development in 1980 because of technical concerns.[35] After a year of continued development and characterization, the 7 remaining forms were given four different assessments by a DOE evaluation team: evaluation at a DOE defense waste site,[34] a peer review evaluation,[36-38] a product performance evaluation,[39-41] and a processability analysis.[42-45] The evaluation process, performance indices, weighting factors, and processability analysis are described by Hench et al.[45] and the ranking of the seven waste form materials in the four screening processes is shown in Table 4.7.

Table 4.7 Ranking of candidate waste form materials for geologic disposal of HLW[a]

Waste form	First input, preproduction and process evaluation, DOE sites	Second input, alternate waste peer review panel	Third input, performance evaluation by SRL	Fourth input, processability evaluation by Du Pont
Borosilicate glass	1	1	2	1
SYNROC	2	2	1	3
Tailored ceramic	2	4	1	3
High-silica glass	2	3	2	2
Special concrete		6	3	1
Coated sol-gel		5	1	4
Glass marbles		7	3	2

[a]Ranks in numerical order, 1 being the best and 7 the worst.

From HLW immobilization forms by L. L. Hench, D. E. Clark, and J. Campbell, *Nucl. Chem. Waste Manage.*, vol. 5, p. 150, 1984. Copyright 1984 by Pergamon Press Ltd. Reprinted by permission.

The bases for selecting borosilicate glass as the HLW form material are as follows:

1. Borosilicate glass demonstrated acceptable product performance properties.
2. Borosilicate glass was ranked as the preferred form by the Alternative Waste Form Peer Review Panel.
3. Borosilicate glass was consistently selected as the preferred form by the DOE defense sites and rated the highest in the commercial waste form evaluations.
4. The process for fabricating the borosilicate glass waste form is the simplest and least expensive of all those considered.

The bases for selecting the crystalline ceramic forms as alternative materials are as follows:

1. The crystalline ceramic forms, SYNROC and tailored ceramic, ranked highest in the product performance evaluation. They also have ability to incorporate large volumes of waste (high waste loading).
2. The SYNROC form, ranked second by the Alternative Waste Form Peer Review Panel, was judged to be the best-characterized and understood of the forms other than borosilicate glass.
3. Ceramic waste forms consistently ranked high in DOE's defense site evaluations.
4. The ceramics have generally better high-temperature leaching characteristics than borosilicate glass; in particular, titanates and phosphates demonstrated long-term resistance to leaching and radiation damage.[32]
5. To a large extent the crystalline ceramic forms have the properties of their mineral analogs, which have proved to be extremely durable in nature and have excellent thermal and mechanical stability. Thus the long-term leach behavior can be predicted from both laboratory studies and observation of the geologic record.[31,33,46]

With the selection of glass as the final waste form for defense and commercial wastes in the United States, no formal development program on the ceramic alternative exists except for the processing work at Rockwell International Science Centre[33,47] in this country. Development work is continuing in Australia and Britain. The French CEA group in 1981 also selected borosilicate glass for the solidification of fission product solution at the La Hague reprocessing facility, based in part on the successful operation of the PIVER pilot installation since 1973 and the AVM prototype plant in Marcoule since 1978.

Nuclear waste glass leach behavior. The primary issue of concern regarding glass and other waste forms is their long-term stability in contact with hot repository ground water in the event of a breached canister. Thus, relative leach performance has been given the highest weighting factor in evaluation studies. Factors that are important in controlling leaching within the storage system are glass composition, type of surface film formed, waste percentage and type, leachant composition, temperature, ratio of surface area of glass to volume of leachant, flow rate and leachant resident time, and canister overpack and backfill interactions. Reasonably well-developed theories and models describe the leach behavior of nuclear waste glasses in terms of these variables. Research on the basic mechanisms and controlling variables is summarized by Hench et al.[45] Figures 4.13 and 4.14 are taken from this reference. They show the effects of flow on waste glass leaching and the silicon release rate as a function of glass hydration energy, respectively. Most research on glass and HLW form corrosion has used static testing, such as that specified in MCC-1.[48] It is possible that under certain conditions ground water will flow through a geologic repository and react with its contents. Flow rates of up to several hundred liters per year have been reported, although a few liters per year is most probable. To evaluate the effects of flow on leaching, MCC-4 was developed by the Materials Characterization Center, using a single pass of the solution through the leaching vessel. As shown in Figure 4.13, the rate of leaching increased as the flow rate was increased from 0.1 to 10 ml/hr, and little difference was observed between the static test and the test with 0.1 ml/hr. At high rates, corrosion products as well as potential surface-passivating species are removed from the leaching vessel, thus reducing the beneficial effects on both solution saturation and protective surface film formation. These results

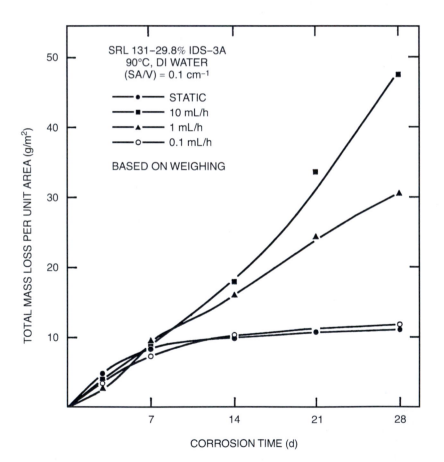

Figure 4.13 Effects of exposure time and flow rate on glass corrosion. From HLW immobilization forms by L. L. Hench, D. E. Clard, and J. Campbell, *Nucl. Chem. Waste Manage.*, 1984, vol. 5, p. 149. Copyright 1984 by Pergamon Press Ltd. Reprinted by permission.

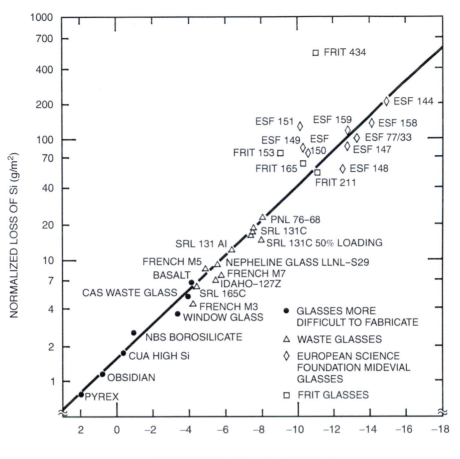

Figure 4.14 Release of structural silicon as a function of glass hydration energy. From HLW immobilization forms by L. L. Hench, D. E. Clard, and J. Campbell, *Nucl. Chem. Waste Manage.*, 1984, vol. 5, p. 150. Copyright 1984 by Pergamon Press Ltd. Reprinted by permission.

suggest that under the low-flow conditions expected in the repository, saturation will prevail and the leach rate of the glass will be limited by the rate of transport of corrosion products from the repository.[45] Similar conclusions were reached by Macedo et al.[49] using a quasi-low-flow test, in which the flow rates were varied in a recirculatory mode, with the same solution passed through the leaching vessel multiple times.[50]

The long-term leaching rate, which cannot be measured directly, is important for analyses of the safety of isolating HLW in a repository because of the potential for leaching long-lived radionuclides through ground water. This rate must be predicted from mathematical models.[51] A number of such studies have been reported.[51–55] In addition to mechanistic considerations and kinetic equations, thermodynamic aspects of the chemical composition of the waste form have been used. Newton and Paul[56] demonstrated a relationship between the free energy of hydration of glass and its durability, considering the glass to be a physical mixture of orthosilicates (Na_2O-SiO_2, $MgO-SiO_2$, etc.) and uncombined oxides. The free energy of hydration of the glass can thus be determined by multiplying the mole fraction of each silicate it contains by the respective free energy and summing. The more negative the value, the less durable the glass. Plodinec et al.[57] applied this concept to a number of natural and synthetic glasses, including simulated radwaste glasses (Figure 4.14). These approaches should eventually permit the development of a unified theory of glass corrosion based on thermodynamics, mechanisms, kinetics, and surface film formation.

In addition to the waste form, canister, and overpack, the backfill and/or buffer materials are often used as part of the engineered barrier to isolate long-lived radionuclides from the biosphere. Bentonite is one of the most promising candidates for the buffer material. Chemical analysis of this type of clay material was also undertaken, as corrosion of the canister as well as radionuclide migration subsequent to breaching of the canister would be largely influenced by chemical species, especially anions, contained in the buffer material.[58] It was shown that migration of the anions is considerably retarded in compacted bentonite.

4.3.3 Interim Storage

Solidified HLW may be stored on site immediately after solidification and placed in storage before transfer to a disposal site. Two generic interim storage options for waste canisters are used: air-cooled and water-cooled storage.[59] An example of the former is shown in Figure 4.15, which is the reference storage concept of SRP. The water-cooled option would be very similar to those used for spent fuel (wet) storage. The canister centerline and wall temperatures during both types of storage as a

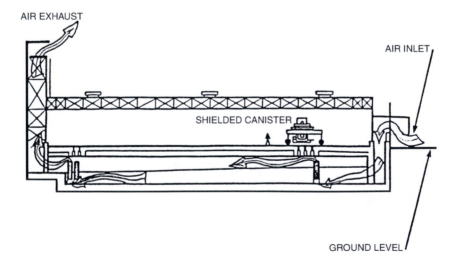

Figure 4.15 Reference (air-cooled) storage concept for Savannah River Plant. From the Storage and Disposal of Radwaste as Glass in Canisters, DOE Report PNL-2764, 1978, Pacific Northwest Laboratory. Reprinted by permission.

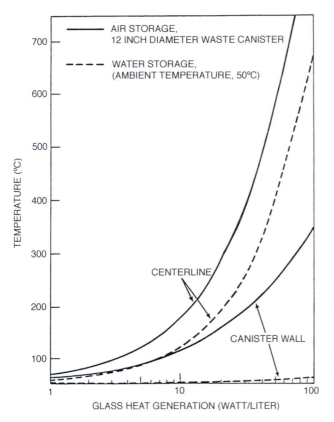

AIR STORAGE,
12 INCH DIAMETER WASTE CANISTER

WATER STORAGE,
(AMBIENT TEMPERATURE, 50°C)

CENTERLINE

CANISTER WALL

TEMPERATURE (°C)

GLASS HEAT GENERATION (WATT/LITER)

Figure 4.16 Canister centerline and wall temperatures during storage. From High-level radioactive waste by J. L. McElroy and M. S. Hanson, in *Radioactive Waste Technology*, edited by A. A. Moghissi et al., ASME, 1986, p. 649. Copyright 1986 by American Society of Mechanical Engineers. Reprinted by permission.

function of specific decay heat load in the glass are shown in Figure 4.16 at a canister diameter of 12 in.[60] The centerline temperature should not exceed 500°C for any length of time, to avoid crystalline growth in the glass.

4.4 PACKAGING OF HIGH-LEVEL WASTE

4.4.1 Performance Objectives of HLW Packages

The engineered waste package, which is defined as including the waste form and all components that enclose the waste form out to the near-field host rock, has the primary objective of limiting the release of radionuclides to the accessible environment. The performance has been categorized into two time periods: (1) the thermal period, which is currently defined as up to 1000 years after permanent closure of the repository, and (2) the postthermal period.

During the thermal period emphasis is placed on radionuclide containment; after that the radionuclide release rate to the geologic medium is controlled to an acceptably low value. In addition to providing containment, the engineered waste package must be retrievable during the early years after initial repository operation in the event of unforeseen circumstances. As mentioned in Section 3.5 regarding spent fuel packaging, although the package designs for different geologic media being considered for repositories are similar, the package must have suitable material and dimensions to meet such conditions as lithostatic and hydrostatic pressures, corrosion environment, and groundwater travel time in the expected geologic environments. The design of engineered waste packages should comply with federal regulations such as 10CFR20, 10CFR50, 10CFR60, 10CFR71, 30CFR57, 40CFR191, and 49CFR173.

4.4.2 Waste Package Environment

The environment can be categorized according to groundwater composition and geotechnical parameters. Table 4.8 gives the groundwater compositions near three different geologic media and Table 4.9 shows their respective geotechnical parameters.

4.4.3 Package Designs for Various Repository Media

The package designs for each repository medium are described briefly for salt, basalt, and tuff.

Salt repository packages. For the purposes of package description, three waste forms of HLW are considered: defense high-level waste (DHLW), commercial high-level waste (CHLW), and consolidated spent fuel (CSF). The last one was described in Section 3.5. Two basic design approaches are presented for each waste form (except for packages for the basalt repository): a *reference package* for borehole emplacement and an *alternate package* (self-shielded) for tunnel emplacement.

Salt reference package The borehole package is based on the principle of placing the containers of waste assemblies in vertical boreholes in the floor of a mined repository tunnel. The crushed salt backfill and the surrounding geology provide radiation shielding

Table 4.8 Package environment

	Groundwater composition (mg/L)			
	Salt waste package			
Species (ion)	Inclusion brine	Dissolution brine	Basalt waste package	Tuff waste package (Jackass Flats well)
Na^+	42,000	115,000	363	51
K^+	30,000	15	3.4	4.9
Mg^{2+}	35,000	10	0.03	2.1
Ca^{2+}	600	900	2.8	14
Cl^-	190,000	175,000	310	7.5
SO_4^{2-}	3,500	3,500	173	22
F^-			33.4	2.2
HCO_3^-	700	10	110	61

Table 4.9 High-level waste package environment: summary of geotechnical parameters

Parameter	Salt	Basalt	Tuff[a] Below water table	Tuff[a] Above water table
Depth of repository (m)	Variable	1128	884	300
Design basis earthquake (g)	0.30	0.23	TBD	TBD
Lithostatic pressure at repository level (MPa)	15.4	29.0	TBD	TBD
Hydrostatic pressure at repository level (MPa)	7.0	11.3	TBD	TBD
Maximum design pressure (MPa)	15.4	11.3	3.0	0.1
Density of basalt, salt, or tuff (g/cm^3)	2.19	2.77	2.18	2.12
Porosity of basalt, salt, or tuff (%)	—	0.1	26	17
Regional hydraulic gradient	—	10^{-3}	—	—
Hydraulic conductivity of medium (m/sec)	—	3.2×10^{-10}	1.2×10^{-4}	TBD
Water constituents	b	b	b	b
Thermal conductivity of medium at 20–300°C (W/m K)	3.99	2.16	1.85	1.8
Thermal expansion of medium at 20–300°C (K^{-1})	—	6.33×10^{-6}	12.0×10^{-6}	10.7×10^{-6}
Specific heat capacity of medium at 20–300°C (J/g K)	0.91	0.93	1.75	1.0
Ambient rock temperature at depth (°C)	—	59	—	—

[a] TBD, To be determined.
[b] See Table 4.8.

while other emplacement and repository operations in the area are in progress. All operations involving waste packages, before and during emplacement, require shielded or remotely controlled equipment. Figure 4.17 shows the reference package design for borehole emplacement. The designs for the other waste forms differ from this only in dimensions, and the different dimensions for various waste forms are also given in the figure.[61] As a primary containment barrier, the overpack is designed for at least a 1000-year corrosion lifetime, which requires a relatively thin-walled titanium alloy (Ti-code-2) container. However, the actual thickness is increased from this requirement of 0.05 cm to approximately 0.25 cm (0.1 in.) to ensure stability during manufacture and handling. To minimize the material costs, the container is reinforced internally with a thick carbon steel overpack reinforcement that contains the waste form. After the container is placed in the borehole, the space between the top of the package and the floor of the tunnel is filled with crushed salt to provide shielding.

Salt alternate package In the alternate (self-shielded) package approach, each package has its own shielding and there is no need for an elaborate shielded packaging facility. This design also eliminates the need to have individual emplacement boreholes, as the packages can simply be placed horizontally on the floor of a tunnel (Figure 4.18). Again, the designs

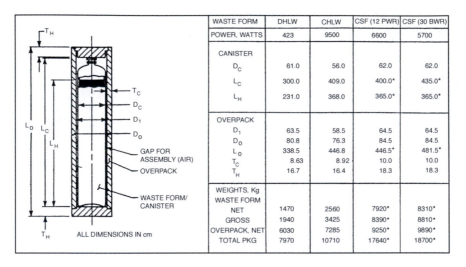

WASTE FORM	DHLW	CHLW	CSF (12 PWR)	CSF (30 BWR)
POWER, WATTS	423	9500	6600	5700
CANISTER				
D_C	61.0	56.0	62.0	62.0
L_C	300.0	409.0	400.0*	435.0*
L_H	231.0	368.0	365.0*	365.0*
OVERPACK				
D_1	63.5	58.5	64.5	64.5
D_0	80.8	76.3	84.5	84.5
L_0	338.5	446.8	446.5*	481.5*
T_C	8.63	8.92	10.0	10.0
T_H	16.7	16.4	18.3	18.3
WEIGHTS, Kg				
WASTE FORM				
NET	1470	2560	7920*	8310*
GROSS	1940	3425	8390*	8810*
OVERPACK, NET	6030	7285	9250*	9890*
TOTAL PKG	7970	10710	17640*	18700*

Figure 4.17 Salt reference HLW package and package dimensions for various waste forms. From Waste Package Reference Conceptual Designs for a Repository in Salt by Westinghouse Waste Technology Services Division, WTSD-TME-001, Rev. A, Waste Technology Services Division, Westinghouse Electric Corporation, Madison, Pa., 1984. Copyright 1984 by Westinghouse Electric Corporation. Reprinted by permission.

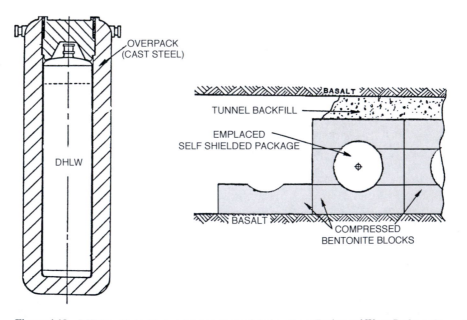

Figure 4.18 DHLW self-shielded package conceptual design. From Engineered Waste Package Conceptual Design, DHLW (Form 1), CHLW (Form 1), and SNF (Form 2), Disposal in Salt by W-AESD, AESD-TME-131, Advanced Energy Systems Division, Westinghouse Electric Corporation, Large, Pa., 1982. Copyright 1982 by Westinghouse Electric Corporation. Reprinted by permission.

for the other waste forms are similar except for dimensional changes to accommodate the different waste form geometries and radiation levels. Features of some self-shielded package (SSP) designs are given in Table 4.10. These designs have a primary containment barrier of a cast steel overpack of sufficient thickness to attenuate the nuclear radiation at the surface. After the waste form is placed in the casting, a top shield plug is welded into place to form a totally leakproof barrier. Cast steel is the reference material, but gray cast iron and ductile (nodular) cast iron may have considerable economic advantages. The primary

Table 4.10 Summary of self-shielded package designs in salt

Parameters	DHLW	CHLW
Design parameters		
Waste form		
Diameter (cm)	61	45.7
Length (m)	3	4.1
Weight of waste (kg)	1470	595
Waste package		
Cross-sectional geometry	Round	Round
Outside dimension (cm)	124.5	116.0
Length (m)	3.9	4.0
Emplacement weight (MT)	31.7	28.8
Heat load (W)	423	2210
Repository		
Package pitch (m)	1.25	2.48
Tunnel height (m)	2.5	2.5
Tunnel width (m)	5	4.3
Performance parameters		
Radiation		
Surface of overpack (mrem/hr)	70	100
Corrosion		
Depth at 100 yr (cm)	0.5	2.5
Overpack		
Required wall (cm)	9.5	6.4
Actual wall (cm)	30.5	40.5
Required head (cm)	18.0	12.4
Actual head (cm)	30.5	40.5
Peak temperature		
Waste form centerline (°C)	101	293
Overpack (°C)	85	205
Salt (°C)	85	205
Local areal load (W/m^2)	15.2	37.5
Package-related cost		
Cost/unit of waste (cast steel; 1981 $/kg)	48	134
Cost/unit of waste (gray cast iron; 1981 $/kg)	29	79

From Engineered Waste Package Conceptual Design by W-AESD, DHLW (Form 1), CHLW (Form 1), and SNF (Form 2), Disposal in Salt, AESD-TME-3131, Advanced Energy Systems Division, Westinghouse Electric Corporation, Large, Pa., 1982. Copyright 1982 by Westinghouse Electric Corporation. Reprinted by permission.

requirement—containment for 1000 years—is satisfied by the thick cast overpack, whose corrosion is limited to an insignificant depth by the very small amount (tens of liters) of brine expected to be available for each package. (Even with unlimited anoxic brine, the maximum corrosion depth is less than a few centimeters after 1000 years.) The reference borehole package (Figure 4.17) for a DHLW waste form of 61 cm (24 in.) diameter is 81 cm (32 in.) in diameter and 339 cm (133.5 in.) long and has a total weight of 8.0 MT and an overpack reinforcement thickness of 8.6 cm (3.4 in.). For the same waste form, the SSP is 125 cm (49.2 in.) in diameter and 390 cm (153.5 in.) long, weighs 32 MT, and has a reinforcement thickness of 30.5 cm (12 in.), resulting in a surface radiation level of 70 mrem/hr. Similarly, the reference borehole package for a CHLW waste form is 76.3 cm (30 in.) in diameter and 447 cm (176 in.) long, weighs 10.7 MT, and has an overpack reinforcement 8.9 cm (3.5 in.) thick. The corresponding SSP design is 116 cm (45.7 in.) in diameter and 400 cm (157.5 in.) long, weighs 28.8 MT, and has an overpack reinforcement 40.5 cm (15.9 in.) thick. To improve the system economics, consideration is given to increasing the amount of waste per package, that is, using a larger diameter waste form. A DHLW canister with an 81 cm (32 in.) diameter could contain nearly twice as much waste as the reference 61-cm (24-in.) canister. Although temperatures are slightly higher with the larger canister, all design requirements are satisfied with adequate margins.

Performance evaluation The borehole design configurations and material selections should satisfy all specified criteria. The performance is summarized in Table 4.11, and

Table 4.11 Salt reference waste package performance

Performance parameter	DHLW	CHLW	Spent fuel
Radiation level (mrem/hr)			
Package surface			
Neutron (reflected)	4.0×10^1	7.7×10^3	2.7×10^3
Gamma	3.3×10^5	1.6×10^6	1.8×10^5
Total	3.3×10^5	1.6×10^6	1.9×10^5
Tunnel floor, emplaced	1.9 (gamma)	1.6 (neutron)	1.8 (neutron)
Initial heat load (W)	423	9500	6600
Areal load (W/m^2)	20	15	14.83
Temperature (°C)			
Peak temperature			
Waste form (limits 500/375)	110	480	348
Overpack (limit above 250)	98	230	175
Salt (limit 250)	98	230	175
Waste form at 1000 yr (limits 100/none)	45	60	99
Corrosion penetration at 1000 yr (cm)			
Expected brine	0.04	0.14	0.07
Unlimited brine	0.95	1.92	2.30
Corrosion allowance (cm)	0.95	1.92	2.30

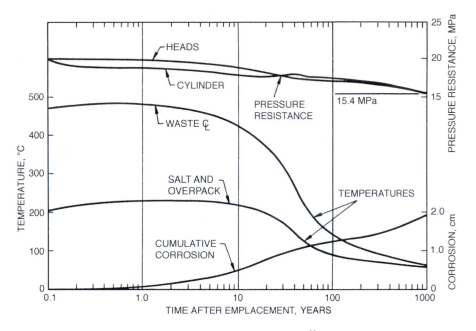

Figure 4.19 CHLW reference waste package performance in salt.[61] From Waste Package Reference Conceptual Designs for a Repository in Salt by Westinghouse Waste Technology Services Division, WTSD-TME-001, Rev. A, Waste Technology Services Division, Westinghouse Electric Corporation, Madison, Pa., 1984. Copyright 1984 by Westinghouse Electric Corporation. Reprinted by permission.

the typical temperature and pressure resistance of a CHLW reference package are shown in Figure 4.19.[61]

The SSP concepts have many advantages over the borehole design. The need for elaborate, shielded hot-cell facilities and shielded surface storage facilities can be reduced or eliminated. Package emplacement operations are greatly simplified and there is no need to drill boreholes. Concerns about the effects of radiolysis of the ground water as well as radiation effects on the host rock are eliminated. Retrieval of the waste packages would also be greatly simplified. From system-wide economic viewpoints, the SSP could be used as "system package" for interim storage, shipping, and disposal. It would make the total system safer by minimizing handling of unshielded material and simplifying recovery from accidents.

A major conclusion reached during the design of salt repository waste packages was that backfill is not necessary with engineered waste packages. The only practical use of a bentonite backfill is to provide additional radionuclide containment during the thermal period. However, in salt geology the corrosion barriers are expected to perform adequately even with an unlimited amount of anoxic brine. Furthermore, because of the lack of a significant quantity of brine and the low permeability of the salt, any radionuclides released from the waste form in the vicinity of the package would be retained.

Basalt repository package.[62] The waste packages for basalt disposal must meet the same containment design standards as the packages developed for salt repositories.

The conceptual designs developed for the Basalt Waste Isolation Project (BWIP) have specific features for each of the three waste forms (DHLW, CHLW, and CSF) to be emplaced in basalt. The package described in this section is designed to facilitate emplacement in a horizontal borehole and is based on a containment barrier made of carbon steel with a wall thickness sufficient to allow for the predicted corrosion depth and the structural capability to preclude buckling during the thermal period. A package backfill is provided to assist the waste form in meeting the long-term radionuclide release criterion of 1 part in 10^5 per year because the data on basalt ground water are insufficient to conclude that the waste form can meet this criterion by itself.

Conceptual design of basalt waste package The basic design to be used is shown in Figure 4.20, which is for all waste forms except for dimensional differences. Variations in dimensions for different waste forms are listed in Table 4.12 (features of the spent fuel package are included for comparison). The carbon steel overpack, which provides the 1000-year containment barrier, consists of a hollow cylindrical main body to which a cap is welded on each end. The top end cap has a lifting pintle for package handling. The wall and end-cap thicknesses are the sum of the thickness required to preclude buckling when subjected to the external design pressure of 13 MPa (1885 psi) plus the

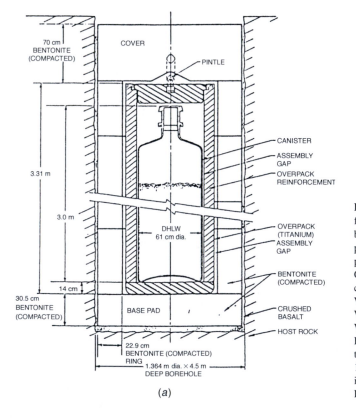

(a)

Figure 4.20 HLW packages for borehole emplacement in basalt. (a) Reference DHLW package. (b) Reference CHLW package. From Waste Package Concepts for Use in the Conceptual Design of the Nuclear Waste Repository in Basalt by W-AESD, AESD-TME-3142, Waste Technology Services Division, Westinghouse Electric Corporation, Madison, Pa., 1982. Copyright 1982 by Westinghouse Electric Corporation. Reprinted by permission.

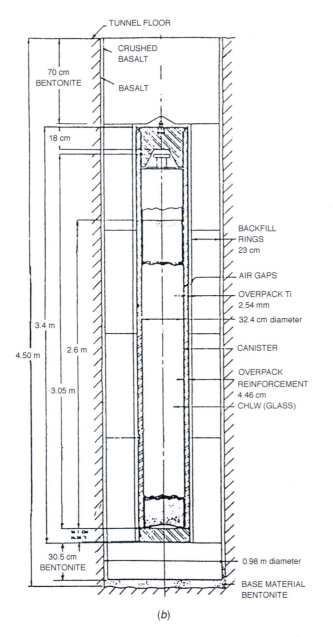

TUNNEL FLOOR

CRUSHED
BASALT

70 cm
BENTONITE

BASALT

18 cm

BACKFILL
RINGS
23 cm

AIR GAPS

OVERPACK Ti
2.54 mm

32.4 cm diameter

3.4 m

2.6 m

4.50 m

CANISTER

3.05 m

OVERPACK
REINFORCEMENT
4.46 cm

CHLW (GLASS)

30.5 cm
BENTONITE

0.98 m diameter

BASE MATERIAL
BENTONITE

(*b*)

Figure 4.20 (continued)

predicted 1000-year corrosion allowance. The backfill is incorporated in the overpack to assist the waste form in meeting the long-term radionuclide release criterion during the postthermal period. The backfill consists of a mixture of bentonite clay and crushed basalt (25% and 75% by weight) compacted into cylindrical rings and disks having a final dry density of 2.1 g/cm^3 (131 lb/ft^3) and a thickness prescribed by BWIP of 15.2 cm (6 in.). The shield plug, which is installed after insertion of the rest of the

Table 4.12 Summary of waste package design features in basalt

Conceptual design feature[a]	DHLW	SF2[b]	CHLW
Waste form diameter [cm (in.)]	61.0 (24)	32.5 (12.8)	32.5 (12.8)
Waste form length [cm (in.)]	300 (118)	417 (164)	300 (118)
Waste form weight [kg (lb)]	1932 (4260)	2210 (4870)	840 (1850)
Waste content per package [kg (lb)]	1470 (3260)	1380 (3040)[c]	595 (1310)[d]
Backfill thickness [cm (in.)]	15.2 (6)	15.2 (6)	15.2 (6)
Backfill outside diameter [cm (in.)]	94.0 (37)	66.0 (26)	66.0 (26)
Backfill weight, kg (lb)	2550 (5620)	2065 (4550)	1580 (3480)
Overpack inside diameter [cm (in.)]	96.5 (38)	68.6 (27)	68.6 (27)
Overpack wall thickness [cm (in.)]	13.0 (5.1)	9.9 (3.9)	9.7 (3.8)
Overpack outside diameter [cm (in.)]	122.5 (48.2)	88.4 (34.8)	88 (34.6)
Overpack head thickness [cm (in.)]	28.2 (11.1)	20.8 (8.2)	20.6 (8.1)
Overpack length [cm (in.)]	394 (155)	496 (195)	379 (149)
Overpack empty weight [MT (t)]	17.0 (18.8)	10.7 (11.8)	8.2 (9.0)
Overpack loaded weight [MT (t)]	21.5 (23.7)	15.0 (16.5)	10.7 (11.7)
Package heat load (W)	423	1650	2210
Borehole pitch [m (ft)]	2.87 (9.4)	5.33 (17.5)	10.4 (34)
Tunnel pitch [m (ft)]	30 (98)	55 (180)	55 (180)

[a]All components are provided with a 1.27-cm (0.5-in.) radial clearance gap and a 2.54-cm (1.0-in.) longitudinal clearance gap.

[b]Dimensions shown for PWR rods. Length, weight, and waste loading will be slightly different for BWR rods.

[c]Kilograms U (pounds U).

[d]This amount of CHLW results from the reprocessing of 2280 kg U (5022 lb U).

From Waste Package Concepts for Use in the Conceptual Design of the Nuclear Waste Repository in Basalt by W-AESD, AESD-TME-3142, Table 8-1, Summary of Waste Package Design Features, Westinghouse Electric Corporation, July 1982. Copyright 1982 by Westinghouse Electric Corporation. Reprinted by permission.

package into the borehole, consists of a hollow steel shell filled with crushed basalt. It attenuates the radiation from the waste at a level in the mined tunnel of less than 1 mrem/hr.

Performance evaluation Performance parameters for the waste package for each of the three waste forms are given in Table 4.13 (the spent fuel package performance is included for comparison). The typical temperature and pressure resistance of the CHLW reference waste package are shown in Figure 4.21.[62]

Tuff repository packages.[63] The conceptual waste packages described in this section are those designed for the Nevada Nuclear Waste Storage Investigations (NNWSI) Project. Four waste forms have been specified for disposal in tuff: DHLW-Form 1, DHLW-Form 2, Spent Fuel-Form 2 (SF2), and CHLW. The DHLW-1 is borosilicate glass that is poured into a stainless steel container while molten. The container, 61 cm in diameter by 3 m long, is filled to only 85% of its capacity to preclude overfilling, and the maximum thermal output is 423 W per canister. The stainless steel canister is

Table 4.13 Summary of waste package concept performance in basalt

Performance parameter	DHLW	SF2 (PWR)	CHLW
Radiation[a] (mrem/hr)	<1.0	<1.0	<1.0
Required structural thickness [cm (in.)]			
Wall	12.4 (4.9)	8.9 (3.5)	8.9 (3.5)
Heads	27.7 (10.9)	19.8 (7.8)	19.8 (7.8)
Corrosion depth at 1000 yr [cm (in.)]	0.2 (0.1)	0.6 (0.2)	0.3 (0.1)
Overpack wall thickness [cm (in.)]			
Required	12.6 (5.0)	9.5 (3.7)	9.2 (3.6)
Actual	13.0 (5.1)	9.9 (3.9)	9.7 (3.8)
Overpack head thickness [cm (in.)]			
Required	28.0 (11.0)	20.4 (8.0)	20.1 (7.9)
Actual	28.2 (11.1)	20.8 (8.2)	20.6 (8.1)
Peak temperature[b] [°C (°F)]			
Initial (dry) conditions			
Waste form	215 (419)	342 (648)	404 (759)
Backfill	210 (410)	238 (460)	327 (620)
Overpack	204 (400)	193 (379)	220 (428)
Basalt	203 (397)	185 (365)	199 (390)
Waste package cost per unit of waste[c] ($/kg U)	30	23	12[d]

[a]Radiation dose rate at the surface of the tunnel shield plug.

[b]Design limits are 500, 375, 300, 430, and 300°C for the glass waste forms, spent fuel, backfill, overpack, and basalt, respectively.

[c]Shield plug cost not included. The cost of the shield plug adds approximately $1.00 per unit of waste, assuming one package per emplacement hole.

[d]Unit cost per kilogram of CHLW glass is $44. This is because 1 kg of CHLW results from reprocessing 3.83 kg U.

From Waste Package Concepts for Use in the Conceptual Design of the Nuclear Waste Repository in Basalt, W-AESD, AESD-TME-3142, Table 8-2, Summary of Waste Package Concept Performance, Westinghouse Electric Corporation, July 1982. Copyright 1982 by Westinghouse Electric Corporation. Reprinted by permission.

suitable for DHLW-1 handling and short-term surface storage; it is considered inadequate as a containment barrier for long-term disposal based on current regulations because of the high glass-filling temperatures and difference in coefficients of expansion. The DHLW-2 is a ceramic (SYNROC-D) waste form with a canister identical in shape to that of DHLW-1. Two hot, isostatically pressed ceramic waste forms are stacked in the stainless steel canister before sealing. Thus the canister does not become sensitized, except possibly around the weld area, and the filling procedure avoids problems related to hoop stress. However, the void space in this canister precludes its use as a containment barrier under expected hydrostatic conditions. Thus, for disposal below the water table in tuff, this waste form must be overpacked. The SF2 waste package was discussed in Section 3.5. The CHLW form is borosilicate glass that is poured into stainless steel canisters while molten. As with DHLW-1, these canisters are filled to about 85% of capacity to preclude overfilling. The canister material was chosen to facilitate handling and interim storage, which for CHLW may include a period of water cooling.

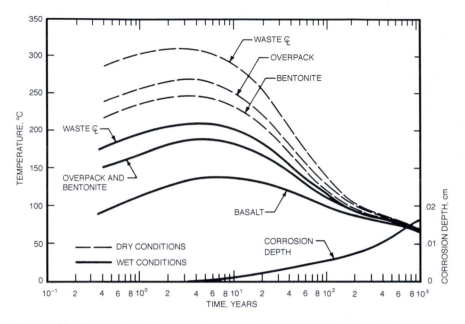

Figure 4.21 Thermal response of CHLW reference waste package in a basalt repository. Initial conditions: 868 W/package at 5.7 W/m². From Waste Package Concepts for Use in the Conceptual Design of the Nuclear Waste Repository in Basalt by W-AESD, AESD-TME-3142, Waste Technology Services Division, Westinghouse Electric Corporation, Madison, Pa., 1982. Copyright 1982 by Westinghouse Electric Corporation. Reprinted by permission.

Like DHLW-1, the CHLW canister is unsuitable for repository containment and must be overpacked.

Tuff waste package design Like the waste package designs for salt, designs for tuff have centered on packages for emplacement in boreholes and on a self-shielded package for tunnel emplacement. The waste forms are assumed to be sufficiently resistant to leaching to limit the release of radionuclides to a rate that meets the eventual NRC criterion (currently defined as 1 part in 10^5 per year) under expected conditions.

The borehole package has a relatively thin, corrosion-resistant titanium alloy overpack, backed by a carbon steel reinforcement (structural member), to provide 1000-year containment and allow for package retrieval. The surrounding geology provides radiation shielding while other emplacement and repository operations are in progress in the area. This design is based on the generic borehole package illustrated in Figure 4.17 and is configured of materials that should perform adequately in a repository below the water table in tuff. The sealed Ti-Code-2 was chosen as overpack material because data indicate that it will be corrosion resistant in the expected tuff ground water. The thin-walled (0.25-cm) overpack is wrapped tightly around the overpack reinforcement to minimize gaps between the overpack and its reinforcement, with welded ends of shaped Ti-Code-2 plates. The carbon steel overpack reinforcement consists of a hollow, cylindrical main body with welded end caps; their thicknesses are designed to preclude buckling when subjected to an external pressure of 3 MPa. Crushed tuff is backfilled remotely after the package is emplaced in the borehole to attenuate radiation levels in the tunnel to less than 2 mrem/hr. Information from Lawrence Livermore National Laboratory indicates that the repository may be above the water table, in which case the design for disposal below the water table would be overly conservative and would probably undergo some changes. The dimensions and weights of this design for three tuff packages are given in Table 4.14, which includes a summary of performance parameters for these packages and costs estimated on the basis of cost per unit weight of waste.[63] The costs listed are those directly associated with the package and related activities, that is, the purchase and shipping costs for fabricated overpack components, prorated repository costs for receiving the wastes, costs for assembling the packages, and mining and emplacement costs. These costs are suitable for comparisons of designs and design approaches.

The self-shielded package has a thick (typically 30–47 cm, or 11.8–18.5 in.) overpack of moderately corrosion-resistant ferrous material that provides containment and its own shielding, thereby eliminating the need for an elaborate shielded packaging facility. The potential benefits of SSPs were discussed earlier in this section. Table 4.15 presents the design parameters of SSPs for DHLW and CHLW, using cast steel or cast gray iron. The flexibility afforded by the casting process allows the use of package dimensions that can optimize waste loadings. Also included in Table 4.15 are performance parameters and costs estimated per unit weight of waste. To be more cost effective the SSP for CHLW in tuff is designed to contain multiple waste canisters; the resulting triangular cross section minimizes the amount of cast metal required for constant-thickness walls around the waste forms.

Bentonite as a backfill material. Unlike the packages for salt repositories, backfill is used in basalt or tuff. Batch sorption studies using sodium and calcium bentonite in reference basalt and tuff ground waters indicate strong retardation of strontium, cesium, and americium and moderate retardation of uranium and neptunium. Similar measurements with saturated brines indicate that plutonium and americium are well retarded and europium, cesium, and strontium are moderately retarded. Such data suggest that a properly chosen backfill (0.3 m thick) could delay breakthrough of plutonium and americium for perhaps 10,000 years.[64] Corrosion data for oxidizing reference basalt and tuff ground water indicate that, using a linear assumption, a corrosion allowance of 0.5 in. of cast iron or steel for 1000 years is satisfactory. When bentonite comes in contact with an aqueous solution it swells and, if confined, reduces the flow. With all three kinds of ground water, diffusion will be the controlling mechanism.

4.4.4 Handling Packages or Casks

Several handling operations have to be performed on packages or casks at both the shipping and receiving locations[65]:

1. Inspecting the package or cask.
2. Radiation surveying.

Table 4.14 Summary of HLW borehole designs in tuff

	DHLW 1 reference	DHLW 2 reference	CHLW reference
Design parameters			
Waste form			
Diameter (cm)	61	61	32.4
Length (m)	3	3	3.05
Weight of waste (kg)	1470	2400	595
Total waste form weight (kg)	1940	3650	845
Waste package			
Outside diameter (cm)	71	71	39.5
Length (m)	3.3	3.3	3.38
Empty weight (kg)	2640	2910	925
Loaded weight (kg)	4580	6560	1770
Heat load (W)	423	1472	2210
Repository			
Number of rows	2	1	
Package pitch (m)	2.51	2.6	5.9
Tunnel height (m)	4.6	4.6	4.6
Tunnel width (m)	6		4.6
Borehole diameter (cm)	76	76	44
Borehole depth (m)	4.6	4.9	4.93
Performance parameters			
Radiation			
Surface of overpack (mrem/hr)	4.3×10^6	1.5×10^7	4.2×10^7
Surface of tunnel (mrem/hr)	<2	1.4	<2
Corrosion			
Depth at 1000 yr (cm)	0.007	0.03	0.03
Allowance (cm)	0.25	0.25	0.25
Overpack reinforcement			
Required wall, (cm)	2.3	2.3	1.3
Actual wall (cm)	3.5	3.5	2
Required head (cm)	7	7	3.8
Actual top head (cm)	18	18	20
Actual bottom head (cm)	9	9	6
Peak temperature			
Waste form (°C)	133	253	360
Overpack (°C)	116	222	272
Tuff (°C)	110	208	239
Tunnel (°C)	97	140	105
Local areal load (W/m^2)	13.5	22.6	15
Package-related costs			
Component cost ($)	19,150	19,150	8,900
Total cost ($)	48,900	56,450	53,050
Cost/unit of waste ($/kg)	33	24	23

From Conceptual Waste Package Designs for Disposal of Nuclear Waste in Tuff, by W-AESD, AESD-TME-3138, Tables 1-1, 1-2, and 1-4, Westinghouse Electric Corporation, December 1982. Copyright 1982 by Westinghouse Electric Corporation. Reprinted by permission.

Table 4.15 Summary of tuff self-shielded package designs

	DHLW		CHLW	
	Reference cast steel	Reference gray cast iron	Reference cast steel	Reference gray cast iron
Design parameters				
Waste form				
Diameter (cm)	61	61	32.4	32.4
Length (m)	3	3	3.05	3.05
Canisters/package	1	1	3	3
Weight of waste (kg)	1470	1470	1785	1785
Total waste form weight (kg)	1940	1940	2535	2535
Waste package				
Cross-section geometry	Round	Round	Triangular	Triangular
Outside dimension(s) (cm)	124.5	129.5	161 × 161 × 161	146 × 146 × 146
Length (m)	3.75	3.83	4.06	4
Empty weight (MT)	29.7	29.7	60	39
Loaded weight (MT)	31.7	31.7	62.6	41.4
Heat load (W)	423	423	6630	6630
Repository				
Package pitch (m)	1.25	1.3	13.25	13.25
Tunnel height (m)	2.5	2.5	2.5	2.5
Tunnel width (m)	5	5	5	5
Performance parameters				
Radiation				
Surface of overpack (mrem/hr)	70	70	100	100
Corrosion				
Depth at 1000 yr (cm)	1.2	1.2	3.1	N3.1
Allowance (cm)	23.5	26	32.2	24.7
Overpack				
Required wall (cm)	2.7	2.7	13.3	13.3
Actual wall (cm)	30.5	33	45.5	38
Required head (cm)	7	7	8.3	8.3
Actual top head (cm)	45.8	50	68.3	57
Actual bottom head (cm)	30.5	33	45.5	38
Peak temperature				
Waste form (°C)	127	127	495	N495
Overpack (°C)	106	106	345	N345
Tuff (°C)	106	106	345	N345
Tunnel (°C)	106	106	345	N345
Local area load (W/m^2)	13.5	13	20	20
Package-related costs				
Component cost ($)	55,900	32,000	126,700	53,000
Total cost ($)	66,550	42,650	166,700	93,000
Cost/unit of waste ($/kg)	45	29	24	14

3. Gas sampling.
4. Leak checking.
5. Loading/unloading.
6. Transferring between workstations.
7. Performing routine maintenance and decontamination where necessary.

At most nuclear power plants, contact operation with a shielded loading/unloading facility is used. For a large-throughput facility, contact operations result in exposure of workers to high radiation levels, which is in conflict with ALARA goals. Thus, there is interest in either handling machines or robotics.

4.5 TRANSPORTING HIGH-LEVEL WASTE

4.5.1 Responsibilities and Regulations

In the United States, transporting commercially generated HLW (essentially the spent fuel) is the responsibility of the federal government as specified in the NWPA of 1982. According to the NWPA,[66] the government (through the Department of Energy) will in time take the title for spent fuel at civilian nuclear power reactor sites and transport the spent fuel to federally owned and operated storage. Such transportation will be subject to licensing and regulation by the NRC and the Department of Transportation (DOT), and DOE may make the expenditure for such operation from the Waste Fund as established by the Act. Thus, DOE is providing the technical or physical development as well as full institutional development of the transportation system. Shipments of substantial quantities of SNF and/or HLW to monitored retrievable storage or a federal repository were originally scheduled to start in year 1998. Under DOE, Transportation Operations and Traffic Management (TOTM) is responsible for these operations. Development work on shipping containers was reported by Allen.[65] Similar development activities in Europe include studies on mechanical requirements for accident-proof packaging radwaste (West Germany) and large transport containers for nuclear reactor decommissioning waste (United Kingdom).

In addition to the licensing and regulations issued by the NRC and DOT, bills have been put before Congress regarding the transportation of HLW. For example, HR 761 amended the Hazardous Materials Transportation Act (HMTA) to prohibit transportation of radioactive material through populous areas; HR 4297 amended the 1980 NRC Authorization Act to require advance notice of HLW shipment; and HR 4850 put further restrictions on spent fuel transportation.

4.5.2 Economic Considerations

Comparisons are made between the truck and rail shipping of radwaste in Table 4.16.[67] In addition to the unit shipping cost, dollars per load weight, which is a function of distance shipped, a security cost is charged per shipment. It can be shown that the total

Table 4.16 Transportation and security costs

One-way distance (miles)	Truck shipping[a]			Rail shipping		
	Loaded leg[b]	Empty leg[b]	Security cost ($/shipment)	Loaded leg[b]	Empty leg[b]	Security cost ($/shipment)
500	3.30	2.10	910	6.20	5.80	1840
1000	5.10	4.10	1820	9.30	8.70	3850
1500	7.60	6.10	2730	11.80	11.10	3940
2000	10.10	8.20	3640	14.00	13.10	4040

[a]For a legal weight truck. The shipping cost of an overweight truck will be higher than what is shown here.
[b]Costs in $/100-lb load.
From Transportation of radioactive waste by R. M. Burgoyne, in *Radioactive Waste Technology*, edited by A. A. Moghissi et al., ASME/ANS, New York, 1986. Copyright 1986 by American Society of Mechanical Engineers. Reprinted by permission.

cost of truck shipping is lower than that of rail shipping as long as the distance is less than 1500 miles each way and the waste load does not exceed 14,000 lb, since the cost advantage of rail shipping is at long distance and high shipping load.

4.6 COMPUTER CODES

There are many computer codes used for different aspects of high-level waste management. Some of the more important codes are listed here, but this is not a complete list of all the computer codes used in any system designed to solidify, package, handle, and transport high-level wastes. The following codes are a good starting place for references to other codes that might be of interest. These codes were selected from a software catalog maintained by the U.S. Department of Energy, Office of Scientific and Technical Information, Energy Science and Technology Software Center, P.O. Box 1020, Oak Ridge, TN 37831. The catalog can be accessed on the Internet at URL doe.gov/waisgate/estsc.html. Descriptions of several other computer codes are provided elsewhere in this book.

CAN. Canister model system analysis. This package provides a computer simulation of a system model for packaging nuclear waste in canisters.

COMRADEX4. Accidental released radiological dose. This was developed to evaluate potential radiological doses in the near environment of radioactive releases.

DCHAINV1.3. Radioactive decay and reaction chain calculations. This code calculates the time-dependent daughter populations in radioactive decay and nuclear reaction chains.

FRA, Fuzzy risk analyzer. This is a general-purpose code for risk analysis using fuzzy, not numeric, attributes.

KEFFMGBSTGAN. Nuclear criticality safety. This package includes three codes used to insure criticality safety.

ORMONTE. Monte Carlo sensitivity analysis code. This code produces a probability histogram for each output variable of interest so that the risk associated with the attainment of a given deterministic value can be assessed.

RSAC-5.1. Radiological safety analysis code. This code calculates the consequences of the release of radionuclides to the atmosphere.

TOAD. Processing of analyzer gamma-ray spectra. This code is used to process and analyze gamma-ray spectra.

4.7 DISCUSSION QUESTIONS AND PROBLEMS

1. Consider the following options and try to reach consensus on the best way to handle radioactive wastes in the United States. Most of these wastes are being stored at the generator sites and it appears unlikely that this will change for a long time.
   (a) Should the waste continue to be stored in this manner indefinitely?
   (b) Should the waste be packaged and sent to a central storage or disposal facility as it is generated? How many storage facilities should exist?
   (c) Should the waste continue to be stored at the generator sites until such sites become inactive and then shipped to a central storage or disposal facility?

2. There has been much discussion and many billions of dollars spent on design and construction of disposal systems for nuclear wastes, yet no wastes have been disposed of. The construction of two facilities, the Waste Isolation Pilot Plant (WIPP) and the Yucca Mountain Disposal Facility, have both been completed and are ready to accept wastes, but no real wastes are allowed in them. Both facilities have conducted extensive testing with simulated wastes, with good results, but use of real wastes has not been approved. What will happen or what will it take to gain approval of a disposal system for nuclear wastes?

3. Determine the radioactive fission products which contribute 90% of the radioactivity of the nuclear wastes from reactor fuels at time of discharge from the reactor. Estimate the volume of these wastes generated by a 1000-MW plant in 1 year. Calculate the number of curies in these fission products, and then calculate the number of curies in these same fission products 300 years later.

4. Determine the total amount of uranium-238 and uranium-235 in a typical 1000-MW PWR plant and calculate the number of curies before the reactor goes critical.

5. Determine the annual volume of HLW for a 1000-MW PWR plant for the case where all spent fuel has been reprocessed. Compare this volume with the volume of HLW with unreprocessed spent fuel.

REFERENCES

1. King, C. L., Nuclear waste disposal can be done safely, *Stanford Observer*, p. 5, October 1985.
2. Pigford, T. H., The National Research Council study of the isolation system for geologic disposal of radioactive wastes, in *Scientific Basis for Nuclear Waste Management VII*, p. 461, Elsevier, New York, 1984.
3. American Nuclear Society, AIChE task force says disposal can be safe, *Nucl. News*, p. 84, March 1986; AIChE, The Role of Nuclear Power in the Energy Future of the U.S., Nuclear Engineering Division,

AIChE, April 1988; Richmond, C. R., Population exposure from the nuclear fuel cycle: Review and future direction, in *Proc. Topical Meeting Population Exposure from the Nuclear Fuel Cycle*, Oak Ridge, Tenn., September 1987; American Nuclear Society, NRC expresses confidence in eventual storage (HLW), *Nucl. News*, p. 86, October 1984.

4. Shapiro, F. C., *Radwaste*, Random House, New York, 1981.

5. U.S. Department of Energy, Integrated Data Base Report-1996, U.S. Spent Nuclear Fuel and Radioactive Waste Inventories, Projections, and Characteristics, DOE/RW-0006, Rev. 13, December 1997.

6. Taylor, W. J., High Level Waste Information Request for the 1997 Integrated Data Base Report, 97-WDD-115, DOE Richlands Operations Office, July 11, 1997.

7. Millet, C. B., IDB Data Spread Sheet, Lockheed Martin Idaho Technologies Co., September 2, 1997.

8. Hester, J. R., DOE Integrated Data Base, Westinghouse Savannah River Co., September 2, 1997.

9. Hollinden, J. J., Submittal of High-Level Waste Information for the 1997 Integrated Data Base Report, West Valley Nuclear Services Co., July 23, 1997.

10. Washington State Department of Ecology, U.S. Environmental Protection Agency, U.S. Department of Energy, Hanford Federal Facility Agreement and Consent Order, EPA Docket Number 1089-03-040120, Ecology Docket Number 89-54, Richland, Wash., May 1989.

11. U.S. Department of Energy, Environmental Impact Statement, Management of Commercially Generated Radioactive Waste, DOE/EIS-0046F, Washington, D. C., 1980.

12. Benedict, M., T. H. Pigford, and W. H. Levi, *Nuclear Chemical Engineering*, 2d ed., McGraw-Hill, New York, 1981.

13. Bebbington, W. P., The reprocessing of nuclear fuels, *Sci. Am.*, vol. 235, no. 6, p. 30, 1976.

14. Study Group on Nuclear Fuel Cycle and Waste Management, Report to the American Physical Society, *Rev. Mod. Phys.*, vol. 50, no. 1. pt. II, 1978.

15. Powell, J., M. Steinberg, H. Takahaski, P. Grand, T. Bolts, and H. J. C. Kouts, The APEX Accelerator Cycle for Transmutation of Long-Lived Fission Wastes, Report BNL-28282, Brookhaven National Laboratory (CONF-800743-11), 1980.

16. Chipman, N. A., and T. E. Carleson, HLW volume reduction using the modified Zirflex fuel dissolution process, in *Spectrum '86, Proc. ANS Int. Topical Meeting, Waste Management and Decontamination and Decommissioning*, J. M. Pope et al., eds., Niagara Falls, New York, p. 1156, 1986.

17. International Energy Association Ltd., Study of the Potential Uses of the Barnwell Nuclear Fuel Plant (BNFP), IEAL Report IEAL-141A, Fairfax, Va., 1980.

18. Bemden, E. V., Fabrication of plutonium enriched fuel at Belgonucleare, *Nucl. Technol.*, vol. 53, p. 186, May 1981.

19. Rochlin, G. I., *Plutonium, Power and Politics*, University of California Press, Berkeley, Calif., 1979, p. 39.

20. Nelson, R. L., N. Parkinson, and W. C. L. Kent. U.K. development toward remote fabrication of breeder reactor fuel, *Nucl. Technol.*, vol. 53, p. 196, May 1981.

21. Garmon, L., and I. Peterson, If it's AVM, this must be France, *Sci. News*, vol. 121, p. 60, January 1982.

22. Belter, W. G., Present and future programmes in the treatment and ultimate disposal of HLW in USA, in *Treatment and Storage of HLWs, Proc. Symp. on Treatment and Storage of HLWs*, International Atomic Energy Agency, Vienna, 1963.

23. Oertel, G. K., J. L. Crandall, and J. L. McElroy, Treatment and conditioning of wastes from nuclear fuel reprocessing plants in the U.S.A., in *Proc. Int. Conf. on Radwaste Management, Seattle, 1983*, vol. 2, p. 209, IAEA, Vienna, March 1984.

24. Chapman, C. C., Design preferences for a slurry-fed ceramic melter suitable for vitrifying West Valley wastes, in *Proc. Int. Symp. Ceramics in Nuclear Waste Management*, vol. 8, p. 149, 1983.

25. Jouan, A., C. Ladirat, and J. P. Moncouyoux, Present status of the French continuous fission product vitrification process, in *Nuclear Waste Management II, Advances in Ceramics*, vol. 20, p. 105, American Ceramic Society, Westerville, Ohio, 1986.

26. Giraud, J. P., J. P. Conord, and P. M. Saverot, Conceptual design for vitrification of HLW at West Valley using a rotary calciner/metallic melter, in *Proc. 2nd Int. Symp. Ceramics in Nuclear Waste Management*, Chicago, April 1983.

27. Thiry, H. B., J. P. Laurent, and J. L. Ricaud, French experience and projects for the treatment and packaging of radioactive wastes from reprocessing facilities, in *Proc. Int. Conf. on Radwaste Management, Seattle, May 1983*, vol. 2, p. 219, IAEA, Vienna, 1984.

28. Morris, J. B., and B. E. Chidley, Preliminary experience with the New Harwell inactive vitrification pilot plant, in *Proc. Symp. Management of Radwaste from the Nuclear Fuel Cycle*, vol. 1, p. 241, IAEA/NEA, Vienna, 1976.

29. Ringwood, A. E., *Safe Disposal of HLW: A New Strategy*. Australian National University Press, Canberra, 1978.

30. Reeve, K. D., et al., The development and testing of SYNROC C as a HLW form, in *Scientific Basis for Nuclear Waste Management IV*, p. 99, North-Holland, New York, 1982.

31. Reeve, K. D., and A. E. Ringwood, The SYNROC process for immobilizing HLW, in *Proc. Int. Conf. on Radwaste Management, Seattle, May 1983*, vol. 2, p. 307, IAEA, Vienna, 1984.

32. Reeve, K. D., D. M. Levins, J. L. Woodlfrey, and E. J. Ramm, Immobilization of HLRW in SYNROC, in *Proc. 2nd Int. Symp. on Nuclear Waste Management, Advances in Ceramics*, vol. 8, p. 200, American Ceramic Society, Westerville, Ohio, 1983.

33. Harker, A. B., and J. F. Flintoff, Hot isostatically pressed ceramic and glass forms for immobilizing Hanford HLW, in *Proc. 2nd Int. Symp. on Nuclear Waste Management, Advances in Ceramics*, vol. 8, p. 222, 1983.

34. U.S. DOE, Environmental Assessment: West Form Selection for Savannah River Plant HLW, DOE/EA-0179, Washington, D. C., 1982.

35. Bemadzikowski, T. A., ed., The Evaluation and Selection of Candidate HLW Forms, Report DOE/TIC-11611, DOE Savannah River Operations Office, 1982; available from National Technical Information Service (NTIS), Springfield, Va.

36. U.S. DOE, The Evaluation and Review of Alternative Waste Forms for Immobilization of HLW, Report DOE/TIC-10228 (August 1979) and Report DOE/TIC-11219 (June 1980). Alternative Waste Form Peer Review Panel; available from NTIS, Springfield, Va.

37. U.S. DOE, The Evaluation and Review of Alternative Waste Forms for Immobilization of HLW, Report Number 3 DOE/TIC-11472, Alternative Waste Form Peer Review Panel, July 1981; available from NTIS, Springfield. Va.

38. Stone, J. A., S. T. Goforth, Jr., and P. K. Smith, Preliminary Evaluation of Alternative Forms for Immobilization of Savannah River Plant HLW, DOE Report DP-1545, E. I. du Pont de Nemours, SRL, Aiken, S. C., 1979.

39. Schulz, W. W., et al., Preliminary Evaluation of Alternative Forms for Immobilization of Hanford HLW, DOE Report RHO-ST-32, Rockwell Hanford Operations, Hanford, Wash., 1980.

40. Post, R. G., Independent Evaluation of Candidate Alternative ICPP HLW, DOE Report ENICO-1088, Exxon Nuclear Idaho Co., 1981.

41. Staples, B. A., H. S. Cole, and J. C. Mittl, Evaluation of Ceramic Materials Developed to Immobilize ICPP Zirconia Calcine, DOE Report ENICO-1095, Exxon Nuclear Idaho Co., 1981.

42. Wald, J. W., et al., Comparative Waste Forms Study, DOE Report PNL-3516, Pacific Northwest Laboratory, Richland, Wash., 1980.

43. E. R. Johnson Associates, Preliminary Evaluation of Alternative Waste Form Solidification Processes, Vol. II: Evaluation of the Processes, DOE Report PNL-3477, Pacific Northwest Laboratory, Richland, Wash., 1980.

44. U.S. DOE, A Method for Product Performance Evaluation of Candidate Waste Forms for Immobilization of HLW, Report DOE/TIC-11612, DOE Interface Working Group on HLW Form Selection Factors, 1982; available from NTIS, Springfield, Va.

45. Hench, L. L, D. E. Clark, and J. Campbell, HLW immobilization forms, *Nucl. Chem. Waste Manage.*, vol. 5, p. 149, 1984.

46. Jantzen, C. M., et al., Leaching of polyphase nuclear waste ceramics: Microstructural and phase characterization, *J. Am. Ceram. Soc.*, vol. 65, no. 6, p. 292, 1982.

47. Harker, A. H., et al., Formulation and processing of polyphase ceramics for HLW, in *Scientific Basis for Nuclear Waste Management IV*, p. 567, Elsevier, New York, 1983.

48. Pacific Northwest Laboratory, Materials Characterization Center (MCC) Test Methods, Preliminary Version, PNL-3990, Richland, Wash., 1981.

49. Macedo, P. B., A. Barkatt, and J. H. Simmons, A flow model for the kinetics of dissolution of nuclear waste glasses, *Nucl. Chem. Waste Manage.*, vol. 3, p. 13, 1982.

50. Clark, D. E., H. Christensen, H. P. Hermansson, and S. B. Sundvall, Effects of flow on corrosion and surface film formation on an alkali borosilicate glass, in *Nuclear Waste Management Symposium, Advances in Ceramics*, vol. 8, p. 19, American Ceramic Society, Columbus, Ohio, 1984.

51. Bradley, D. J., Basic research for evaluating nuclear waste form performance, *Nucl. Technol.*, vol. 51, p. 111, 1980.

52. Isard, J. O., A. R. Allnett, and P. J. Melling, An improved model of glass dissolution, *Phys. Chem. Glasses*, vol. 23, p. 185, 1982.

53. Banba, T., and T. Murakami, The leaching behavior of a glass waste form—Part II: The leaching mechanism, *Nucl. Technol.*, vol. 70, p. 243, 1985; T. Banba, T. Murakami, and H. Kimura, The leaching behavior of a glass waste form—Part III: The mathematical leaching model, *Nucl. Technol.*, vol. 76, p. 84, 1987.

54. McGrail, B. P., A. Kumar, and D. E. Day, Sodium diffusion and leaching of simulated nuclear waste glass, *J. Am. Ceram. Soc.*, vol. 67, p. 463, 1984.

55. Sullivan, T. M., and A. J. Machiels, Modeling chemical interactions in the hydrated layers of nuclear waste glasses, in *Scientific Basis for Nuclear Waste Management II*, p. 597, North-Holland, New York, 1984.

56. Newton, R. G., and A. Paul, A new approach to predicting the durability of glasses from their chemical compositions, *Glass Technol.*, vol. 21, p. 307, 1980.

57. Plodinec, M. J., G. G. Wicks, and N. E. Bibler, An Assessment of Savannah River Borosilicate Glass in the Repository Environment, DOE Report DP-1629, Savannah River Laboratory, Aikens, S. C., 1982.

58. Takahaski, M., M. Muron, A. Inoue, M. Aoki, M. Takizawa, K. Ishigure, and N. Fujita, Properties of bentonite clay as buffer material in HLW geological disposal. Part I: Chemical species contained in bentonite, *Nucl. Technol.*, vol. 76, p. 221, 1987.

59. Mendel, J. E., The Storage and Disposal of Radwaste as Glass in Canisters, DOE Report PNL-2764, Pacific Northwest Laboratory, Richland, Wash., 1978.

60. McElroy, J. L., and M. S. Hanson, High-level radioactive waste, in *Radioactive Waste Technology*, A. A. Moghissi et al., eds., chapter 17, p. 649, ASME/ANS, New York, 1986.

61. Westinghouse Advanced Energy Systems Division, Waste Package Reference Conceptual Designs for HLW Repository in Salt, WTSD-TME-001, Rev. A, Madison, Pa., 1984.

62. Westinghouse Advanced Energy Systems Division, Waste Package Concepts for Use in the Conceptual Design of the Nuclear Waste Repository in Basalt, AESD-TME-3142, Madison, Pa., 1982.

63. Westinghouse Advanced Energy Systems Division, Conceptual Waste Package Designs for Disposal of Nuclear Waste in Tuff, AESD-TME-3138, Madison, Pa., 1982.

64. Kircher, J. F., and D. J. Bradley, NWTS Waste Package Design and Materials Testing, in *Scientific Basis for Nuclear Waste Management VII*, p. 383, Elsevier, New York, 1983.

65. Allen, G. C., Advanced Transportation System Options for Spent Fuel and HLRW, SAND 84-0826C TTC-0491, Sandia National Laboratories, 1984.

66. U.S. 97th Congress, *Nuclear Waste Policy Act of 1982*, Public Law 97-425, January 1983.

67. Burgoyne, R. M., Transportation of radioactive waste, in *Radioactive Waste Technology*, A. A. Moghissi et al., eds., chapter 11, ASME/ANS, New York, 1986.

CHAPTER FIVE

DISPOSAL OF SPENT FUEL
AND HIGH-LEVEL WASTE

5.1 INTRODUCTION

The technical community is in general agreement that the technology exists to safely dispose of nuclear wastes. The preferred method has always been deep geologic disposal, but other methods have also been studied such as deep-sea burial and placing the wastes in outer space. Many scientists also believe that long-term engineered storage is also acceptable. Early studies of deep geologic disposal centered on placing the wastes in salt deposits because they have been stable for very long periods of time, no water has been associated with the deposits for very long periods of time, and salt is plastic and so tends to flow under pressure, thus sealing any openings over time. These studies were later expanded to include hard rock geologies. The U.S. program included salt, basalt, tuff, and granite. Granite was dropped from the program early since most of the potentially acceptable deposits were in the eastern U.S., where the population densities were high. In 1987 Congress passed the NWPAA of 1987, redirecting the nuclear waste program to phase out site specific activities at all sites other than tuff at the Yucca Mountain site in Nevada. The development of WIPP in a salt deposit near Carlsbad, New Mexico, has continued. The WIPP facility is intended to eventually provide for deep geologic disposal of TRU wastes.

5.2 SPENT NUCLEAR FUEL AND HIGH-LEVEL WASTE DISPOSAL METHODS

The hazard of HLW declines dramatically with time as it undergoes natural radioactive decay. One way to quantify this change is to compare the combined hazard potential of

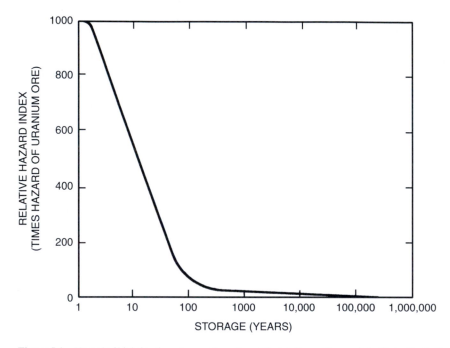

Figure 5.1 Hazard of high-level waste over time. From Electric Power Research Institute, Geologic disposal of nuclear waste, *EPRI J.* (Palo Alto, Calif.), May 1982. Copyright 1982 Electric Power Research Institute. Reprinted by permission.

all the radioactive products in the waste with the hazard presented by an equal volume of uranium ore as it is mined from the ground. The ratio of these two hazards can be considered as a relative hazard index. Figure 5.1 shows how such an index for HLW changes with time. The HLW in its first year of storage is about 1000 times as hazardous as natural uranium ore, but after 1000 years of storage, radioactive decay has brought the hazard down to about 1% of the original figure.[1] Thus, it may be advantageous to store the HLW for a period of time before final disposal. Table 5.1 provides the chronology of initiatives taken on the disposal of HLW. As shown in this table, burial of HLW in bedded salt deposits was first recommended by the National Research Council over 40 years ago. Several independent studies since that time have also concluded that HLW can be safely disposed of in a deep geologic repository,[1–3] and that view is shared by countries such as France, West Germany, Belgium, Canada, Japan, Sweden, and the United Kingdom. France is considering primarily salt, granite, and clay as the geologic media. West Germany is considering primarily salt at its Gorleben site as the geologic medium. Japan is investigating mined repositories in granite, diabase, shale, zeolitic tuff, limestone, slate, and schist, with a target date of 2000 for trial disposal.[2] In the United States, three candidate geologic media were selected for the first repository and approved for site characterization: basalt, bedded salt, and volcanic tuff. However, in December 1987, Congress amended the NWPA of 1982 by selecting the Yucca Mountain (tuff) site for characterization and called for work stoppage at the other two candidate sites.

Table 5.1 Chronology of initiatives taken on the disposal of high-level wastes[a]

1957	National Academy of Sciences (NAS)–National Research Council (NRC) recommends that HLW be buried in bedded salt deposits
1957–1961	Studies made on the feasibility of above proposal
1961	Results of the study are reviewed by NAS–NRC and further study recommended
1963–1967	Studies made on salt vault (eventually to be located near Lyons, Kansas)
1963	Calcining demonstrated
1965	Atomic Energy Commission (AEC) begins to convert liquid wastes to salt cake
1966	NAS–NRC committee continues to advocate deep salt burial
1968	Committee on Radioactive Waste Management (CRWM) established by NAS–NRC to review programs for solidification and disposal
1969–1970	Conceptual design developed for a prototype facility to bury HLW in salt
1970	Panel on disposal in salt mines meets to hear presentation on radwaste burial
1970	AEC declares that all commercially generated HLW must be solidified (within 5 years) and delivered to a federal repository (within 10 years)
1970	AEC tells state of Idaho that the waste stored there will be removed by 1980 (to Lyons facility)
1971	Congress directs AEC to stop Lyons project until its safety can be certified
1972	AEC abandons Lyons project
1976	Energy Research and Development Administration informs governors of plans to conduct field investigations to have a deep geologic repository in salt available by 1985
1978	DOE issues draft report to Task Force for Review of Nuclear Waste Management
1978	DOE conducts a series of local hearings on its proposed Waste Isolation Pilot Project (WIPP)
1980	Carter Administration scraps the WIPP; President Carter announces the interim strategy in a statement to Congress
1981	DOE reinstates WIPP for a repository for government-produced transuranics and, for limited demonstration purposes, a small quantity of unreprocessed defense spent fuel
1982	Congress passes the NWPA of 1982, which becomes Public Law 97-425
1983	NRCI completes a geologic disposal study of HLW
1985	President Reagan approves the three candidates for repository selection as recommended by DOE
1987	Congress passes the NWPAA of 1987, redirecting the nuclear waste program to phase out site-specific activities at all candidate sites other than the Yucca Mountain site and establishing an MRS Review Commission

[a]Source (up to 1980): *Radwaste: A Reporter's Investigation of a Growing Nuclear Menace* by Fred C. Shapiro, Random House, New York. Copyright 1981 Random House, Inc. Reprinted by permission.

5.2.1 Geologic Repositories

The isolation mechanisms of geologic waste disposal systems consist in delaying the ingress of groundwater, slowing the dissolution of radionuclides, increasing the groundwater travel time, delaying sorption of radionuclides in the geologic medium, dispersion action, and dilution of radionuclides by surface water. Thus an ideal geologic formation for the containment of HLW should have minimal permeability, maximal flow dispersion, minimal chance of forming apertures, minimal thermal and mining disturbance, and maximal ion retention capacity. As stated before, salt, clay, and crystalline rock (granite) have been universally considered as candidate geological formations for the containment of HLW.[4] Table 5.2 compares these three types of formation.

Table 5.2 Comparison of three types of geologic formation for the containment of HLW

Formation	Advantages	Disadvantages
Salt (evaporite)	Absence of groundwater flow	Small inclusions of brine might migrate and corrode the waste canisters
	Relatively good heat conductor	Being a resource itself, it may contain valuable resources
	Able to "creep"; thereby any cracks tend to seal themselves	Groundwater would dissolve salt rock
Clays and claylike formation (argillite)	Capacity to absorb radmaterial	Lack of information about the response to heating
	Low permeability	Possible need to use supporting structure
	Ion retention capacity very good	Morphologies of clays may change
Hard rocks (crystalline)	Unfissured state (no cracks) permits very little water movement (i.e., can be very impermeable)	Not as good a heat conductor as salt
	Good absorber of materials leached out of the glass block	Requires increased spacing of canisters or longer cooling time prior to disposal

From Survey of foreign terminal radioactive waste storage programs by K. M. Harmon, in *Proc. 1983 Civilian Radwaste Management Information Meeting*, p. 229, DOE OCRWM, February 1984. Reprinted by permission.

5.2.2 Repository Siting

As specified in the NWPA of 1982, the DOE has the responsibility for the siting policy and guidelines for the repository, recommendations on sites, and detailed site characterization, which includes testing to obtain technical information supporting a licensing application. During the selection process, DOE is to prepare the environmental impact statement (EIS) and site selection report before submitting the recommendation to the president, who in turn submits to Congress the selection for the first repository. Along the way, DOE is to hold a series of public hearings in the vicinity of the sites. Figure 5.2 shows geologic sites that were actively considered for the disposal of radwaste.[6] From the start of issuance of a site characterization plan to the final site designation, a site investigation can take 4–5 years; this time is needed to obtain state and local permits for drilling exploratory shafts and boreholes, construct such shafts and develop underground test facilities, and conduct surface and subsurface tests to support site qualification.

5.2.3 Repository Design

In addition to being in compliance with the performance objectives of the NRC and EPA (as specified in Section 5.2.2), the repository design should be consistent with and complementary to the natural geologic settings and represent a comprehensive treatment of the total system with conservative margins. It should also be sufficiently transparent to

Figure 5.2 Specified geologic sites under active consideration for the disposal of radwaste.[5] From NWPA—The salt geologic waste repository project, Presented at the AIF Fuel Cycle Conference '84. Atlanta, Ga., April 1984. Copyright 1984 Atomic Industrial Forum. Reprinted by permission.

withstand independent scrutiny. Possible designs for a final repository are shown in Figure 5.3. The following are important design considerations:

1. Human-induced phenomena. These include improper waste emplacement, undetected past intrusion (such as mine shafts), inadvertent future intrusion (e.g., archeological exhumation, weapon testing, resource mining), or intentional intrusion (e.g., war, sabotage).
2. Perturbation of the groundwater system (e.g., irrigation, reservoirs, chemical liquid waste disposal) and biosphere alteration (e.g., establishment of population centers, climate modification).
3. Shaft seal failure. As shown in Figure 5.4, a number of repository seals are used, including borehole seals, shaft seals, and tunnel seals. Failure of shaft seals would cause leakage between the repository and the surface (biosphere).

Worldwide geologic disposal and repository design concepts are summarized in Table 5.3. Figures 5.5 and 5.6 show conceptual repositories in salt, and Figure 5.7 shows a conceptual repository in basalt.

5.2.4 Performance Assessment

Operation of the repository will have short- and long-term effects. The former include temperature rise, brine migration, and changes in moisture content. The temperature profiles,

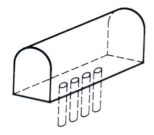

(a)

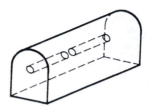

(b)

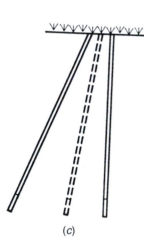

(c)

(d)

Figure 5.3   Possible designs for a final repository. (a) Disposal of the waste in shallow boreholes drilled in tunnels excavated at great depth. (b) Disposal of the waste in tunnels excavated at great depth (horizontal boreholes). (c) Disposal of the waste in very deep boreholes drilled from the surface of the ground. (d) Disposal of the waste in an isolated body of rock that is sealed off from its surroundings by an impervious material.

both vertical and horizontal, peak at the center of disposal area. The spacing of the waste canisters in the mine is designed to limit peak temperatures in the waste to a value below 200°C (392°F). The vertical temperature profile measured in Project Salt Vault is shown in Figure 5.8.[7] The heat source was placed 1000 ft below the surface with an initial heat generation rate of 2 kW per canister and an areal spacing of 65 kW/hectare. The profile is symmetrical until the surface heat flux peaks (estimated at $q_s = 0.11$ Btu/hr ft^2) at 700 years after the disposal. Brine migration is affected primarily by the temperature-dependent solubility of the salt. Thus the brine trapped in the salt formation tends to migrate up the thermal gradient. Moisture may also be contained in formations and is driven out of minerals into the pores. This could cause the pore pressure to increase and change the geologic and structural character of the formation.

The long-term effects consist in deformation of rocks and transport of radionuclides; the latter is of most concern because of its impact on public safety. Figure 5.9

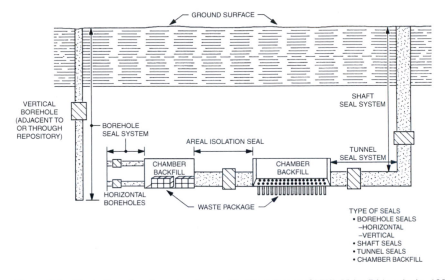

Figure 5.4 Illustration of various repository seals. From Report ONWI-55 by D'Appolonia, 1980, DOE ONWI, Battelle Memorial Institute. Reprinted by permission.

shows a summary of deformations in and around the center pillar of the Project Salt Vault experimental area.[7] The solid lines indicate the room outlines, planes of measurement, and baselines for measured deformations, and the dashed lines represent deformation, on a significantly larger scale, in horizontal and vertical directions. The deformation can be estimated by extending the rock mechanical analyses, and its possible consequences are rock fracture (resulting from induced stresses) and disturbance of the overlying groundwater regime. The uncertainties in predicting the potential release and transport of radionuclides through the medium to the biosphere are the major concern in terms of public safety and are largely responsible for the unwillingness of most localities to accept a waste repository. However, experience with fossil fission reactors and natural ore bodies such as those discovered near Oklo in Zaire indicates that any reasonable geology gives assurance of excellent radionuclide containment.[2] Conservative requirements have been set by the NRC in 10CFR60,[8] which specifies 300- to 1000-year containment by the waste package (during the thermal period) and then restriction of outleakage to 1 part in 100,000 of the 1000-year inventory for 10,000 years after the containment period. At the same time, the groundwater travel time to the accessible environment must be over 1000 years. The EPA also intended, through 40CFR191,[9] to require the individual dose of the waste, before repository closure, to be less than 25 mR (whole-body dose) and the release to the accessible environment, after repository closure, not to exceed limits conservatively estimated by EPA to produce 1000 health effects in 10,000 years from irradiation by 100,000 MTHM fuel.

The current assessment of performance for various types of geologic disposal is given in an NRC study[10] in which calculations were made of the long-term isolation and environmental releases for conceptual repositories in salt, basalt, granite, and tuff. The waste disposal technology identified in the study is predicted to be more than

Table 5.3 Worldwide geologic disposal and repository design concepts[a]

Country	Waste type	Cooling time (yr)	Earliest time available	Geologic media	Design concept	Media selected for further study
Belgium	HLW, ILW	50	Mid-1990s	Clay	In-floor disposal (−225 m)	Clay
Canada	HLW/SF	50	2010+	Granite/gabbro	In-floor disposal	Salt as backup
Denmark	HLW	40	2040	Salt	Deep boreholes (−1200 to −2500 m)	Salt
Finland	HLW/SF	TBD	2020	Granite	In-floor disposal	Granite
France	HLW	TBD	1993	Granite	In-floor disposal	Clay as backup
Germany	HLW	30	2000	Salt	In-floor disposal	Salt
India	HLW	30	TBD	Granite/gneiss	—	Granite/gneiss
Italy	HLW	50	TBD	Clay	—	Granite as backup
Japan	HLW	30	2020	TBD	In-floor disposal	Granite, clay
Netherlands	HLW	10	TBD	Salt	Gallery/deep boreholes	Salt
Spain	SF	10	TBD	Salt/granite	—	Salt/granite
Sweden	HLW/SF	30	2020	Granite/gneiss/gabbro	In-floor disposal (−500 m)	—
Switzerland	HLW	35	2020	Granite	In-floor disposal (to −2500 m)	Granite
United Kingdom	HLW	50	2040	Granite	—	Granite (salt, clay as backup)
United States	HLW	10	2000	Salt, basalt, tuff, granite	In-floor disposal	Salt, basalt, tuff, granite

[a]TBD, To be determined.

Adapted from Survey of foreign terminal radioactive waste storage programs by K. M. Harmon, in *Proc. 1983 Civilian Radwaste Management Information Meeting*, p. 229, DOE OCRWM, February 1984. Reprinted by permission.

adequate for isolation of radioactive waste from the biosphere and protection of public health and safety, although more accurate prediction of long-term performance of the waste disposal system is necessary before a detailed design of a repository can be completed. Except for possible human intrusion or a major and unexpected natural diversion of groundwater, all radionuclides can be contained in a repository in salt. Most radionuclides can be contained in basalt, granite, or tuff. Only small amounts of long-lived radionuclides are predicted to be released to the surrounding media in contaminated groundwater. The study concluded that all candidate geologies can meet the performance criteria if there is surface water into which contaminated groundwater can discharge before being used by humans, that release of radionuclides to the biosphere by groundwater transport may not occur for tens of thousands of years or longer after the waste is emplaced, and that some important radionuclides may not appear until a

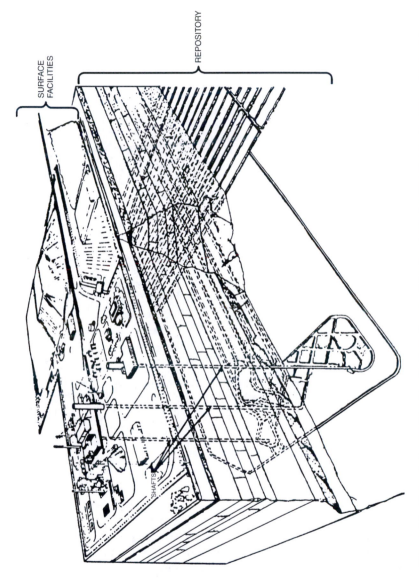

SURFACE
FACILITIES

REPOSITORY

SHAFTS

Figure 5.5 Conceptual repository in salt. From Survey of foreign terminal radioactive waste storage programs by K. M. Harmon, in *Proc. 1983 Civilian Radioactive Waste Management Information Meeting*, DOE OCRWM, 1984. Reprinted by permission.

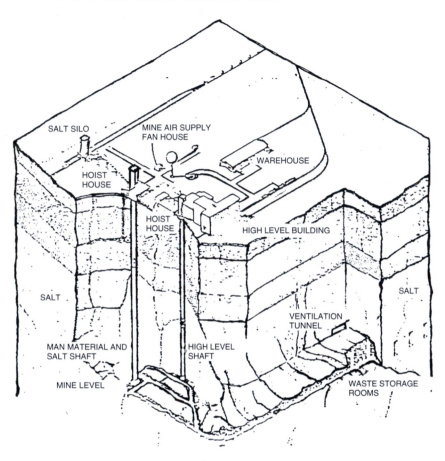

Figure 5.6 Schematic view of pilot plant waste repository in bedded salt formation. From Survey of foreign terminal radioactive waste storage programs by K. M. Harmon, in *Proc. 1983 Civilian Radioactive Waste Management Information Meeting*, DOE OCRWM, 1984. Reprinted by permission.

few hundred thousand or a few million years later.[10] Recognizing the large uncertainties in calculating long-term releases of radioactivity from the repository, the study recommended a continuing program of system analysis to estimate probable radioactivity releases and future doses from repositories and to evaluate and reduce uncertainties in these estimates. The program should include experiments to test the predicted dissolution rates limited by solubility and by diffusion and convection in the groundwater surrounding the waste forms, to determine the effects of repository heating on the dissolution rate, and to determine the long-term release rate, under repository conditions, of cesium from spent fuel and borosilicate glass and of carbon and iodine from spent fuel. Experiments and analyses should also be carried out to determine the probable failure modes of waste packages under repository conditions, with attention to stresses induced during resaturation of a sealed repository and the risks from unexpected events.

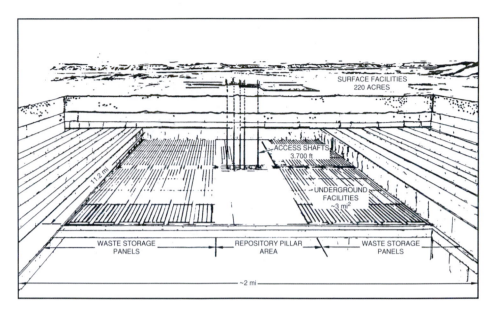

Figure 5.7 Conceptual cutaway and perspective of a nuclear waste repository in basalt. From Westinghouse Electric Corporation, Waste Package Concepts for Use in the Conceptual Design of the Nuclear Waste Repository in Basalt by W-AESD, AESD-TME-3142, Madison, Pa., July 1982. Copyright 1982 Westinghouse Electric Corporation. Reprinted by permission.

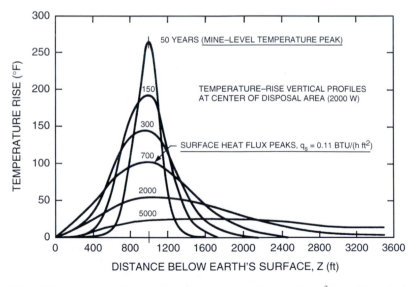

Figure 5.8 Temperature-rise vertical profiles at center of disposal area.[7] From Disposal of radwaste in bedded salt formations by W. C. McClain and A. L. Boch, *Nucl. Tech.*, 1974, vol. 24, p. 398. Copyright 1974 American Nuclear Society. Reprinted by permission.

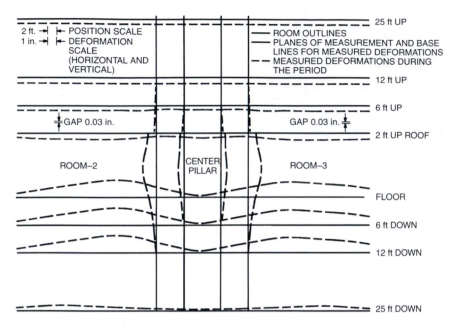

Figure 5.9 Summary of deformations in and around center pillar of Project Salt Vault experimental area. From Disposal of radwaste in bedded salt formations by W. C. McClain and A. L. Boch, *Nucl. Tech.*, 1974, vol. 24, p. 398. Copyright 1974 by American Nuclear Society. Reprinted by permission.

Computer codes are being developed and validated by experimental data to predict the transport and behavior of radionuclides in repositories. The code MISER, used by Cheung et al.,[11] models radionuclides from leaching of the waste form to subsequent groundwater transport to the biosphere. Their results describe the effect of repository system design on a probabilistic framework in both near field and far field. Uncertainties in waste form release rate, package properties, and geotechnical data are accounted for with Monte Carlo techniques. The postclosure risk assessment thus obtained shows that for both generic layered salt and basalt repositories, the limiting individual peak dose rate is less than 0.1 rem (background radiation) for both alkali borosilicate glass and polyphase ceramic waste forms, and the difference in performance between these waste forms is insignificant.

5.2.5 Repository Development

Development of performance assessment. The performance assessment during the thermal (containment) and postthermal periods was described in the preceding section. For developing the repository, the performance should also be assessed during the period of excavation that leads to site confirmation or rejection and the operational period when the waste is emplaced in the repository but can be retrieved if necessary. Nevertheless, the long-term isolation and environmental releases are of primary importance. Earlier analyses of mass transfer of radionuclides from waste packages into surrounding backfill and rock considered diffusive–convective transport of dissolved species in groundwater in the rock but assumed no flow of groundwater through it. To assess the possible contribution of backfill water flow to radionuclide release, the local flow field in layers of backfill and rock surrounding a waste package was analyzed.[12]

Figure 5.10 shows some results from a potential-flow theory developed by Chambré et al. to predict the steady flow streamlines in backfill and porous rock surrounding a spherical or cyclindrical waste package for different ratios α of the permeability of backfill to that of the rock, with backfill thickness $b = 0.25$ m, waste radius $r_0 = 0.25$ m, and the waste form assumed impermeable. Refraction of streamlines at the backfill/rock interface is noted. Adjacent streamlines form stream tubes, with a constant discharge rate through the flow cross section of each stream tube. Thus, for low-permeability backfill ($\alpha = 0.1$), the

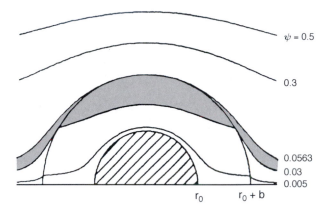

(a) $\alpha = 0.1$

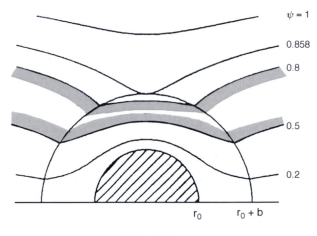

(b) $\alpha = 10$

Figure 5.10 Flow of groundwater around buried waste. Streamlines $\psi = $ cons for $\alpha = 0.1$ and 10. From Flow of groundwater around buried waste, by P. L. Chambré, C. H. Kang, and T. H. Pigford, *Trans. ANS*, vol. 52, p. 77, 1986. Copyright 1986 American Nuclear Society. Reprinted by permission.

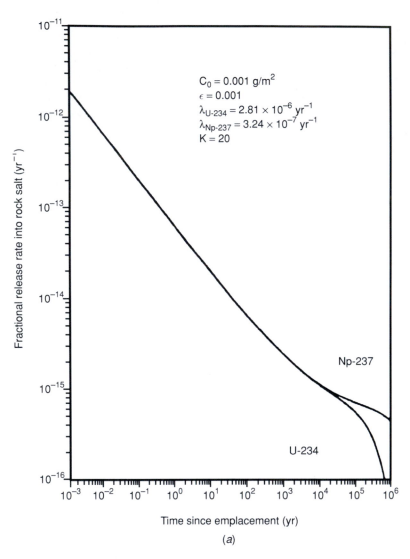

$$C_0 = 0.001 \text{ g/m}^2$$
$$\epsilon = 0.001$$
$$\lambda_{U\text{-}234} = 2.81 \times 10^{-6} \text{ yr}^{-1}$$
$$\lambda_{Np\text{-}237} = 3.24 \times 10^{-7} \text{ yr}^{-1}$$
$$K = 20$$

Np-237

U-234

Time since emplacement (yr)

(a)

Figure 5.11 Fractional release rates from spent fuel waste package in a salt repository. (a) Fractional release rates of ^{234}U and ^{237}Np, based on a 1000-year inventory. (b) Fractional release rates of ^{135}Cs, ^{137}Cs, and ^{129}I, based on initial inventory. From Release rates from waste packages on a salt repository by P. L. Chambré et al., *Trans. ANS/AO*, 1987, vol. 55, p. 131. Copyright 1987 by American Nuclear Society. Reprinted by permission.

stream tube expands in the backfill, indicating reduced Darcy velocity within the backfill and locally increased Darcy velocity in the adjacent rock. If backfill permeability is greater than that of the rock ($\alpha = 10$), stream tubes narrow and local velocities increase in the backfill. By integrating over the backfill cross section, the volumetric flow rate through the backfill per unit length of cylinder can be calculated as a function of the permeability ratio.[12]

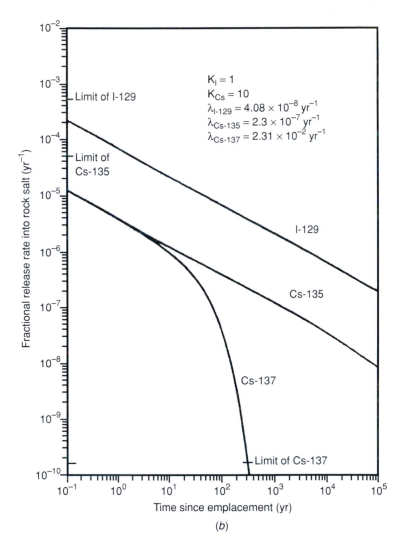

Figure 5.11 (continued)

Analytical solutions for the rates of dissolution of solubility-limited species from solid waste embedded in saturated porous salt have been presented.[13,14] In other cases where dissolution may be rapid, the rate of release of the soluble species from the waste package would be limited by the rate of mass transfer of the dissolved species (e.g., fission products in the fuel/cladding gap when exposed to groundwater) through the surrounding porous media.[15,16] Figure 5.11a shows the fractional release rates of ^{234}U and ^{237}Np from a spent fuel waste package based on a 1000-year inventory, assuming a spherical spent fuel waste package 0.72 m in radius, diffusion coefficient 10^{-7} cm^2/sec, salt porosity $\varepsilon = 0.001$, uranium retardation coefficient $K = 20$, and a uranium solubility of 0.001 g/m.^{3}

Figure 5.11b shows the time-dependent fractional release rates, normalized to initial inventories, of readily soluble nuclides, assuming that the metallic container fails and stagnant brine reaches the spent fuels; retardation coefficients and half-lives are as shown in the figure.[16] Thus, for the selected parameters and for containment times greater than 300 years, release rates from individual waste packages in salt can meet the NRC's requirement. If the container fails shortly after emplacement, the release rate for ^{137}Cs is exceeded for up to 300 years (Figure 5.11b). The release rates of ^{234}U and ^{237}Np normalized to their 1000-year inventory are below the NRC limit of 10^{-5}/year for all times on the graph (Figure 5.11a).

Other mathematical models have been reported in the literature:

1. Models of radionuclide transport in porous media were adapted to the analysis of transport in jointed porous rock.[17] Radionuclide transport in jointed porous rock can be approximated as occurring in an equivalent porous medium.
2. A chemical transport model, CHEMTRN, includes advection, dispersion/diffusion, complexation, sorption, precipitation or dissolution of solids, and dissolution in water.[18] The transport, mass action, and site constraint equations were expressed in a differential/algebraic form with the sorption process modeled by either ion exchange or surface complexation.
3. The effects of several variables and model assumptions have been assessed in the calculation of radionuclide discharge from hypothetical repositories in tuff and bedded salt.[19] The repository sites were modeled in a way consistent with the current understanding of the characteristics of the geologic environments being studied by the DOE.
4. As part of a program to develop a methodology for use in assessing the long-term risk of disposal of radwaste in deep geologic formations, the dynamic network (DNET) model was developed to investigate waste/near-field interactions associated with the disposal of radwaste in bedded salt formations.[20]
5. The coupled thermomechanical, thermohydrologic, and hydromechanical processes were studied with a numerical code, ROCMAS, for a radwaste repository in a fractured rock medium.[21]
6. A probabilistic source-term code, AREST, has been developed to provide a quantiative assessment of the performance of the engineered barrier system relative to the regulatory requirements.[22]
7. Using uncertainty analysis techniques to quantify the level of confidence and to assess the long-term risk, probabilistic systems assessment codes (PSACs) are being developed in several member countries of the Nuclear Energy Agency (NEA) within the PSAC user group.[23] The work carried out on the application of PSAC in these countries is reported in reference.[23]

In considering the time period of concern for judging the long-term hazard of geologic disposal, Merz[24] recommended that analyses be divided into two categories according to the following time frame:

1. Up to 10,000 years, for which the consequence analysis technique provides reliable quantitative results in the form of release data and radiation exposure values.

2. From 10,000 years to the maximum of 10^6 years, where safety analyses are performed in a qualitative manner without exposure calculations, since they would become highly speculative. An absolute time cutoff is applied for time frames greater than 10^6 years.

Although this selection is largely arbitrary, adoption of the 10,000-year limiting value seems to be scientifically justifiable.[24]

A way to increase confidence in long-term performance predictions has been sought through natural analogs. Although there are no exact analogs of a radioactive waste disposal system, studying natural analogs of the most important components or subsystems of a repository can overcome the necessity of demonstrations over extremely long time periods. Research has been initiated on igneous intrusions into proposed repository-type host rock, on uranium ore body analogs, on actinide solubility as a function of natural complexing agents in closed basin lakes, and on field migration of radionuclide species at disposal sites for LLW and uranium mill tailings.[25] As the evaluation of repository performance will require the use of complex numerical models for comparison with the NRC and EPA standards, three difficult problems will be encountered:

1. Predicting for thousands of years or more without a record of equal or greater length.
2. Predicting interactions of man-made elements (e.g., Np, Pu, and Tc) in the repository.
3. Scaling up complex processes from the laboratory to repository dimensions.

Thus, predictions based on natural processes that do have a long (1000 years), well-defined, and stable record can be used as a data base. In addition, laboratory studies must be conducted to describe the chemistry of the radionuclides, particularly the thermodynamics, kinetics, and retardation mechanisms. Analogous elements that are present in the geologic environment can be studied to obtain bounding estimates for important processes (for time scales up to 40 years). Uncomplicated natural systems that can be described by relatively simple models are better than complex analogs with many uncontrolled variables.[25]

Performance criteria. The criteria should be specified in the following two categories, on which the experimental evaluation may be based:

1. Near field: (a) Physical and chemical effects of heat and radiation, and (b) cavern stability.
2. Far field: (a) Thermomechanical effects, and (b) human actions and natural phenomena.

Regarding the phenomena relevant in making licensing decisions about the disposal of HLW, a major issue is the validity of the mathematical models that will be used in predicting the performance of HLW repositories. In addressing this issue, several avenues of inquiry have been pursued, including examinations of available scientific results that can be used in predicting HLW repository performance, investigations of laboratory tests that are analogous to or dynamically similar to selected aspects of expected situations, and examinations of natural geologic processes and events whose sequence of occurrence is analogous to some aspects of phenomena expected to affect the release and transport of radionuclides from emplaced HLW.[26] Thus, a combination of data from field observations, field experiments, and laboratory experiments on phenomena and conditions analogous and dynamically similar to those of an HLW repository will be used in NRC's HLW research program to test the applicability of models and portions of models used to assess, understand, and predict repository performance. Because of the reliance on mathematical models to support claims of compliance with regulatory criteria, the NRC attempts to resolve questions about the validity of assumptions underlying models used to predict the performance. To this end, the NRC held a workshop on the validity of mathematical models applied to HLW disposal, the findings of which have been published.[27]

5.3 MAJOR DEVELOPMENT PROGRAMS IN THE UNITED STATES

5.3.1 West Valley Demonstration Project

The HLW generated at West Valley, New York, resulted from the operation of the world's first commercial fuel reprocessing plant for 7 years by the original owner, Nuclear Fuel Services, Inc. The source of radioactivity in the HLW was fission products from the spent fuel from 10 nuclear reactors, which was reprocessed during 27 separate campaigns from 1966 to 1972. The variations in kinds of fuels and the first use of the chop/leach head-end process, where the cladding is not removed chemically but is cut up and the product is dissolved or leached out, led to the use of a variety of chemicals and materials.[28] The WVDP is to demonstrate the solidification and preparation of HLW for permanent disposal, including solidification of liquid HLW in a form suitable for transportation and disposal, development of containers suitable for permanent disposal, disposal of LLW and TRU waste produced thereupon, and decontamination and decommissioning of tanks, facilities, material, and hardware used in the solidification process.

The flow sheet for HLW processing in the project is shown in Figure 5.12. The slurry-fed ceramic melter (SFCM) is shown in Figure 5.13, and the physical properties of the glass waste form produced from the SFCM are shown in Table 5.4. The process rate of the WVDP SFCM (150 L/hr) is four times that of the Radioactive Liquid Feed Ceramic Melter (RLFCM) used in the Tokai Plant in Japan and of the Pamela Plant in West Germany, which is 30–35 L/hr. The vitrification operation will last 1–1 1/2 years and produce a total of 300 low-heat cans.

5.3.2 In Situ Testing: Climax Stock and Colorado School of Mines

The Climax spent fuel test facility at the Nevada Test Site (NTS) undertook experiments to (1) simulate the thermal environment of a panel of a full-scale repository, (2) determine the effects of heat in combination with intense ionizing radiation (^{60}Co source) on the

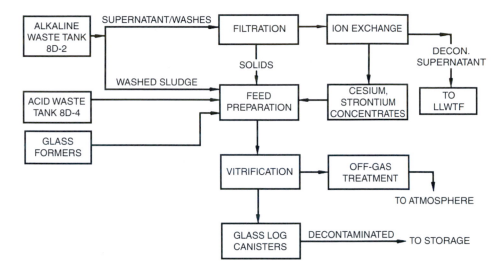

Figure 5.12 West Valley process flow diagram. From Conceptual design of HLW vitrification process at West Valley using a SFCM by G. M. Hughs, et al., in *Proc. 2nd International Symposium on Ceramics in Nuclear Waste Management, Advances in Ceramics*, vol. 8, p. 145, American Ceramic Society, 1983. Copyright 1983 American Ceramic Society. Reprinted by permission.

canister environment, and (3) measure thermal and thermomechanical responses of the facilities and compare them with model calculations.[30]

The Colorado School of Mines (CSM) participated in the U.S.–Canada Joint Radioactive Waste Management Program by contributing the excavation technology experience from their experimental mine. The work involves (1) assessing effects of blasting on the rock mass, (2) determining constitutive relationships for crystalline rocks, and (3) evaluating the heated flat-jack test as a method for obtaining the mechanical properties of jointed rock masses for input to thermomechanical models. Extensometer and leveling pins were installed during the construction of the experimental mine. In situ stress, fracture parameter, and permeability measurements will be made. In situ tests are also being carried out in other countries, some of them under a cooperative program with the U.S. government. These experimental programs are described in Section 5.4.

5.3.3 Nuclear Waste Terminal Storage (NWTS) Program

The NWTS program considered three different geologic environments: salt, basalt, and tuff. The geologic and geochemical conditions for these media are sufficiently different that some variations in design approaches and materials and component choices are dictated.

Salt studies. National, regional, and area surveys resulted in activities in three identified areas, namely the Paradox basin, the Permian basin, and the Gulf Coast basin.[31] The Paradox basin covers an area of about 28,500 km^2 (11,000 square miles) including southeastern portions of Utah and southwestern portions of Colorado. It is composed of

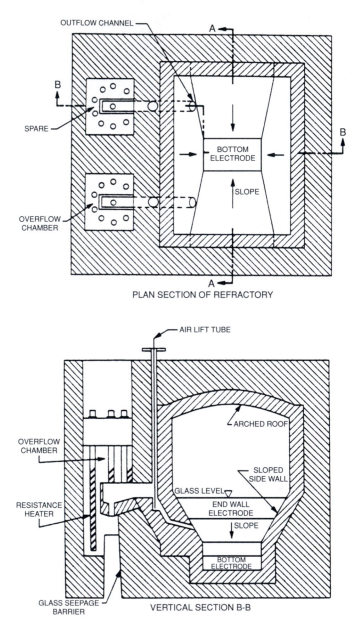

Figure 5.13 Schematic of slurry-fed ceramic melter. From Design preferences for a slurry-fed ceramic melter suitable for vitrifying West Valley wastes by C. C. Chapman, in *Proc. International Symposium on Ceramics in Nuclear Waste Management, Advances in Ceramics,* vol. 8, p. 149, American Ceramic Society, 1983. Copyright 1983 American Ceramic Society. Reprinted by permission.

Table 5.4 Physical properties of glass waste forms

Property	Value
Thermal conductivity (100°C)	0.55 Btu/hr ft °F (0.95 W/m °C)
Heat capacity (100°C)	0.22 cal/g °C (0.22 Btu/lb °F)
Fractional thermal expansion[a]	1.22×10^{-5}/°C (0.68×10^{-5}°F)
Young's modulus[b]	9×10^6 psi (6.3×10^5 kg/cm²)
Tensile strength	9×10^3 psi (6.3×10^2 kg/cm²)
Compressive strength	1×10^5 psi (7.0×10^3 kg/cm²)
Poisson's ratio[c]	0.2
Density (100°C)[a]	2.75 g/cm³ (172 lb/ft³)
Softening point[a]	502°C (936°F)

[a]Experimentally determined for Frit 131 glasses.
[b]Young's modulus, or the modulus of elasticity, measures the stiffness of the material.
[c]Poisson's ratio is equivalent to the ratio of equatorial to axial strain under an applied axial stress.
  From Survey of foreign terminal radioactive waste storage programs by K. M. Harmon, in *Proc. 1983 Civilian Radwaste Management Information Meeting*, p. 229, DOE OCRWM, February 1984. Reprinted by permission.

roughly 30 individual salt beds with depth varying from 122 m (400 ft) to 2720 m (8920 ft) and thickness from 46 m (150 ft) to 2900 m (9505 ft). The Permian basin covers an area of approximately 310,680 km² (120,000 square miles) including portions of Kansas, Oklahoma, Colorado, Texas, and New Mexico. The depth of the basin varies from 137 to 1160 m (450–3810 ft), and thickness varies from 9 to 260 m (30–853 ft). The Gulf Coast basin contains both salt beds and salt domes, covering an area of about 298,000 km² (115,000 square miles); it encompasses parts of Alabama, Louisiana, Mississippi, and Texas and extends southward beneath the Gulf of Mexico. Thickness of the salt beds is estimated to be 305–5334 m (1000–17,500 ft). The number of salt domes is estimated to be more than 300, with diameters varying from 1.6 to more than 6.4 km (1–4 miles) and depths ranging from 20 to 3050 m (65–10,000 ft).[32] One site (Deaf Smith location) in the Permian basin was nominated for recommendation. Repository conceptual design and waste package design in salt have also been studied.[6]

Basalt Waste Isolation Project (BWIP). Established in 1976 as part of the NWTS program, the BWIP is to assess the feasibility of siting, constructing, and operating a repository in the basalt, of meeting all the specific regulatory criteria, and of providing the necessary engineering technology. The Hanford site in southeast Washington was one of the sites recommended for the first geologic repository for commercial HLW. Site screening and characterization work included the drilling of more than 100 small-diameter boreholes to better understand the hydrology and geology of the Hanford site and the Pasco Basin. A key technical issue is a determination of the key hydrologic parameters for groundwater motion. Of four candidate basalt horizons, all of which are below 914 m (3000 ft), a preferred horizon (the Cohassett Flow) at 1036 m (3400 ft) was identified. The conceptual design of a repository that can hold 70,000 MTHM was completed. Several repository field tests were initiated: Drillability tests were started at the Colorado School of Mines to evaluate efficient methods of mining basalt. Overcoring tests provided a measure of the basalt stress levels in situ and demonstrated the ability to drill coreholes from a shaft 1.8 m (6 ft) in diameter in the presence of high-pressure, high-temperature water. Initial data have been obtained from the jointed block test in the Near-Surface Test Facility as well as field hydrofracturing tests in deep boreholes.[33]

Nevada Nuclear Waste Storage Investigations Project (NNWSI). This project investigated the possible location of a geologic repository in the volcanic tuff formation. Through systematic screening, the Yucca Mountain site was selected for recommendation as a repository horizon. Hydrology and geology studies included surface and fracture mapping, drilling, permeability and porosity measurements, and study of saturated and unsaturated zones and microfractures. The age of the underlying basalt was determined to evaluate the chances of volcanic disruption, and faults were characterized for type, age, and pattern of occurrence.[34] Waste package and repository design concepts for tuff were also developed. As a result of a congressional amendment of December 1987, this site was selected for characterization.

5.4 DEVELOPMENT PROGRAMS IN OTHER COUNTRIES

The development of methods for safe disposal of HLW and reliable performance assessment has been actively pursued in countries such as the United Kingdom (AERE), Canada (AECL), Sweden (SKB), West Germany (GSF), Belgium (Mol), France, Japan, Finland, and Switzerland. It was realized that the level of sophistication needed to carry out comprehensive assessments could best be met by sharing experience and coordinating R & D internationally. Several international or bilateral cooperative projects are described briefly below.

5.4.1 The Stripa Project

This project has as participants Canada, Finland, France, Japan, Sweden, Switzerland, and the United States. Granite rock formations have attracted considerable attention in many NEA member countries because they have properties that may be suitable for the isolation of HLW and because of their relatively common existence in these countries, with ages of many hundreds of millions of years. The physical and chemical properties of hard crystalline rocks are complex, particularly in terms of groundwater movement, fracture networks, and chemistry for a repository environment. To investigate the behavior of hard rocks in such circumstances, in situ experiments and measurements must be performed at a depth similar to that of a conceptual repository. At the Stripa mine in Sweden such depths can be reached to appropriate comparable geologies. The project had two phases. Phase one (1980–1983) included hydrologic and geochemical investigations in deep boreholes, rock mechanical investigations, tests of migration of injected tracers in fractures, and a buffer mass test for the temperature, pressure, and water uptake into compacted bentonite around electrical heaters. Phase two, started in 1983, involves the development of cross-hole geophysical investigation techniques, sealing of

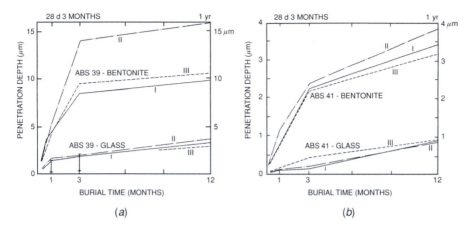

Figure 5.14 Corrosion depth (micrometers) in glass versus burial time: (a) ABS 39 and (b) ABS 41. Upper curves: glass in contact with bentonite. Lower curves: glass–glass contact. Graphs: I, inner edge of Ca-rich "plateau"; II, onset of unchanged B concentration ("uncorroded bulk"); III, position halfway up the plateau concentration of Al. From Analysis of one year in situ burial of nuclear waste glasses in stripa by L. L. Hench, A. Lodding, and L. Werme, in *Proc. 2nd International Symposium on Ceramics in Nuclear Waste Management, Advances in Ceramics,* vol. 8, p. 321, American Ceramic Society, 1983. Copyright 1983 by American Ceramic Society. Reprinted by permission.

boreholes, shafts, and tunnels, a large-scale three-dimensional tracer migration test, and fracture hydrology tests.[35] Before the international project was established in 1977–1980, investigations were conducted in the Stripa mine by Lawrence Berkeley Laboratory (LBL) in cooperation with the Swedish Nuclear Fuel Supply Company (SKBF/KBS). These experiments were aimed at developing techniques to determine near-field rock mechanics and far-field hydrologic, geochemical, and geophysical parameters at potential repository sites. Figure 5.14 is an example of the in situ experiment at the Stripa mine to test possible synergistic interactions of the materials in a nuclear waste storage system. Two nuclear waste glasses containing 9% simulated waste, alkali borosilicate (ABS) 39 and ABS 41, were used for the burial experiment because their compositions (Table 5.5) were close to that to be used for commercial solidification

Table 5.5 Compositions of two waste glasses used in the Stripa burial experiment

Glass	SiO$_2$	B$_2$O$_3$	Al$_2$O$_3$	Na$_2$	Fe$_2$O$_3$	Zn	Li$_2$O	UO$_2$	Simulated fission product (%)
ABS39	48.5	19.1	3.1	12.9	5.7	0	0	1.7	9
ABS41	52.0	15.9	2.5	9.9	3.0	3.0	3.0	1.7	9

From Analysis of one year in-situ burial of nuclear waste glasses in Stripa by L. L. Hench, A. Lodding, and L. Werme, in *Proc. International Symposium on Ceramics in Nuclear Waste Management, Advances in Ceramics,* vol. 8, p. 311, Table 1, American Ceramic Society, 1983. Copyright 1983 by American Ceramic Society. Reprinted by permission.

operations in France.[36] The thickness of the reaction layer formed at 90°C during 1 year's burial in Stripa for glass–glass interfaces is calculated to be 1×10^{-6} m $(3.9 \times 10^{-5}$ in.) for ABS 41 and 3×10^{-6} m for ABS 39. Thus an alkali–boron depletion zones for glass–glass interfaces at the end of the thermal period of storage (300 years) would be no more than 0.3 mm (0.12 in.) for glass ABS 41 and 0.9 mm (0.035 in.) for glass ABS 39, even if water breached the canister immediately after burial. Corrosion zones at glass interfaces with bentonite may be two to five times as thick.[36]

5.4.2 Asse Salt Mine Project

The HLW Test Disposal Project at the Asse Salt Mine in West Germany has been carried out by the Institut für Tieflagerung (IfT) of the Gesellschaft für Strahlen und Umweltforschung mbH Munchen (GSF). The experiments at the mine are considered to be pilot tests for the future nuclear repository and are carried out in cooperation with the Energy Research Center of the Netherlands (ECN).[37] A significant step with regard to the test disposal of HLW is represented by the Asse brine migration tests, which are part of the joint U.S.–German Cooperative Radiative Waste Management Agreement and were started in May 1983. The following issues were addressed:[38]

1. Brine migration (liquid and vapor).
2. Radiation effects of gamma rays.
3. Gas generation caused by radiation and corrosion.
4. Accelerated corrosion and leaching.
5. Altered properties of salt (effects of heat, radiation, and brine).
6. Effects of heat and radiation on test assemblies, instruments, and various materials exposed to repository conditions.

The tests were designed to have a maximum salt temperature of 210°C (410°F) and a radiation dosage of $(0–3) \times 10^8$ rad/year, simulated with a ^{60}Co source. Figure 5.15 shows the Asse test gallery of four sites with electrical heat input but only two sites containing ^{60}Co gamma sources. The test site cross section is shown in Figure 5.16. Evaluation of the temperature response of the experiment over the first 6 months of operation (Figure 5.17) indicated that (1) the initial heatup of the experiment was very close to the predicted rate and magnitude and (2) except for the initial magnitude of the temperature rise, the shape of the temperature distribution in salt is very close to predictions.[38,39]

Figure 5.18 shows the motions of the salt around the test site, specifically the horizontal closure of the room at site 2. Also shown are two computer predictions, one by DAPROK (a finite-element rock mechanics code used by D'Appolonia) and the other MAUS, a similar code used by IfT. It appears that the MAUS code agrees better with the measured data from reference 39. Results of the thermomechanical experiments performed in the 300-m (984-ft) hole in the Asse mine are reported in reference 40. The experiments were performed under thermal conditions relevant to the present design of a repository. It could be concluded that an initial radial gap of 5 cm (2 in.) between container and borehole was closed because of the thermomechanical behavior of the salt within 20 days. After that, compression builds up on the container. For a repository at a

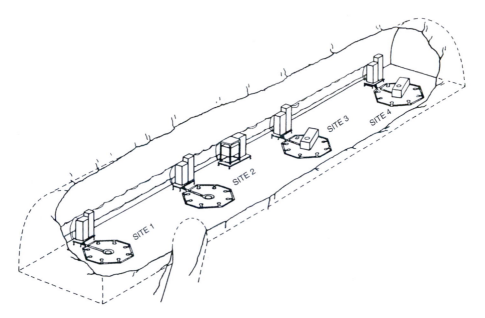

Figure 5.15   Asse test gallery. (Only two sites contain ^{60}Co gamma sources.) From Asse salt mine nuclear waste repository simulation experiments by A. J. Coyle, in *Waste Management '83*, edited by R. G. Post, University of Arizona Press, Tucson, Ariz., 1983. Copyright 1983 by University of Arizona Press. Reprinted by permission.

depth of 1000 m (3280 ft) and a lithostatic pressure of 22 MPa (3190 psi), the maximum compression will be 40 MPa (5800 psi).[40]

5.4.3 Joint Program with Canada

The overall objective of the U.S.–Canada bilateral agreement is to cooperate on mutually agreed topics of interest that are associated with radwaste management, including the following:[41]

1. Preparation and packaging of wastes
2. Decontamination and decommissioning
3. Surface and subsurface storage
4. Geologic characterization and disposal
5. Environmental and safety considerations
6. Public acceptance issues
7. Transportation

The United States is cooperating in the Canadian Underground Research Laboratory (URL) project, which has the following components:

1. Excavation technology and blast damage assessment are provided by CSM in reviewing URL design and test plans and in a technology transfer workshop.

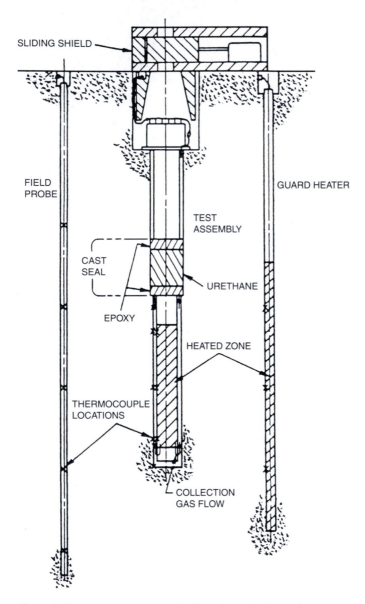

Figure 5.16 Asse test site cross section. From Asse salt mine nuclear waste repository simulation experiments by A. J. Coyle, in *Waste Management '83*, edited by R. G. Post, University of Arizona Press, Tucson, Ariz., 1983. Copyright 1983 by University of Arizona Press. Reprinted by permission.

2. Instrumentation development includes a modified Bureau of Mines gage for long-term stress monitoring, modifications of a block test, and a sonic probe borehole extensometer suitable for displacement measurement during rock excavation.

3. Geohydrology support through hydrologic modeling and analysis of fracture and well test data from the URL site is provided by Lawrence Berkeley Laboratory, which

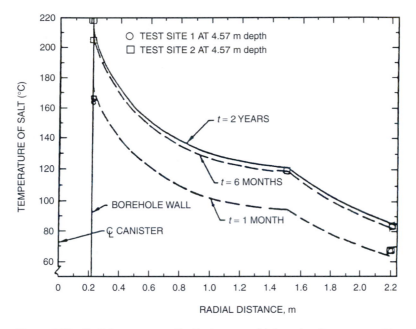

Figure 5.17 Radial temperature distribution around brine migration test assembly at heater midline. Actual readings at sites 1 and 2 coincide. From Westinghouse Nuclear Waste Department, Brine Migration Test for Asse Mine, FRG by W-AESD, AESD-TME-3095, p. 101, May 1981. Copyright 1981 by Westinghouse Nuclear Waste Department. Reprinted by permission.

has developed two- and three-dimensional numerical models to determine whether a fracture system behaves as a continuum and to estimate the size of the representative equivalent volume (the smallest volume of rock that may be treated as a continuum). The application of borehole geophysical characterization of crystalline rock on the URL site was studied by the U.S. Geological Survey (USGS), including resistivity logs, neutron and gamma–gamma logs, and an acoustic bore-hole televiewer.

5.4.4 Clay Disposal Program in Belgium[42]

The Belgian Nuclear Research Establishment (CEN/SCK) has carried out work at Mol, where the Boom clay formation has favorable characteristics for waste disposal. In the first 5-year program (1975–1979), experimental work was carried out on site and in the laboratory, while theoretical studies were devoted to examination of the technical and economic feasibility of disposal at a depth of 220 m (722 ft) in the Boom clay. The main issues of the second 5-year program (1980–1984) were as follows:

1. Construction of an underground laboratory, collection of in situ data, confirmation of laboratory observations, and demonstration of the capabilities for underground structures in clay.

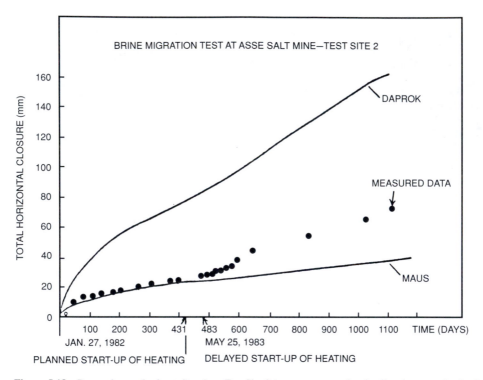

Figure 5.18 Room closure (horizontal) at Asse Test Site 2 (measurements taken by directly measuring horizontal distance between two extensometer stations).[39] From Asse salt mine nuclear waste repository simulation experiments by A. J. Coyle, in *Waste Management '83*, edited by R. G. Post, University of Arizona Press, Tucson, Ariz., 1987; and HLRW Test Disposal Project in the Asse salt mine-FRG by T. Rothfuches and R. Stippler, in *Waste Management '87*, edited by R. G. Post, University of Arizona Press, Tucson, Ariz., 1987. Copyright 1987 by University of Arizona Press. Reprinted by permission.

2. Continuation of laboratory work on samples to study the physicochemical equilibria in the natural clay and the effects of clay/waste package interactions.
3. Installation of an extended network of hydrologic observation wells and modeling of the groundwater flow regime.
4. Evaluation of the performance and safety of the disposal concept.

Figure 5.19 shows the scheme of the as-built underground experimental facility at Mol. The experience gained over the past several years has provided valuable information for the demonstration and pilot phase of the project. The site-specific investigations on the Boom clay confirm the suitability of a deep clay formation for a radwaste repository. The recent experiences narrow down the variability of the parameters for the design, feasibility, and safety studies. In situ experiments in the underground laboratory related to the excavation, lining, and construction of underground facilities provide data needed for the construction of an underground repository in deep clay.

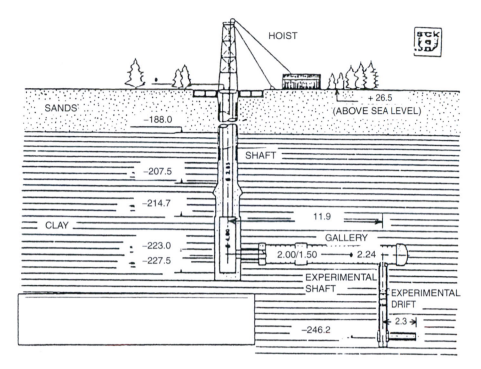

Figure 5.19   Scheme of the as-built underground experimental facility at Mol, Belgium. From Disposal of Radwaste in Clay: The Experience Gained in Belgium by A. Bonne and B. Neerdael, Belgium Nuclear Research Establishment (CEN/SEK), IAEA, Geneva, 1986. Reprinted by permission.

5.5 EXPERIMENTAL PROGRAMS FOR DISPOSAL SYSTEMS

Experimental evidence for the safety of deep geologic burial can be obtained in laboratory experiments, field experiments, and in situ experiments. Laboratory experiments usually involve reasonably small specimens and determine specific properties, while field and in situ experiments involve characterization of the local geology and hydrology at a particular site. The difference between field experiments and in situ experiments is in the depth of the test location and whether the configuration closely resembles a repository in terms of waste canister dimensions, spacing, and thermal output as well as burial room characteristics.

Most of the laboratory and field experiments were performed by the Spent Fuel Handling and Packaging Program and are discussed in greater detail in Chapter 3. Most of these experiments, however, also provide information that is valuable to understanding the disposal system. Technology development testing on PWR spent fuel elements was carried out at the Engine-Maintenance Assembly and Disassembly (E-MAD) facility at the Nevada Test Site, using isolated drywells and concrete casks. In addition to the maximum temperatures of fuel clad, canister, and liner, the axial and radial fuel assembly temperatures under varying gas coolant conditions inside the canister were measured with external thermal impedance conditions representing a variety of system types. The maximum fuel cladding temperature was well below the 380°C imposed on these tests. This value was calculated conservatively as a temperature to which the fuel cladding could be exposed for an infinite time without clad rupture due to creep. Figure 5.20 shows the typical fuel clad temperature test results at these conditions. For a specified lifetime (time before rupture) of cladding under storage conditions with air

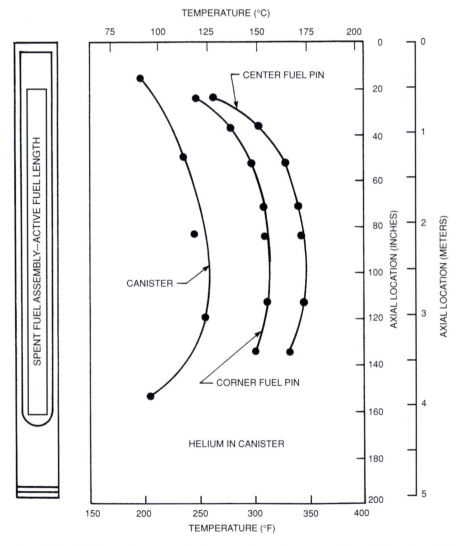

Figure 5.20 Fuel temperature test results for imposed drywell storage conditions. Test performed in the hot cell with spent fuel. From Heat transfer associated with dry storage of spent LWP fuel by G. E. Driesen et al., in *Heat Transfer in Nuclear Waste Disposal*, ASME HTD vol. 11, p. 16, ASME, 1980. Copyright 1980 by American Society of Mechanical Engineers. Reprinted by permission.

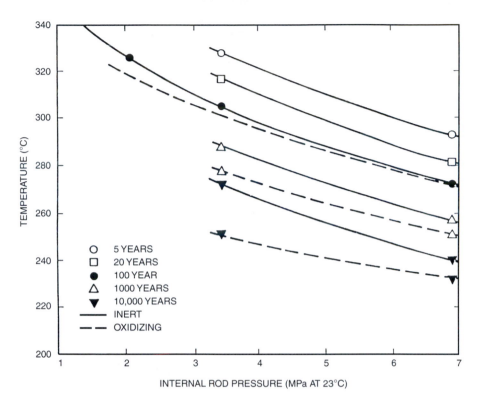

Figure 5.21   Isothermal storage temperature for designed lifetime of cladding. From Low-temperature stress rupture behavior of PWR spent fuel rods under dry storage conditions by R. E. Einziger and R. Hohli, Presented at the Commercial Spent Fuel Management Program Annual Technical Review Meeting, Richland, Wash., October 1983. Reprinted by permission.

or inert gas in the canister, the isothermal storage temperature corresponding to room temperature and internal rod pressure have been calculated. The results are shown in Figure 5.21. Both in situ and field experiments have contributed to our knowledge of thermal effects, rock mechanics, hydrology, and excavation technology, and no serious discrepancies have been found between observations and predictions. Brief descriptions of Project Salt Vault and the Avery Island experiments follow.

5.5.1 Project Salt Vault[43,44]

This was the first in situ test of the disposal of HLW; it was done in an inactive salt mine in Lyon, Kansas, in 1965–1967. Spent fuel assemblies from Idaho Falls National Engineering Laboratory supplemented with electric heaters were used, and data on the properties and behavior of salt in situ were collected as the heat sources were located at an array of holes in the floor of newly mined rooms at a depth of about 1000 ft. Rock mechanical and thermal properties of the salt formation were measured, including thermal conductivity, heat capacity, thermal expansion coefficients, stress–strain characterstics, elastic and plastic constants, and creep behavior. The accomplishment was recognized in the statement that "Project Salt Vault provided one of the first demonstrations of the effects of thermal loading imposed on a salt mine. It is of interest to note that the results have been reproduced analytically through specific independent computer simulations."[45] Negligible radiation effects were detected.

5.5.2 Experiments at Avery Island[46,47]

At Avery Island, Louisiana, in situ tests were done to confirm the Project Salt Vault results for dome salt deposits. Electric heaters were used to simulate the thermal characteristics of radwaste. Data were obtained on brine migration, temperature distribution and stress in salt, stability of salt rock, and corrosion of candidate sleeve materials. The volume of brine predicted to flow into the heater borehole was in agreement with the amount of moisture measured in the in situ experiment. The computation at low temperatures is sensitive to the interpolation of the temperature-dependent proportionality constant in the equation for the rate of brine inclusion movement, which was derived on the basis of theories and measurements for bedded salt.

5.6 SOCIOECONOMIC IMPACTS AND INSTITUTIONAL ISSUES

5.6.1 Socioeconomic Impacts

An important aspect of the disposal of HLRW relates to its socioeconomic effects. In addition to effects that result from unique features and public perceptions of a repository, there are also effects that result from development-related growth. These effects can be categorized as follows.[48]

1. Demographic changes resulting from the arrival of a new population and the characteristics in which the immigrants differ from the host population.
2. Economic impacts from the promotion of new business, raised cost of living due to higher wages, economic downturn after repository closure, and effects on tourism and agriculture.
3. Community-service impacts, including housing and community facilities and services.
4. Social impacts due to changes in age distribution, ethnic composition, male-to-female ratio, lifestyles, cultural traditions, and political views, and increased social problems caused by rapid growth, land acquisition, and public concerns.
5. Fiscal impacts, such as increases in tax revenue and funds for financial assistance.

5.6.2 Institutional Issues[48]

The institutional issues that could impede the implementation of the NWPA include the following:

1. Failure to reach or to implement a consultation and cooperation agreement and to develop full participation capabilities.

2. State or local laws that are in conflict with federal laws or otherwise incompatible with DOE's responsibilities. Such conflict may occur if the state or local government imposes substantive or procedural requirements that prevent the DOE from fulfilling its responsibilities under the NWPA in a timely manner.

3. State or tribal notice of disapproval of a site selected for a repository as provided by the NWPA. Such a notice of disapproval would eliminate that site from further consideration for a repository unless both houses of Congress override the notice by passing a joint resolution that approves the selection of the site. For instance, Mississippi is concerned about the impact on its relatively small land holdings and its economy tied to farming, small manufacturing, and forest products industries; Texas worries about the effects on the water source (Ogallala Aquifer) and the marketability of food products grown nearby; and Utah is concerned that the potential siting would affect national parks and other nearby state parks and tourism.

4. Impediments to the transportation of waste. Regardless of where facilities for the storage or permanent isolation of waste are eventually sited, spent fuel from reactors will be transported through several states. Concerns about this may attract the political involvement of elected officials in a large number of states and of Indian tribes that may be affected by transportation activities. Among the issues that may arise are routing, travel time, prenotification of states, escorts and safeguards, emergency response capability, and preparedness. (Further discussion may be found in Chapter 11.)

REFERENCES

1. Electric Power Research Institute, Geologic disposal of nuclear waste, *EPRI J.* (Palo Alto, Calif.), May 1982.
2. Crandall, J. L., H. Krause, C. Sombret, and K. Uematou, High level waste processing and disposal, *Trans. ANS/AO*, vol. 48, p. 106, 1985.
3. Heath, C. A., Review panel comments on geomedia specific research, in *Scientific Basis for Nuclear Waste Management VII*, p. 455, Elsevier, New York, 1984.
4. Harmon, K. M., Survey of foreign terminal radioactive waste storage programs, in *Proc. 1983 Civilian Radioactive Waste Management Information Meeting*, December 1983.
5. Chipman, N. A., and T. E. Carleson, HLW volume reduction using the modified Zirflex fuel dissolution process, in *Spectrum '86, Proc. ANS Int. Topical Meeting, Waste Management and Decontamination and Decommissioning*, J. M. Pope et al., eds., p. 1156, Niagara Falls, New York, September 1986.
6. Goldsmith, S., NWPA—The salt geologic waste repository project, Presented at the AIF Fuel Cycle Conference '84, Atlanta, April 1984.
7. McClain, W. C., and A. L. Boch, Disposal of radwaste in bedded salt formations, *Nucl. Technol.*, vol. 24, p. 398, 1974.
8. Nuclear Regulatory Commission, Disposal of High Level Radioactive Waste in Geologic Repositories, Title 10, Part 60, U.S. Code of Federal Regulations, June 1983.
9. U.S. Environmental Protection Agency, Environmental Standards for the Management and Disposal of Spent Nuclear Fuel, High Level and Transuranic Wastes, Title 40, Part 191, U.S. Code of Federal Regulations, 1982.
10. Pigford, T. H., The National Research Council study of the isolation system for geologic disposal of radioactive waste, in *Scientific Basis for Nuclear Waste Management VII*, p. 482, Elsevier, New York, 1984.
11. Cheung, H., et al., Nuclear Waste Form Risk Assessment for U.S. Defense Waste at SRP, Annual Report FY1982 UCRL-53188-82, Lawrence Livermore Laboratory, 1982.

12. Chambré, P. L., C. H. Kang, and T. H. Pigford, Flow of groundwater around buried waste, *Trans. ANS*, vol. 52, p. 77, 1986.
13. Pigford, T. H., and P. L. Chambré, Reliable Predictions of Waste Performance in a Geologic Repository, LBL-20166, Lawrence Berkeley Laboratory, 1985.
14. Chambré, P. L., T. H. Pigford, W. W.-L. Lee, J. Ahm, S. Kajiwara, C. L. Kim, H. Kimura, H. Lung, W. J. Williams, and S. J. Zavoshy, Mass Transfer and Transport in a Geologic Environment, LBL-19430, Lawrence Berkeley Laboratory, 1985.
15. Kim, C. L., P. L. Chambré, and T. H. Pigford, Mass-transfer-limited release of a soluble waste species, *Trans. ANS*, vol. 52, p. 80, 1986.
16. Chambré, P. L., Y. Hwang, W. W.-L. Lee, and T. H. Pigford, Release rates from waste packages on a salt repository, *Trans. ANS/AO*, vol. 55, p. 131, 1987.
17. Erickson, R. L., Approximations for adapting porous media radionuclide transport models to analysis of transport in jointed, porous rock, in *Scientific Basis for Nuclear Waste Management VI*, p. 473, Elsevier, New York, 1983.
18. Miller, C. W., Toward a comprehensive model of chemical transport in porous media, in *Scientific Basis for Nuclear Waste Management VI*, p. 481, Elsevier, New York, 1983.
19. Siegel, M. D., M. S. Chu, and R. E. Pepping, Compliance assessments of hypothetical geologic nuclear waste isolation systems with the draft EPA standard, in *Scientific Basis for Nuclear Waste Management VI*, vol. 15, p. 497, Elsevier, New York, 1983.
20. Cranwell, R. M., Modelling of waste/near field interactions for a waste repository in bedded salt: The dynamic network (DNET) model, in *Scientific Basis for Nuclear Waste Management VI*, p. 507, Elsevier, New York, 1983.
21. Tsang, C. F., J. Noorishad, and J. S. Y. Wang, A study of coupled thermomechanical, thermohydrological and hydromechanical processes associated with a nuclear waste repository in a fractured rock medium, in *Scientific Basis for Nuclear Waste Management VI*, p. 515, Elsevier, New York, 1983.
22. Liebetran, A. M., et al., AREST: A probabilistic source-term code for waste package performance analysis, in *Waste Management '87*, R. G. Post, ed., vol. 1, p. 535, University of Arizona, Tucson, Ariz., 1987.
23. Carlyle, S. G., The activities, objectives and recent achievements of the NEA probabilistic system assessment codes (PSAC) users group, in *Waste Management '87*, R. G. Post, ed., vol. 1, p. 661, University of Arizona, Tucson, Ariz., 1987.
24. Merz, E. R., Time period of concern for judging the long-term hazards of geological disposal of radwaste, in *Waste Management '87*, R. G. Post, ed., vol. 2, p. 181, University of Arizona Press, Tucson, Ariz., 1987.
25. Birchard, G. F., and D. H. Alexander, Natural analogues—A way to increase confidence in predictions of long-term performance of radioactive waste disposal, in *Scientific Basis for Nuclear Waste Management VI*, p. 323, Elsevier, New York, 1983.
26. Randall, J. D., and F. A. Constanzi, Results from and issues raised by NRC's HLRW research program's integrated experimental modeling and model validation work, in *Waste Management '87*, R. G. Post, ed., vol. 1, p. 153, University of Arizona Press, Tucson, Ariz., 1987.
27. Randall, J. D., et al., eds. Validation of Mathematical Models for Waste Repository Performance Assessment—Confidence Building through Synthesis of Experiments and Calculations, NUREG Report, Nuclear Regulatory Commission, Washington, D.C., 1987.
28. Duckworth, J. P., and P. Burn, Generation of high level waste at West Valley, N.Y., in *Proc. 1983 Civilian Radwaste Management Information Meeting*, Washington, D.C., December 1983.
29. Office of Nuclear Waste Isolation, Battelle Memorial Institute, Preliminary Thermomechanical Analysis of a Conceptual Nuclear Waste Repository at Four Salt Sites, BMI/ONWI-512, Battelle Memorial Institute, Columbus, Ohio, 1985.
30. U.S. DOE, Mission Plan for the Civilian Radioactive Waste Management Program, DOE/RW0005, vol. 1, pt. 1, 1985.
31. U.S. DOE, Briefing on the National Waste Terminal Storage Program—Salt Studies, Energy Committee, Southern Legislative Conference, Savannah, Ga., 1983.
32. Lefond, S. J., *Handbook of World Salt Resources*, Plenum Press, New York, 1969, pp. 1, 21, 25.
33. Ash, E. B., The Basalt Waste Isolation Project—Status and issues, Presented at the Atomic Industrial Forum Fuel Cycle Conference, Atlanta, Ga., 1984.

34. Spaeth, M. E., et al., Overview of Nevada Nuclear Waste Storage Investigations Project activities, Presented at the Atomic Industrial Forum Fuel Cycle Conference, Atlanta, Ga., 1984.

35. NEA, Summary and conclusions, in *Proc. Workshop on Geological Disposal of Radwaste, In-Situ Experiments in Granite*, Stockholm, 1982.

36. Hench, L. L., A. Lodding, and L. Werme, Analysis of one year in-situ burial of nuclear waste glasses in Stripa, in *Proc. 2nd International Symposium on Ceramics in Nuclear Waste Management, Advances in Ceramics,* vol. 8, p. 310, American Ceramic Society, 1983.

37. Rothfuches, T., and R. Stippler, HLRW test disposal project in the Asse Salt Mine—FRG, in *Waste Management '87*, R. G. Post, ed., vol. 1, p. 89, University of Arizona Press, Tucson, Ariz., 1987.

38. Coyle, A. J., Asse Salt Mine nuclear waste repository simulation experiments, in *Waste Management '83*, R. G. Post, ed., University of Arizona Press, Tucson, Ariz., 1983.

39. Eckert, J. L., H. N. Kalia, and A. J. Coyle, Quarterly Brine Migration Data Report, January–March 1985, Nuclear Waste Repository Simulation Experiments Technical Report BMI/ONWI-650, 1987.

40. Prij, J., Results of the thermomechanical experiments performed in the 300 m hole in the Asse Mine (FRG), in *Waste Management '87*, R. G. Post, ed., vol., 2, p. 187, University of Arizona Press, Tucson, Ariz., 1987.

41. Robinson, R. A., and N. Bulut, U.S. joint program with Sweden and Canada: Crystalline rock, in *Proc. 1983 Civilian Radioactive Waste Management Information Meeting*, p. 227, 1983.

42. Bonne, A., and B. Neerdael, Disposal of Radwaste in Clay: The Experience Gained in Belgium, Belgium Nuclear Research Establishment (CEN/SEK), IAEA, Geneva, 1986.

43. Hench, L. L., D. E. Clark, and J. Campbell, HLW immobilization forms, *Nucl. Chem. Waste Manage.*, vol. 5, p. 149, 1984.

44. Jantzen, C. M., et al., Leaching of polyphase nuclear waste ceramics: Microstructural and phase characterization, *J. Am. Ceram. Soc.*, vol. 65, no. 6, p. 292, 1982.

45. Harker, A. H., et al., Formulation and processing of polyphase ceramics for HLW, in *Scientific Basis for Nuclear Waste Management IV*, p. 567, Elsevier, New York, 1983.

46. Pacific Northwest Laboratory, Materials Characterization Center (MCC) Test Methods, Preliminary Version, PNL-3990, Richland, Wash., 1981.

47. Macedo, P. B., A. Barkatt, and J. H. Simmons, A flow model for the kinetics of dissolution of nuclear waste glasses, *Nucl. Chem. Waste Manage.*, vol. 3, p. 13, 1982.

48. U.S. DOE, Mission Plan for the Civilian Radioactive Waste Management Program, vol. 1, pt. II, DOE/RW0005, June 1985.

CHAPTER SIX

TRANSURANIC WASTE

6.1 TRANSURANIC WASTE. DEFINITIONS, SOURCES, AND INVENTORIES

6.1.1 Transuranic Elements

The elements of atomic numbers greater than 88 form a transition group called the actinide series. The elements of greatest interest within the actinide series are those with atomic numbers greater than 92 (uranium), which are called transuranic (TRU) elements. These elements and their half-lives are shown in Table 6.1; many of the radioisotopes are alpha emitters.

6.1.2 Definition of Transuranic Waste

TRU waste is defined in the United States as radwaste that is not classified as HLW but contains an activity of more than 100 nCi/g from alpha-emitting TRU isotopes with half-lives greater than 20 years. No consensus has been reached worldwide on the numerical limit used here; the definition varies in different countries from 0.3 to 1000 nCi/g. What is common to all countries is that they prefer and are planning to use deep geologic repositories for final disposal of TRU waste.[1]

The U.S. definition of TRU waste given above is used in 40CFR191.[2] Even in the United States, the limit of 100 nCi/g was changed from an earlier (1970) number of 10 nCi/g as a threshold value for TRU waste.[3,4] The justification can be found in the proceedings of a workshop on the management of alpha-contaminated waste that was held in 1982.[5] The program committee of the workshop concluded that, for purposes of classification, radwastes not classified as HLW and having concentrations greater than

Table 6.1 TRU elements and alpha emitters

Atomic number	Element name	Chemical symbol	Nuclide	Half-life (yr)	α-Emitter specific activity (Ci/g)
93	Neptunium	Np	Np-237	10^6	7.1×10^{-4}
94	Plutonium	Pu	Pu-238 to Pu-244	87 to 10^7	1.8×10^{-5} to 182
95	Americium	Am	Am-241 to Am-243	150 to 7×10^3	0.2 to 10.3
96	Curium	Cm	Cm-243 to Cm-250	18 to 10^7	9.3×10^{-5} to 83
97	Berkelium	Bk	Bk-247	1.4×10^3	1.1
98	Californium	Cf	Cf-249 to Cf-252	2.6 to 9×10^2	1.6 to 540
99	Einsteinium	Es	—	—	—
100	Fermium	Fm	—	—	—
101	Mendelevium	Md	—	—	—
102	Nobelium	No	—	—	—
103	Lawrencium	Lr	—	—	—
104	Rutherfordium	Rf	—	—	—
105	Hahnium	Ha	—	—	—

100 nCi/g of long-lived alpha-emitting radionuclides should be designated as TRU wastes. Those containing long-lived alpha-emitting radionuclides at concentrations below this level may be classified as LLW and destined for near-surface disposal.

Before 1970, most TRU waste in the United States was buried directly, often in concrete containers. Since 1974, TRU waste has been placed in interim storage (20-year retrievable) in steel, concrete, or wooden boxes on surface storage pads. Although a few daughter products have energetic gamma emission, most TRU waste can be handled with the shielding provided by the waste package itself. This waste is classified as contact-handled TRU waste (CHTRU). Less than 3% of the retrievably stored TRU waste inventory contains enough beta, gamma, and neutron emitters to produce a dose of greater than 200 mrem/hr at the package surface, thus requiring that the package be handled remotely. Such waste is designated as remotely handled TRU waste (RHTRU).

6.1.3 Sources, Inventories, and Characteristics of TRU Waste

The principal sources of TRU are fuel reprocessing and waste immobilization facilities, fuel fabrication for recycle of plutonium, weapons material (defense) production, and decommissioning of nuclear reactors or fuel cycle facilities. Lakey et al. reviewed the TRU waste-producing countries throughout the world and listed their sources(s) of such waste, as shown in Table 6.2.[1] Of 26 countries that have nuclear reactors, 10 have reported TRU waste generation. As facilities begin to be decommissioned, these wastes will increase.

An accurate assessment of the buried volumes of TRU waste at the Department of Energy (DOE) sites is difficult because early burial practices were not governed by the current requirement for identification and segregation. Before 1970, transuranic waste

Table 6.2 Countries producing transuranic wastes

Country with reactor	Spent fuel reprocessing	Mixed-oxide fuel fabrication	Weapons material production
Argentina	—	—	—
Belgium/Eurochemie	X	X	—
Brazil	—	—	—
Bulgaria	—	—	—
Canada	—	—	—
China (PRC)	X	—	X
Czechoslovakia	—	—	—
Finland	—	—	—
France	X	X	X
East Germany	—	—	—
West Germany	X	X	—
Hungary	—	—	—
India	X	X	X
Italy	X	X	—
Japan	X	X	—
Korea (South)	—	—	—
Netherlands	—	—	—
Pakistan	—	—	—
Spain	—	—	—
Sweden	—	—	—
Switzerland	—	—	—
Taiwan	—	—	—
United Kingdom	X	X	X
United States	X	X	X
USSR	X	X	X
Yugoslavia	—	—	—

From Management of transuranic waste throughout the world by L. T. Lakey, et al., *Nucl. Chem. Waste Manage.*, 1983, vol. 4, p. 35. Copyright 1983 by Pergamon Press Ltd. Reprinted by permission.

was buried on-site in shallow DOE landfills. Since 1970, TRU waste has been placed in retrievable storage until it can be transported to a permanent disposal facility. This waste is called "retrievably stored" transuranic radioactive waste. Table 6.3 and Figure 6.1 show the calculated inventories at the end of fiscal year 1996 and projections through fiscal year 2033 of retrievably stored TRU waste from defense and commercial activities.[20]

Most transuranic waste is solid waste consisting of protective clothing, equipment, or cleaning cloths used in areas where spent fuel is reprocessed. Some sludges remaining after Pu or other transuranic elements are removed during reprocessing are also TRU wastes. Approximately half of TRU waste generated to date is thought to include hazardous wastes governed by the Resource Conservation and Recovery Act (RCRA).[17] As a result, a TRU repository must also be designed to confine some hazardous wastes.

Table 6.3 Summary of as-generated waste-form volumes (m³) of retrievably stored and projected mixed and nonmixed waste

Sites	CHTRUW			RHTRUW			Grand Total at end of FY2033
	Stored at end of FY1996	Projected waste during FY1997–2033	Total at end of FY2033	Stored at end of FY1996	Projected waste during FY1997–2033	Total at end of FY2033	
Major sites							
ANL-E	80.6	166.9	247.5	0.0	0.0	0.0	247.5
Hanford	11,008.0	6,270.0	17,278.0	203.0	1,660.0	1,863.0	19,140.9
INEEL	64,760.2	550.0	65,310.2	62.0	0.0	62.0	65,372.2
LANL	8,610.1	6,218.6	14,828.7	93.2	33.8	127.0	14,955.7
LLNL	239.6	645.6	885.1	0.0	0.0	0.0	885.1
Mound	235.8	18.0	253.8	0.0	0.0	0.0	253.8
NTS	618.2	8.0	626.2	0.0	0.0	0.0	626.2
ORNL	921.1	370.0	1,291.1	1,283.0	200.0	1,483.0	2,774.1
RFETS	1,889.2	3,218.1	5,107.2	0.0	0.0	0.0	5,107.2
SRS	6,033.5	8,348.7	14,382.3	0.57	0.0	0.57	14,382.8
Subtotal	94,396.2	25,813.9	120,210.0	1,641.7	1,893.8	3,535.5	123,745.6
Small-quantity sites							
Subtotal	25.43	170.44	195.87	30.63	393.50	424.13	620.00
Commercial site							
WVDP	37.41	143.64	181.05	483.63	28.56	512.19	693.27
Grand total	94,459.0	26,128.0	120,586.9	2,156.0	2,315.9	4,471.8	125,058.9

Derived from Integrated Data Base Report-1996: U.S. Spent Nuclear Fuel and Radioactive Waste Inventories, Projections, and Characteristics by U.S. DOE, Report DOE/RW-0006, Rev. 13, 1996, U.S. Department of Energy, Washington, D.C., 1997.

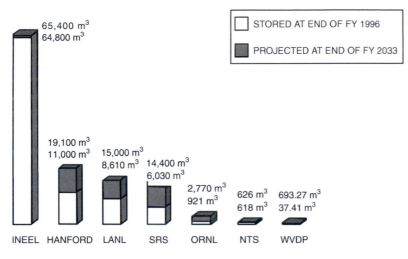

Figure 6.1 Current and projected inventories of stored TRU waste at DOE sites to 1996 (open bars) and 2033 (shaded bars). Data for figure from Integrated Data Base Report-1996: U.S. Spent Nuclear Fuel and Radioactive Waste Inventories, Projections, and Characteristics by U.S. DOE, Report DOE/RW-0006, Rev. 13, 1996, U.S. Department of Energy, Washington, D.C., 1997.

6.2 PROCESSING, STORAGE, AND TRANSPORTATION OF TRANSURANIC WASTE

Transuranic waste is currently being disposed of at the Waste Isolation Pilot Plant (WIPP) near Carlsbad, New Mexico. Any TRU waste sent to WIPP must meet strict waste acceptance criteria. As a result, those waste acceptance criteria must be considered when the TRU waste is packaged. This section outlines waste acceptance criteria and then describes TRU packaging, storage methods, and transportation.

6.2.1 Waste Acceptance Criteria[8]

Requirements in the WIPP Waste Acceptance Criteria are organized into six categories:

1. Container properties
2. Radiologic properties
3. Physical properties
4. Chemical properties
5. Gas generation properties
6. Data package contents

All TRU waste disposed at WIPP must meet these requirements. Each DOE site that generates TRU waste must have detailed plans and procedures that guarantee that its shipments of TRU waste comply with the waste acceptance criteria. Furthermore, the site must send to WIPP paperwork certifying that a shipment of TRU waste complies with the criteria *before* the shipment is sent.

Details of the WIPP Waste Acceptance Criteria can be found in reference 8. A general description of the types of requirements covered by each of the six categories follows.

1. *Container properties*: (a) types of containers allowed, (b) container weights, (c) removable surface contamination limits, (d) container identification markings, (e) types of filters used in vents if a container is vented.

2. *Radiologic properties*: (a) radionuclides present, (b) concentration of ^{239}Pu (fissile material), (c) alpha activity of transuranic isotopes, (d) radiation dose rate at container surfaces and 2 m from container.

3. *Physical properties*: (a) amount of free liquid allowed, (b) limits on the size of sealed containers.

4. *Chemical properties*: (a) limits on pyrophoric materials, (b) limits on hazardous waste, (c) no explosives, corrosives, or compressed gases, (d) all containers must be sampled for volatile organic compounds (VOCs), (e) limits on concentration of polychlorinated biphenyl concentration, (f) presence of asbestos must be disclosed.

5. *Gas generation properties*: (a) decay heat limits, (b) hydrogen gas generation rate limits, (c) flammable VOC concentration limits, (d) length of time retrievably stored drums must be vented before shipment.

6. *Data package contents*: (a) characterization of the waste, (b) bill of lading, (c) land disposal restriction notification.

6.2.2 Packaging and Storage

WIPP did not begin accepting TRU waste until 1999. Between the early 1970s and 1999 TRU waste was stored for later retrieval and disposal. A variety of packages are currently in use for retrievable storage of TRU waste. Several sizes of drums, boxes, metal pipes, and concrete cylinders are used. Drums and boxes account for most of the storage volume. The steel drums have a 208-L (55-gal) capacity, and the plywood or steel boxes generally have outside dimensions of 213 cm (84 in.) (length) × 122 cm (48 in.) (width) × 132 cm (52 in.) (height). The height includes a 10-cm (3.9-in.) footing block to facilitate handling. Additional waste is contained in approximately 10 different types of containers. Specifications have been proposed for standardized packaging of TRU-contaminated materials (Table 6.4). The standards would require that all waste be placed in either 208-L metal drums or rectangular metal boxes. Four sizes of rectangular boxes have been approved.[6]

6.2.3 Transportation

Since the transport of radioactive waste is necessary to place it in temporary storage, to provide additional processing of the waste, or to place it in a permanent disposal facility, a design was developed by Sandia National Laboratories for a container that can transport a variety of CHTRU waste forms. The transporter, called TRUPACT (transuranic

Table 6.4 TRU waste packages for retrievable storage and transportation

Waste package	Outside dimensions (cm)	Volume (m³)		Required storage space (m³)	
		Outside	Inside	Rectangular array	Triangular array
Principal waste packages in current use					
Metal drums (nominal 55 gal)	62.2 diam × 90.8 H	0.276	0.21	0.352	0.304
Plywood boxes	213.4 L × 121.9 W × 132 H	3.43	2.92	3.43	
Proposed standardized waste packages[a]					
Metal drums (nominal 55 gal)	62.2 diam × 90.8 H	0.276	0.21	0.352	0.304
Metal boxes					
Box A	189.2 L × 128.0 W × 97.8 H	2.4	2.3	2.4	
Box B	172.7 L × 137.2 W × 97.8 H	2.3	2.2	2.3	
Box C	223.5 L × 137.2 W × 137.2 H	4.2	4.1	4.2	
Box D	284.5 L × 172.7 W × 193.6 H	9.5	9.4	9.5	

[a] Data from reference 9.

From Order 5820.2, Radioactive Waste Management, Department of Energy, Washington, D.C., February 6, 1984. Reprinted by permission.

package transporter), was designed for safe containment of TRU waste and protection from physical and thermal abuse under hypothetical accident conditions. It was a large rectangular box that contained drums or boxes of waste. TRUPACT was suitable for either rail or truck shipment. Figure 6.2 shows the TRUPACT container in a truck configuration with three containers per trailer.

However, in 1980, the Department of Energy decided to ship TRU wastes in Type B packages with certification from the Nuclear Regulatory Commission (NRC). Contact-handled TRU will be shipped in TRUPACT-II containers, while remote-handled TRU will be shipped in RH-72B containers. TRUPACT-II is cylindrical in shape, about 8 ft in diameter and 10 ft high. It holds fourteen 55-gal drums weighing about 500 lb each or two 63-ft³ waste boxes. The waste drums or boxes are placed in a central, double-walled, stainless steel chamber. Surrounding the chamber are 10 in. of polyurethane foam and ½ in. of ceramic fiber. Finally, everything is encased in another stainless steel cylinder. Up to three TRUPACT-II containers can be shipped on one truck. The RH-72B is approximately 12 ft long, longer than the TRUPACT-II. However, it is only 3.5 ft in diameter. The RH-72B also has inner and outer cylindrical vessels. In addition, a 2-in.-thick lead liner is added to the container to provide extra protection against gamma rays.

Both the TRUPACT-II and the RH-72B must satisfy four rigorous tests in order to receive NRC certification as Type B containers: (1) free drop, (2) puncture, (3) thermal, and (4) immersion. The tests are described in Chapter 11.

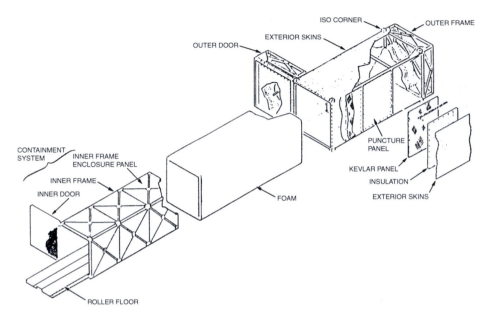

Figure 6.2 The TRUPACT-1 packaging schematic. From Design Team Response to Peer Review of the Preliminary Design for the TRU Transporter, TRUPACT by R. B. Pope et al., Report SAND 82-1493 (TTCO326), Sandia National Laboratory, January 1983; and Transportation of Contact Handled TRU Wastes by R. T. Haelsig, G. J. Ginn, R. A. Johnson, and K. L. Fergeson, *Waste Management '88*, University of Arizona Press, Tuscon, Ariz., 1988. Reprinted by permission.

6.3 REGULATIONS AND STANDARDS FOR TRANSURANIC WASTE

The Waste Isolation Pilot Plant was not required to be licensed by the NRC. However, it was required to meet EPA standards specified in 40CFR191 and 40CFR194. In addition, to accept TRU waste containing hazardous waste, WIPP was required to obtain an RCRA Part B permit. Shipments of TRU waste must also comply with DOT regulations.

6.3.1 EPA Regulations

Regulations promulgated by the U.S. Environmental Protection Agency (EPA) are published as Title 40 of the Code of Federal Regulations (40CFR). Part 191, Environmental Radiation Protection Standards for Management and Disposal of Spent Nuclear Fuel, High-Level and Transuranic Radioactive Wastes (40CFR191), sets limits on the radiation dose the general public can receive from SNF, HLW, or TRU wastes. Subpart A, Section 191.03, states[2] as follows:

(a) Management and storage of spent nuclear fuel or high-level or transuranic radioactive wastes at all facilities regulated by the Commission or by Agreement States shall be conducted in such a manner as to provide reasonable assurance that the combined annual dose equivalent to any member of the public in the general environment resulting from: (1) Discharges of radioactive material and direct radiation from such management and storage and (2) all operations covered by Part 190; shall not exceed 25 millirems to the whole body, 75 millirems to the thyroid, and 25 millirems to any other critical organ.

Subpart B of 40 CFR191 specifies actions the disposal facility operator must take to ensure long-term isolation of the wastes buried in a repository.

Part 194 (40CFR194) is specific to transuranic waste. It is entitled Criteria for the Certification and Re-Certification of the Waste Isolation Pilot Plant's Compliance with the 40CFR Part 191 Disposal Regulations. This regulation was issued in 1996. On May 13, 1998, the EPA certified that WIPP met all applicable federal nuclear waste disposal standards.

Because some of the transuranic waste also contains hazardous waste, WIPP was required to have an RCRA Part B permit. 40CFR270 is the EPA regulation that prescribes the process of applying for and receiving the RCRA permit. However, New Mexico is an "agreement state" and has been given authority to issue RCRA Part B permits. An "agreement state" is one that has incorporated regulations at least as stringent as those in pertinent federal regulations (in this case 40CFR270) into state law, identified a state agency to enforce the regulations, and demonstrated that the agency's personnel are adequately trained to do the job. In New Mexico, the RCRA regulations are in Title 20, Chapter 4, of the state laws, and the responsible agency is the New Mexico Environment Department.

6.3.2 U.S. Department of Transportation Regulations

For shipments and packages, the U.S. Department of Transportation (DOT) regulations in 49CFR173 set forth requirements for the transportation of radmaterials by carriers and shippers (these requirements are in addition to those of 10CFR71), including activity limits and radiation level limitations for packages, transportation rules for all packages, and prescribed tests to be performed on these packages.

6.4 WASTE ISOLATION PILOT PLANT

The Waste Isolation Pilot Plant is constructed in the salt beds of southeast New Mexico. Its mission as authorized by Congress in 1979 was "for the express purpose of providing a research and development facility to demonstrate the safe disposal of radwaste resulting from the defense activities and programs of the U.S. exempted from regulation by the NRC." Thus it is to provide for permanent isolation of TRU wastes generated by defense programs. It was also to provide an "underground laboratory" in which the concepts for safe disposal of defense high-level waste (DHLW) in salt would be validated and demonstrated. However, in later years, DOE was forbidden to bring any high-level waste to WIPP.

The WIPP project issued a geologic characterization report for the site in 1978 and published a final environmental impact statement (EIS) in October 1980. From July 1981 to March 1983, two shafts and about 3.2 km (2 miles) of underground drift were excavated at the proposed facility depth of 2150 ft in order to implement the Site and Preliminary Design Verification (SPDV) program. The underground exploration and construction of prototype TRU waste rooms provided additional confidence in the WIPP site and the safety of the room design. After the publication of the summary of geotechnical studies and the SPDV program in March 1984 and a formal comment period and public hearing, construction of the full WIPP facility began with the sinking of a third shaft in the October 1984.

This section outlines the site evaluation process that led to a final decision to construct WIPP, describes the WIPP facilities, and discusses some legal and regulatory actions between the time WIPP construction was completed in 1989 and the first waste shipment was received in 1999.

6.4.1 Site Evaluation Studies

The following site selection criteria were used:

1. A bed of rock salt (halite) at least 60 m thick, of purity sufficient to minimize chemical complications from brine of complex composition and from water released from hydrous minerals.
2. A depth greater than 300 m to ensure freedom from surface influence.
3. A depth less than 1000 m to ensure acceptably low creep rates in the salt.
4. Approximate horizonality to minimize difficulty in mining operations.
5. Little indication of recent tectonic or igneous activity.
6. Sufficient distance from an exposed edge or underground aquifer where salt dissolution is occurring.
7. An area without a history of resource extraction and sizable economic resources.

The geologic and hydrogeologic studies to evaluate the suitability of the WIPP site progressed to the point that the site was recommended for full facility construction in March 1983. Site evaluation studies began with compilation of relevant geotechnical knowledge by the U.S. Geological Survey (USGS) in the early 1970s. Drilling of initial exploratory holes began in 1974, and the present site was identified for detailed investigation in November 1975. Extensive geologic mapping and correlation of borehole data were done through more than 350 petroleum industry boreholes penetrating the evaporites in a 2300-km² (900 square mile) area centered on the WIPP site and many potash exploration holes penetrating into the middle of the Salado Formation. Structure and isopach maps based on these holes aid in understanding salt-dissolution and salt-deformation issues. The industry borehole data, together with about 1500 line-miles of petroleum industry seismic reflection data and regional gravity and aeromagnetic surveys, were already available before detailed site characterization began. Seismic reflection profiling was most useful, and nearly 200 line-miles was run at the WIPP site to evaluate structure with the evaporites.

A representative geologic cross section through the site is shown in Figure 6.3, where the excavations are approximately centered in the more than 915 m (3000 ft) of evaporites constituting the Salado and Castile Formations. Further geotechnical studies were done at

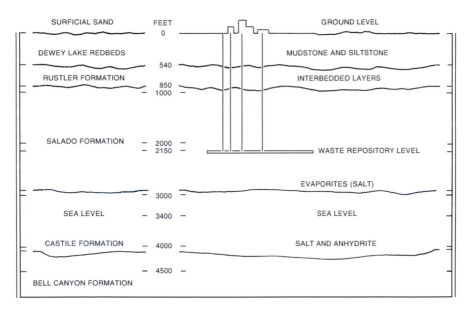

Figure 6.3 Geologic section through the WIPP site. Modified after U.S. Geological Survey cross section for the U.S. Department of Energy, November 1979. From *WIPP, Waste Isolation Pilot Plant* by U.S. DOE, WIPP-DOE 164, p. 5, U.S. Department of Energy, 1984. Reprinted by permission.

the site for three major purposes: better understanding of salt dissolution at and near the site, investigation of mechanisms for deformation of evaporite rock sequences, and improved definition of the hydrology. Studies showed the transport of isotopes via groundwater supplies to be highly improbable. Even if transport occurred, transport time to the nearest water supply used by humans would be on the order of 500,000 years. With regard to potential conflict with natural resources, the WIPP site was selected to minimize the potential conflict with natural resources while fulfilling other site selection criteria. It is not possible, however, to avoid potential conflict with hydrocarbons in the Delaware Basin, and the WIPP site does have some potash resources within its boundary. Natural gas may occur beneath the WIPP site at depths of 3050–4500 m (10,000–15,000 ft). These depths may be explored beneath WIPP without drilling through the salt within the site boundaries by using existing technology for drilling deviated holes. The important issue related to hydrocarbons at the site is not the existence of resources but the possible attractiveness of the geologic setting for future exploration at a time when the existence of WIPP may have been forgotten. It is impossible to guarantee that drilling into or through WIPP cannot happen; however, based on several breach scenarios, the study concluded that the radiologic consequences of such intrusions to humans in the future are not significant. Breaches of the salt barrier by natural processes are considered as not credible.

6.4.2 WIPP Facilities

WIPP is located on a 10,240-acre site near Carlsbad, New Mexico. The facilities at WIPP are classified as surface facilities and subsurface facilities.

Surface facilities. Principal on-site surface facilities are the waste handling building, support building, exhaust/filter building, warehouse and shops, sewage treatment plant, and other auxiliary buildings. Separate areas are provided in the waste handling building for CHTRU and RHTRU handling. The CHTRU waste area has an exterior truck and rail loading and unloading area, a receiving and inspection area, an inventory and preparation area, and an overpack and repair room. The RHTRU area includes the shipping and receiving area, cask preparation/decontamination area, cask unloading room, hot cell for preparing canisters for transfer underground, and a facility cask loading area. Also in the waste handling building are a solid radioactive waste collection/compaction room, HVAC equipment area, mechanical/electrical equipment rooms, and worker changing area. Site-generated waste comes from decontamination of externally contaminated casks and waste packages. Liquid wastes are accumulated and solidified before being packaged for disposal. Critical surface waste handling facilities are designed to withstand a tornado with wind speeds up to 183 mph and a design-basis earthquake imposing 0.1 g on the facilities.

The support building contains offices, radiologic control facilities, change rooms, and personnel equipment storage. The exhaust filter building houses the high-efficiency particulate aerosol (HEPA) filter banks, which can be switched into the exhaust only on command or, as a fail-safe feature, in the event of a power failure. The surface facilities are shown in Figure 6.4.

Subsurface facilities. The underground TRU waste isolation and experimental areas are reached by three shafts—construction and salt handling shaft, waste shaft, and exhaust shaft—all of which are lined through the sedimentary rocks overlying the salt and are unlined below about 850 ft. The construction and salt handling shaft has an inside diameter of 10 ft. With a hoist capacity of 10 tons and an 8-ton skip, it allows a muck removal rate of 100 tons/hr. This shaft also provides for electrical power and signal cable routing and for personnel access. The waste will be moved to the isolation horizon through the waste shaft, which has a diameter of 19 ft and a hoist capacity of 45 tons. The exhaust shaft is 14 ft in diameter and is the exhaust duct for all the underground ventilation air. The subsurface facilities are shown in Figures 6.5 and 6.6.

The underground excavation originally included the TRU waste isolation, experiment, and support areas. The waste rooms will be reached by four parallel entries to allow separation of waste room and construction ventilation. Each TRU waste room is 13 ft high, 33 ft wide, and 300 ft long, and the rooms are separated from each other by 100-ft pillars. CHTRU waste drums and boxes are stacked in the rooms, which are backfilled with salt. As one room is filled with waste, the next room is being excavated and some of the salt muck is used to backfill the voids in the room being filled. The 10-ft-long RHTRU waste canisters are placed 6 ft deep in horizontal holes drilled into the pillars of the CHTRU waste rooms. During the pilot phase, these holes will be lined with steel to facilitate retrievability, and a 5-in. steel plug will be placed in the mouth of the hole to provide adequate shielding from the radiation. The RHTRU waste will be emplaced using a specially designed and shielded facility cask and transporter, while the CHTRU waste will be moved and emplaced by transport and forklift equipment requiring a minimum amount of shielding.

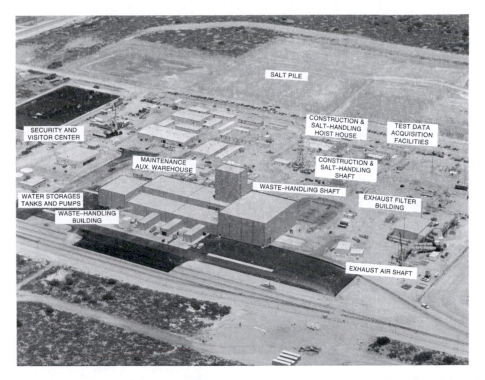

Figure 6.4 Surface facilities and storage areas for WIPP. (1) Sewage treatment plant, (2) switch yard, (3) main substation, (4) waste shaft, (5) construction and salt handling shaft, (6) construction and salt handling shaft head house, (7) construction and salt handling shaft hoist house, (8) exhaust shaft, (9) waste handling building, (10) exhaust filter building, (11) water storage tanks, (12) support building, (13) warehouse/shops building, (14) water pump house, (15) surface salt storage area, (16) saltwater evaporation pond, (17) service building, (18) salt storage area (service), (19) soil stockpile (service), (20) security building. From *WIPP, Waste Isolation Pilot Plant* by U.S. DOE, WIPP-DOE 164, p. 13, U.S. Department of Energy, 1984. Reprinted by permission.

The WIPP has been designed for a nominal operating life of 25 years. For the first 5 years of operation (pilot phase), all CHTRU and RHTRU waste will be retrievable, and operations during this phase will be limited to handling rates of 7080 m^3/year (250,000 ft^3/year) of CHTRU waste and 10 canisters of RHTRU waste. Normal operations would accommodate 14,158 m^3/year (500,000 ft^3/year) of CHTRU and 283 m^3/year (10,000 ft^3/year) or 400 canisters/year of RHTRU waste. The current design capacity for WIPP is about 1.78×10^5 m^3 (6.3×10^6 ft^3) CHTRU and 5100 m^3 (180,000 ft^3) RHTRU.

6.4.3 Legal and Regulatory Actions

Underground construction of WIPP was completed in 1989. However, it would be 10 years before the first shipment of TRU waste arrived at WIPP for permanent disposal. Much of the delay was due to the need to meet a number of legal and regulatory requirements, some of which were imposed after WIPP construction was complete.

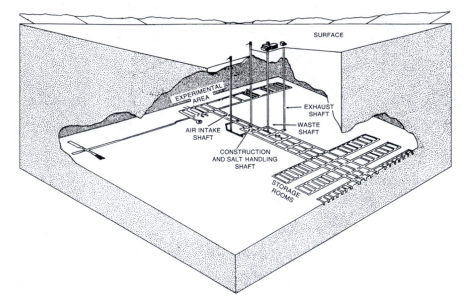

Figure 6.5 Cutaway diagram of the WIPP underground design. From *WIPP, Waste Isolation Pilot Plan* by U.S. DOE, WIPP-DOE 164, p. 12, U.S. Department of Energy, 1984. Reprinted by permission.

A few of these requirements and the associated timelines are described in the following paragraphs. More detail is provided in reference 10.

First, WIPP was located on federal land managed by the U.S. Department of the Interior. DOE formally requested the withdrawal of 10,240 acres from public use and for

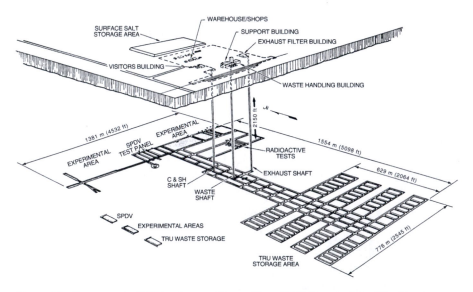

Figure 6.6 Representative WIPP underground layout. From *WIPP, Waste Isolation Pilot Plant* by U.S. DOE, WIPP-DOE 164, 1986, U.S. Department of Energy, p. 8. Reprinted by permission.

exclusive use by WIPP in 1989. In October of 1992, Congress passed the WIPP Land Withdrawal Act (PL 102–579) making the land available for WIPP. In 1996, Congress passed an amendment to this law, which prohibited use of WIPP for experiments on disposal of high-level waste. This guaranteed that no HLW waste would be shipped to WIPP. The amendment confirmed a decision DOE had made in 1993 to conduct the HLW disposal experiments elsewhere.

WIPP was required to comply with 40CFR191,[2] the EPA regulation limiting the radiation dose the public can receive from SNF, HLW, or TRU waste. However, 40CFR194,[11] Compliance Criteria for WIPP Certification, was not issued until 1996. DOE submitted a Compliance Certification Application of 84,000 pages to EPA in October 1996. In May 1998, the EPA certified that WIPP met all applicable federal nuclear waste disposal standards.

As early as 1986, the U.S. EPA asserted that WIPP needed to comply with RCRA requirements for disposal of radioactive waste that also contained hazardous waste. In 1990, EPA recognized New Mexico as an agreement state and authorized the New Mexico Environment Department (NMED) to regulate hazardous and mixed waste disposal in the state. In 1995, DOE submitted an application for an RCRA Part B permit to NMED. The permit was issued in October 1999.

6.5 DISCUSSION QUESTIONS AND PROBLEMS

1. What are the principal sources of TRU waste?

2. What is the definition of TRU waste?

3. What is the name of the container in which CHTRU waste is licensed to be transported? RHTRU?

4. Describe the underground rooms in which CHTRU and RHTRU will be buried and describe how those two types of material will be placed in the rooms.

5. If you were a WIPP representative and asked to speak to residents in areas near WIPP, what questions about WIPP do you think they would have? Suggest four possible questions, and answer them. Choose some questions not answered in this chapter. Additional information on WIPP can be found at the WIPP website http://www.wipp.carlsbad.nm.us/wipp.htm.

REFERENCES

1. Lakey, L. T., H. Christensen, et al., Management of transuranic waste throughout the world, *Nucl. Chem. Waste Manage.*, vol. 4, p. 35, 1983.
2. U.S. EPA, Environmental Standards for the Management and Disposal of Spent Nuclear Fuel, High Level and Transuranic Waste, Title 40, Part 191, U.S. Code of Federal Regulations, June 1983.
3. U.S. NRC, Licensing Requirements for Land Disposal of Radwaste, Title 10, Part 61, U.S. Code of Federal Regulations, 1982.
4. Hollingsworth, K. S., Policy Statement Regarding Solid Waste Burial, AEC Directive IAD No. 0511-21, March 1970.
5. Science Applications, Inc., *Proc. of Alpha-Contaminated Waste Management Workshop*, CONF-820845, Oak Ridge National Laboratory, Gaithersburg, Md., August 1982.

6. U.S. DOE, Integrated Data Base Report-1996, Rev. 13, Report DOE/RW-0006, U.S. Department of Energy, Washington, D.C., 1997.

7. U.S. DOE, Integrated Data Base Report-1993, Rev. 10, Report DOE/RW-0006, U.S. Department of Energy, Washington, D.C., 1994.

8. U.S. DOE, Waste Acceptance Criteria for the Waste Isolation Pilot Plant, Rev., Report DOE/WIPP-069, U.S. Department of Energy, Washington, D.C., 1999.

9. U.S. DOE, Spent Fuel and Radwaste Inventories, Projections and Characteristics, Report DOE/NE-0017-1, 1982.

10. U.S. DOE, WIPP Milestones, U.S. Department of Energy, Carlsbad Area Office, June 1999.

11. U.S. EPA, Criteria for the Certification and Re-certification of the Waste Isolation Pilot Plant's Compliance with the 40CFR Part 191 Disposal Regulations, Title 40, Part 194, U.S. Code of Federal Regulations, July 2000.

ADDITIONAL READINGS

Charlesworth, D. L., Process and mechanical development for the Savannah River TRU waste facility, in *Spectrum '86*, American Nuclear Society, September 1986.

Chaturvedi, L., et al., Performance assessment for a nuclear waste repository—The WIPP experience, in *Waste Management '87*, vol. 1, R. G. Post, ed., University of Arizona Press, Tucson, Ariz., March 1987.

Haelsig, R. T., G. J. Quinn, R. A. Johnson, and K. L. Ferguson, Transportation of Contact Handled TRU Wastes, in *Waste Management '88*, University of Arizona Press, Tucson, Ariz., March 1988.

Jensen, R. T., Inventories and characteristics of TRU waste, *Nucl. Chem. Waste Manage.*, vol. 4, p. 19, 1983.

Lewis, E. L., TRU Waste Certification: Experimental Data and Results, Mound Laboratory Report MLM-3096, Miamisburg, Ohio, 1983.

ORNL, *Proc. 7th International Symposium on Packaging and Transportation of Radmaterials, PATRAM '83*, vol. 1, sponsored by Transportation Technical Center, Sandia National Laboratories and Oak Ridge National Laboratory, May 1983.

TRU Waste Systems Office, TRU Waste Certification Compliance Requirements for Newly-Generated Contact-Handled Wastes for Shipment to the WIPP, WIPP-DOE-114, 1981.

U.S. DOE, Long Range Master Plan for Defense TRU Waste Management, Report DOE-TRU-8201, 1982.

U.S. DOE, Order 5820.1, Radioactive Waste Management, 1983.

U.S. DOT, Shippers General Requirements for Shipments and Packages, Title 49, Part 173, U.S. Code of Regulations, 1984.

CHAPTER SEVEN

LOW-LEVEL WASTE

7.1 INTRODUCTION

Having defined HLW, SNF, and TRU waste in previous chapters, we now consider the remaining radwaste, which belongs to the category of low-level waste (LLW). As indicated in Section 1.2.2, LLW often has relatively little radioactivity and contains practically no transuranic elements. Some LLW, however, may have high enough radioactivity to require special treatment and disposal. Also included in LLW is the category called naturally occurring and accelerator-produced radmaterials (NARM). Because of their unique contents, the tailings from uranium mills constitute another type of waste, also with low-level radioactivity, and are treated separately. Types of LLW include contaminated dry trash, paper, plastics, glass, clothing, discarded equipment and tools, wet sludges, and organic liquids. This chapter covers the management of LLW; that of mill tailings will be discussed in Chapter 8. Sections are presented in this chapter on sources of LLW, different forms and compositions of LLW, historical background, state compacts and regulations, waste treatment processes, packaging and handling LLW, economic evaluations, operational experiences with volume reduction systems, and shallow land disposal and its development.

7.1.1 Sources and Volumes of LLW

Low-level waste is produced by the Department of Energy at many of its facilities and by "commercial" generators, which include nuclear power plants, medical facilities, industry, academic institutions, and non-DOE government facilities. The DOE and its predecessors have been generating LLW in defense and other national programs since the inception of the nuclear weapons programs during World War II. Commercial generation of LLW, primarily by nuclear power plants and medical facilities, began in the early 1960s, and wastes from industrial activities added to the steadily increasing volume of commercial wastes that required proper treatment and disposal.

DOE generates LLW in day-to-day operations and during cleanup of contaminated sites as part of its environmental restoration program. Wastes generated during operation include protective clothing, cleaning cloths, laboratory equipment, environmental monitoring samples, and contaminated hand tools. Cleanup operations typically produce high-volume wastes with low Curie content. Examples are slightly contaminated soils or rubble from demolished buildings where surfaces of the walls had some embedded contamination.

Commercial LLW from nuclear power plants results from activities such as maintenance and replacement of contaminated equipment, collection and analysis of environmental samples, laundering or disposal of protective clothing, and cleanup of small spills. Another type of LLW generated at nuclear power plants is resins used to clean water in spent fuel pools. As pool water is pumped through these resins, they collect spent fuel particles that escape from cracked or pitted fuel rods. The resins are often highly radioactive and must be treated and disposed of separately from other LLW.

Medical facilities generate LLW during diagnoses and treatment of patients using radioactive materials. Academic institutions and non-DOE government facilities generally use small amounts of radioactive material, often as tracers in experiments. Many university laboratories may use radioactive materials, but the total amount of waste is generally small. There are many industrial uses of radioactive materials such as in gages for measuring thickness of paper or metal, sterilization of medical supplies, irradiation of food to kill bacteria, and nondestructive testing. Again, the volumes of LLW are low except in cases when a facility is being decommissioned and all contaminated equipment and structural materials must be disposed of.

Commercial LLW is currently sent to one of three disposal facilities located near Barnwell, South Carolina, Richland, Washington, and Clive, Utah. Four other LLW disposal facilities have been used for various periods in the past. All seven of these facilities will be discussed in more detail later. Figure 7.1 shows the total volume of commercial LLW buried at those sites (except Envirocare near Clive, Utah) between 1962, when the first one opened, and 1999, the latest year for which complete data are available. Envirocare, which opened more recently, accepts high-volume, low-activity waste, primarily large amounts of soil and building rubble, and including it would present a misleading picture.

Superimposed on the graph of buried commercial LLW volume in Figure 7.1 is a graph showing the change in the cost per cubic foot to bury waste at the Barnwell facility. As the cost of LLW disposal at Barnwell has increased by a factor of about 17 between 1986 and 2000,[1] generators have had significant incentive to reduce their waste volumes. Disposal costs vary from facility to facility, and costs are generally lower at the Richland, Washington, and Envirocare facilities. However, since the Richland facility only accepts waste from a few states, and Envirocare only accepts certain types of waste, Barnwell's prices were used.

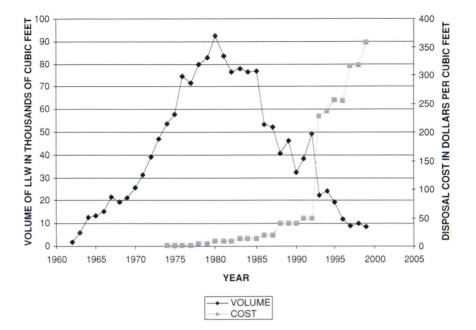

Figure 7.1 Total annual volume of LLW buried at all commercial sites except Envirocare, from 1962 to 1999, and LLW disposal costs between 1974 and 1999. Prepared using information from references 1–3.

Figure 7.2 is a graph showing the Curie content of LLW disposed of at all burial sites except Envirocare between 1963 and 1999. The number of Curies sent to Envirocare to date is so small that adding them to the totals would not change the graph noticeably. While the *volume* of LLW disposed of has decreased significantly in recent years, Figure 7.2 indicates that the *Curie* content has not. The Curie content varies from year to year depending on activities of the generators. For example, the peak in 1978 is probably due to LLW generated during cleanup at the Three Mile Island reactor following a partial meltdown. The peak in 1992 likely occurred because federal law allowed all three LLW disposal facilities operating in 1992 (Barnwell, Richland, and Beatty) to close after December 31, 1992, and many generators made an effort to dispose of as much of their LLW as possible before the end of that year. The peak in 1999 appears to be due to disposal of the reactor vessel from the Trojan nuclear power plant when that plant was decommissioned.

The total volume of commercial LLW buried through 1999 at facilities other than Envirocare was approximately 1.6 million m^3. By contrast, through 1999, the DOE had buried 3.1 million m^3 of LLW at various DOE sites, and DOE expects to bury an additional 1.5 million cubic meters by 2030.[1]

7.1.2 Forms and Compositions of LLW

Of the five categories of commercial LLW generators, utilities often generate the highest volume of waste with the highest Curie content (see Figure 7.3). This section will

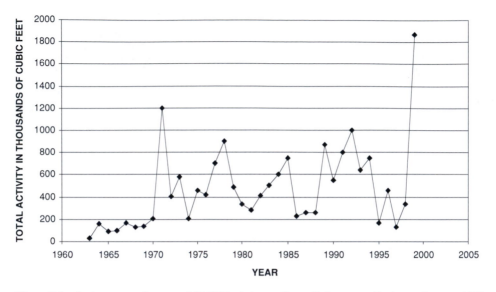

Figure 7.2   Curie content of commercial LLW buried annually at all sites except Envirocare between 1963 and 1999. Prepared using information from references 1 and 2.

therefore focus on the form and composition of LLW from utilities. The LLW generated from nuclear power plants can be in gaseous, liquid, and solid forms.

Gaseous waste   The principal gaseous waste release points from a reactor plant are at the main condenser evacuation system; turbine gland seal system; ventilation system exhaust from containment, auxiliary, turbine, and radwaste buildings; and steam-generator blowdown and steam leakage from the secondary system (pressurized-water reactor, PWR, plants only). Contaminants must be removed from the gaseous waste by filtration,

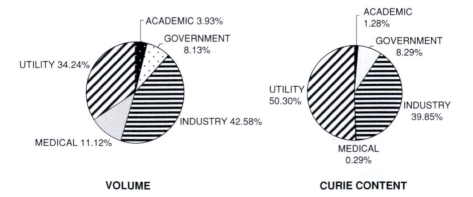

Figure 7.3   Percentage of LLW volume and Curie content generated by each of the five categories of commercial generator. Average for the period 1989–1999. Prepared using information from reference 2.

scrubbing, or absorption to within the allowable limit before the gaseous stream can be discharged to the atmosphere.

Liquid waste Liquids contain relatively low concentrations of solids ($\leq$1% of suspended solid). Examples of liquid wastes are chemical solutions, decontaminated solutions, liquid scintillation fluids, and contaminated oil.

Solid waste Solid wastes consist of two types: wet and dry solids. Wet solid waste usually contains $\geq$10% suspended solid and includes evaporator bottoms, spent ion-exchange media, filter precoat material, expended (loaded) filter cartridges, and research biological waste. Dry solid waste, or dry active waste (DAW), consists of trash, contaminated material, and equipment. The following types of compactible waste are reported by power plants:

1. Plastics: nonhalogenated plastics such as coveralls, protective suits, gloves, hats, bags, and bottles.
2. Polyvinyl chloride (PVC): halogenated plastics, e.g., protective suits, coveralls, boots, gloves, hoses, and bottles.
3. Paper: coveralls, laboratory coats, and absorbent paper cartons.
4. Absorbent materials: hygroscopic materials used to absorb fluids, such as vermiculite and bentonite.
5. Cloth: coveralls, laboratory coats, rags, and gloves.
6. Rubber: boots, hoses, gloves, and sheet.
7. Wood: construction lumber and plywood packing.
8. Noncompactibles: items that are noncompactible but are inadvertently packed with compactible waste. These can include small tools, hardware (nuts, bolts, and screws), or any other noncompactible material.
9. Metals: metallic items that can be compacted such as aerosol cans and paint cans.
10. Filters: high-efficiency particulate air (HEPA) filters and respirator canisters.
11. Glass: bottles, laboratory glassware, faceplates, and viewports.

The following types of noncompactible waste are reported by power plants:

1. Wood: construction lumber and plywood packing.
2. Conduit: tubing, cable, wire, and electrical fittings.
3. Pipe/valves: pipes, tubing, valves, and pipe fittings.
4. Filter frames: wooden or metal frames such as those that surround HEPA filters.
5. Compactibles: compactible material inadvertently or intentionally packed with noncompactible waste.
6. Concrete: debris from scarifying and demolishing concrete structures.
7. Tools: generally hand tools.
8. Dirt: dust, floor sweepings, and similar small particles or large quantities of contaminated dirt or sand.
9. Glass: bottles, laboratory glassware, instrument tubing, and faceplates.
10. Lead: generally shielding material in any configuration.

The specific radioactive isotopes most commonly found in LLW from nuclear power plants are Cs-137 and Sr-90 (fission products) and Co-60, Ni-63, Nb-95, and C-14 (in activated metal).

7.2 HISTORICAL BACKGROUND, STATE COMPACTS, AND REGULATIONS

7.2.1 Historical Perspective on Disposal of LLW

Table 7.1 shows a chronology of initiatives taken on the disposal of LLW,[5] starting from the time when the LLW was disposed of at defense sites. The first commercial disposal

Table 7.1 Chronology of initiatives taken on the disposal of low-level wastes[a]

1940s	Disposal begins in defense sites
1956	Burial of first nuclear engineering laboratory (Argonne) at the top of a hill in Palos Forest (complete nature or volumes of the wastes unknown)
1960	AEC announces that the land burial sites at Idaho National Engineering Laboratory (INEL) and Oak Ridge National Laboratory (ORNL) will be used to dispose of LLW (commercial)
1962	First commercial site (Beatty, Nevada) licensed (U.S. Ecology)
1963	Two more commercial sites, Maxey Flats, Kentucky (U.S. Ecology), and West Valley, New York (NUS), are licensed (agreement states)
1963	AEC sites cease to accept commercial waste
1965	Richland, Washington, becomes licensed site (through the agreement state) (U.S. Ecology)
1967	Sheffield, Illinois, adds to the commercial site (U.S. Ecology)
1970	Sea disposals ended by the AEC in June
1971	Barnwell, South Carolina, licensed by agreement state (Chem. Nuclear)
1975	West Valley site closed (6.57×10^4 m^3 or 2.3×10^6 ft^3 deposited)
1977	Maxey Flats ceases operations (1.36×10^5 m^3 or 4.8×10^6 ft^3 deposited)
1979	Sheffield closed due to capacity exhaustion (8.5×10^4 m^3 or 3×10^6 ft^3 deposited).
1979	Governor Ray closes Washington's commercial site in October and reopens it following action by appropriate federal regulatory agencies; Governor List closes the Beatty site in October and allows resumption in late November; South Carolina also announces a reduction in disposal in Barnwell
1980	Congress enacts the Low-Level Radioactive Waste Policy Act, urges all states to form compacts (with an exclusionary deadline set for January 1986)
1984	Data base management system (DBMS) made available by the DOE's National LLW Management Program.
1985	Congress amends the LLW Policy Act to extend the deadline to end of 1992
1991	Envirocare, a private disposal facility near Clive, Utah, receives a license to dispose of Class A LLW with specified isotopes and concentrations
1992	Beatty site permanently closed (1.37×10^5 m^3 deposited)
1994	Barnwell closes on June 30 to all generators outside the Southeast Compact
1995	South Carolina withdraws from the Southeast Compact and reopens to all states except North Carolina on July 1
1998	Texas Compact approved by U.S. Congress
1998	DOE's National LLW Management Program discontinued
2000	South Carolina, Connecticut, and New Jersey form the Atlantic Compact

[a] Information for dates up to 1981 from *Radwaste: A Reporter's Investigation of a Growing Nuclear Menace* by Fred C. Shapiro, Random House, New York, 1981. Copyright 1981 by Random House, Inc. Reprinted by permission.

site, at Beatty, Nevada, was licensed by what is now the U.S. Ecology, Inc., in 1962. Subsequently, until 1971, five more commercial licensed land disposal sites were added, some operated by U.S. Ecology and others by Chemical Nuclear and Nuclear Service (NUS). After not more than a dozen years of operation, two of them were closed because of operational problems and the Sheffield site was closed because of capacity exhaustion. As a result, a geographic imbalance existed between the location of disposal facilities and the location of LLW generators. Most wastes were generated in the east, while two of the remaining three disposal sites (the Beatty, Richland, and Barnwell sites) were in the west.

In 1979 the two western sites were temporarily closed by their host states in response to problems concerning waste packaging and transportation.[6] In the same year, South Carolina imposed volume restrictions on the Barnwell site, which received approximately 79% of the nation's waste. The annual waste intake at Barnwell was reduced to 1.2 million ft^3 in 1982. Short-term disposal shortage was threatened by these actions. Added to the technical and policy considerations was the general public's apprehension about radioactive materials and waste disposal facilities of any type. Congress therefore enacted the LLW Policy Act of 1980[7] to urge all states to form compacts to handle LLW generated in their region. The law specifies that the disposal of LLW is the responsibility of the states. To force a timely resolution of the problem, Congress set an exclusionary deadline for January 1, 1986, when the commercial disposal sites could refuse to accept waste from outside their compacts. Unfortunately, the progress of state compacts was so slow that in 1985 Congress found it necessary to extend the deadline to 1993 in the Low-Level Waste Policy Amendments Act (LLWPAA) of 1985.[8] The status of such interstate compacts is described in the next section.

7.2.2 Regulations on Disposal and Status of Interstate Compacts

As the LLW can be most safely and efficiently managed on a regional basis, the LLW Policy Act stated that each state is responsible for providing for the availability of capacity either within or outside the state for the disposal of LLW generated within its borders (except for waste generated from defense activities or federal research and development activities). To carry out this responsibility, the states may enter into compacts to provide for the establishment and operation of regional disposal facilities, with the consent of Congress.[7] The law also provided that after January 1, 1986, any such compact may restrict the use of the regional disposal facilities under the compact to the disposal of LLW generated within the region. Because of the slow progress in forming and obtaining congressional consent for compacts, Congress passed the LLWPAA of 1985, which introduced a plan for interregional disposal facility access through 1993. Stepped surcharges could be added by compacts accepting out-of-region wastes after January 1, 1986, and compacts would be required to accept out-of-region wastes only up to a certain volume level.

Figure 7.4 shows the compacts as they were proposed in 1987. Some states chose not to belong to any compact and were referred to as "unaffiliated" states. Between 1985 and 2000, ten compacts and five unaffiliated states spent a total of nearly $600 million[9] in efforts to site and construct LLW disposal facilities. However, in 2000, Envirocare and the facilities near Barnwell, South Carolina, and Richland, Washington, were still the only operating LLW disposal facilities in the United States. Table 7.2 lists the

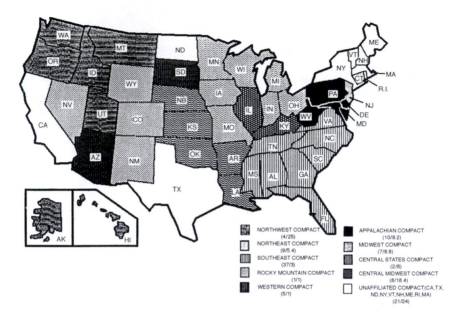

Figure 7.4 U.S. map showing LLW compacts as proposed April 1987. Name of compact is given along with the percent by volume of the national total for each region in 1985 and 1986 (number in parentheses). From State of Washington, Annual Report, 1984, Department of Social and Health Services Division, Office of Radiation Protection, 1986. Copyright 1986 by Office of Radiation Protection, State of Washington. Reprinted by permission.

compacts and their members as of the end of 2000. Each compact had a slightly different approach to providing the required disposal capacity. Some examples follow.

The disposal facility near Richland, Washington, is within the Northwest Compact and was selected as that compact's disposal facility. The Rocky Mountain Compact contracted with the Northwest Compact to use the Richland facility for its waste. In the Southwest Compact, California selected and licensed a disposal facility site in Ward Valley. However, the facility was to be constructed on land owned by the federal government, and as of 2000, the U.S. Department of the Interior had refused to sell or transfer the land to California. Illinois was to be the permanent host state for the Central Midwest Compact's disposal facility, and a volunteer site was identified. However, a three-judge panel disallowed that site, and a search for a new site was never completed. In the Midwest Compact, Michigan was initially selected as the first host state. But Michigan set such strict criteria for its LLW disposal site that no site could be identified, and in 1991, the Midwest Compact revoked Michigan's membership. Ohio became the new host state and proceeded to pass legislation and establish an agency required to site and construct a LLW disposal facility within the state. The Midwest Compact decided to halt the process just before Ohio began its search for a site. The Northeast Compact consisted of two states, Connecticut and New Jersey, each of which sought to site its own LLW disposal facility. By forming the compact, however, they did not have to take LLW from any other state.

The Southeast Compact was one of the most complex. The facility near Barnwell, South Carolina, was to serve as the LLW disposal facility for the Southeast Compact

Table 7.2 LLW compacts and their members in 2000

Appalachian Compact	Northwest Compact	Southwest Compact
Delaware	Alaska	Arizona
Maryland	Hawaii	California
Pennsylvania	Idaho	North Dakota
West Virginia	Montana	South Dakota
Atlantic Compact	Oregon	Texas Compact
Connecticut	Utah	Maine
New Jersey	Washington	Texas
South Carolina	Wyoming	Vermont
Central Compact	Rocky Mountain Compact	Unaffiliated States
Arkansas	Colorado	Massachusetts
Kansas	Nevada	Michigan
Louisiana	New Mexico	New Hampshire
Nebraska	Southeast Compact	New York
Oklahoma	Alabama	North Carolina
Central Midwest Compact	Florida	Rhode Island
Illinois	Georgia	
Kentucky	Mississippi	
Midwest Compact	Tennessee	
Indiana	Virginia	
Iowa		
Minnesota		
Missouri		
Ohio		
Wisconsin		

Compiled using information from the Low Level Waste Forum. Found at http://www.afton.com/llwforum/pubs/mapscharts/06-00 compactmap.gif

until 1992, at which time North Carolina was to become the compact's host state and construct a new facility. North Carolina did not construct a facility, and in 1995, South Carolina withdrew from the Southeast Compact, opening its facility to all states except North Carolina. In 2000, South Carolina joined with Connecticut and New Jersey, the former Northeast Compact, to form the Atlantic Compact. Meanwhile North Carolina withdrew from the Southeast Compact. More details on the activities in each of the compacts are summarized in reference 9.

Several reasons have been cited for the compact's failures to site and build new LLW disposal facilities. First among them were general public and political opposition. Others included decreasing waste volumes due to improved waste management and treatment techniques, continuing access to existing disposal facilities, and the high cost of building and operating new disposal facilities.[9]

7.2.3 Regulations on Processing, Storing, and Shipping LLW

These regulations consist of those issued by the Department of Transportation on transportation of hazardous materials (Title 49 of the Code of Federal Regulations) and by the

Nuclear Regulatory Commission on processing and on design objectives for equipment controlling radioactivity in effluents rules to limit specification to levels "as low as is reasonably achievable" (ALARA), and rules on transportation of radmaterials (CFR Title 10). Table 7.3 lists the regulations under CFR Title 49 concerning the transportation of hazardous materials. These include general regulations for hazardous material and its transportation by shippers and carriers via rail, highway, or aircraft as well as regulations on shipping containers and tank cars. Clearly specified in these regulations are what things should be done and who is responsible. For instance, to comply with packaging requirements, the following factors must be evaluated by the shipper:

1. Radionuclides in the waste.
2. Specific activity of the radionucludes.
3. Physical form and special properties of the waste.
4. Shielding requirements.
5. Proper packaging against accident.

The NRC regulations under CFR Title 10 give the following guidelines:

1. 10CFR20 defines concentration limits on effluents.
2. 10CFR50 sets forth design objectives for equipment to control radioactivity in effluents.
3. 10CFR61 covers licensing requirements for land disposal of radwaste.
4. 10CFR71 establishes requirements for packaging, preparation for shipment, transportation of radmaterials, and the procedures and standards for NRC approval of packaging and shipping procedures for fissile material. Most of the definitions in the DOT regulations are retained in the NRC regulations, but the latter pertain to quantities of LLW with higher activities. Thus, NRC regulations address licensed material that exceeds specified activity levels and must be shipped in NRC-approved packages having an NRC certificate of compliance. Furthermore, NRC requires advance notification of shipments of LLW across state lines.

Off-site doses from all power plant sources, including both release of radioactivity and direct radiation, are limited by the Environmental Protection Agency standard

Table 7.3 Title 49, Transportation Subchapter C—Hazardous materials regulations

49 CFR 171	General information, regulations, and definitions
49 CFR 172	Hazardous materials tables and hazardous materials communications regulations
49 CFR 173	Shippers—general requirements for shipments and packagings
49 CFR 174	Carriage by rail
49 CFR 175	Carriage by aircraft
49 CFR 176	Carriage by vessel
49 CFR 177	Carriage by public highway
49 CFR 178	Shipping container specifications
49 CFR 179	Specifications for tank cars

Table 7.4A Radionuclide and concentrations for waste type classification

Radionuclide	Concentration[a]
^{14}C	8
^{14}C in activated metal	80
^{59}Ni in activated metal	220
^{94}Nb in activated metal	0.2
^{99}Tc	3
^{129}I	0.08
Alpha-emitting transuranic nuclides with half-life greater than 5 yr	100[b]
^{241}Pu	3,500[b]
^{242}Cm	20,000[b]

[a] Units are Ci/m^3 unless otherwise noted.
[b] Units are nCi/g.

Table 7.4B Radionuclide and concentrations for waste type classification

Radionuclide	Concentration (Ci/m^3)		
	Col. 1	Col. 2	Col. 3
All nuclides with less than 5-yr half-life	700	[a]	[a]
^3H	40	[a]	[a]
^{60}Co	700	[a]	[a]
^{63}Ni	3.5	70	700
^{63}Ni in activated metal	35	700	7000
^{90}Sr	0.04	150	7000
^{137}Cs	1	44	4600

[a] No estimated limits.

40CFR190, Environmental Radiation Protection Standards for Nuclear Power Operations. State regulations are summarized in reference 10. These regulations affecting LLW management address mainly the transportation phase and the burial site licenses. The burial site license has an interim status until the interstate compacts, mentioned previously become operative.

Low-level wastes are classified in three classes, A, B, and C, according to the concentration and the radionuclides contained in the waste. Tables 7.4A and 7.4B, taken from 10CFR61.55, provide the values for the classification as follows:

Class A
1. Does not contain nuclides listed in Tables 7.4A and 7.4B.
2. Contains only those in Table 7.4A with concentrations equal to or less than 0.1 times the values in the table.
3. Contains only those in Table 7.4B with concentrations equal to or less than the value in Col. 1 of the table.
4. Combination of cases 2 and 3 above. This class is mostly trash, low-level resins, and biomedical waste that does not require physical stability.

Class B
1. Contains only those in Table 7.4B with concentrations greater than the values in Col. 1 but equal to or less than those in Col. 2.
2. In addition to case 1, contains nuclides in Table 7.4A with concentrations equal to or less than 0.1 times values in the table.

This class largely comprises evaporator concentrates, filter sludges, and spent resins, and the waste must be solidified in a stable matrix material (or high-integrity containers).

Class C
1. Contains nuclides in Table 7.4A with concentrations greater than 0.1 times the values but equal to or less than those in the table. This waste may also contain nuclides in Table 7.4B with concentrations up to those in Col. 3 of the table.
2. Contains only nuclides in Table 7.4B with concentrations greater than those in Col. 2 but equal to or less than those in Col. 3.
3. In addition to case 2, contains nuclides in Table 7.4A with concentrations equal to or less than 0.1 times the value in the table.

This class includes ion exchange resins, sealed sources, and isotope production wastes.

When the nuclide concentration of waste exceeds the values in either Table 7.4A or Col. 3 of Table 7.4B, such waste becomes "above class C," or class C+. This waste is not *generally* acceptable for near-surface disposal.

7.3 TREATMENT AND CONDITIONING PROCESSES

Waste management steps for LLW are shown in Figure 7.5 from waste collection to eventual disposal.[11] The treatment process is a process for separating one stream of radwaste at higher concentration from another at lower concentration so that the latter can be recycled or discharged to the environment. The conditioning process puts the waste into an acceptable concentration and form for packaging and shipment. Table 7.5 shows

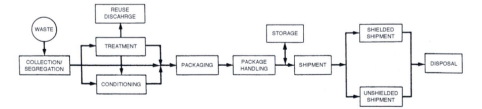

Figure 7.5 Waste management steps for LLW. From Low-Level Radioactive Waste Treatment Technology by EG&G Idaho, Inc., National LLW Management Program, DOE/LLW 13 Tc, 1984, U.S. Department of Energy, 1984. Reprinted by permission.

Table 7.5 Matrix of treatment/conditioning technologies and waste forms

Type of technology	Waste form		
	Liquids	Wet solids	Dry solids
Transfer technologies			
Decontamination	—	—	X
Filtration	X	X	—
Ion exchange	X	—	—
Chemical regeneration	—	X	—
Ultrafiltration	X	—	—
Reverse osmosis	X	—	—
Concentration technologies			
Evaporation	X	—	—
Distillation	X	—	—
Crystallization	X	—	—
Flocculation	X	—	—
Precipitation	X	—	—
Sedimentation	X	X	—
Centrifugation	X	X	—
Drying	—	X	—
Dewatering	—	X	—
Dehydration	—	X	—
Compaction	—	—	X
Baling	—	—	X
Shredding	—	—	X
Integrated systems	X	X	—
Transformation technologies			
Incineration	X	X	X
Calcination	X	X	—
Conditioning technologies			
High-integrity containers	—	X	X
Solidification	X	X	X
Absorption	X	X	—

Adapted from Low-Level Radioactive Waste Treatment Technology by EG&G Idaho, Inc., National Low-Level Radioactive Waste Management Program, DOE/LLW 13 Tc, 1984, U.S. DOE. Reprinted by permission.

the matrix of treatment/conditioning technologies and waste forms.[11] The treatment processes are classified as transfer, concentration, and transformation technologies:

1. *Transfer technologies* are processes that remove radioactive species from a waste stream and transfer them to another medium such as a filtration or ion-exchange medium.
2. *Concentration technologies* are processes that reduce the waste volume, such as evaporation, crystallization, and drying.
3. *Transformation technologies* are processes that concentrate radwaste by changing its physical form, such as incineration, calcination, and compaction.

For the purpose of assessing the treatment methodologies in different applications, these processes are grouped into categories principally according to treatment objectives (Section 7.3.6): dewatering, thermal/physiochemical, sorting/segregation, decontamination, mechanical treatment (volume reduction), and solidification.

7.3.1 Technology Selection Procedure for Liquid LLW

Figure 7.6 shows a technology selection chart for liquid LLW.[11] The basic system for processing liquid radwastes consists of several possible processes or treatment combinations: particulate removal, ionic solid removal, and effluent control.[12] These processes involve such unit operations as evaporation, ion exchange, filtration, and, to a lesser extent, centrifugation and reverse osmosis (Table 7.6).[11]

Ion exchange, demineralization. Ion exchange is the reversible interchange of ions between a solid phase and a liquid phase under conditions such that no permanent change in structure occurs in either phase. The major ion exchange materials are natural and synthetic inorganic polymers, such as aluminosilicates, and synthetic organic polymers (resins). Styrene and divinylbenzene are the most frequently used organic compounds. The ion exchange process may be carried out as a batch or a fixed-bed column operation. Liquid LLW treatment most frequently employs a mixed-bed system, which consists of a stationary bed containing mixed anionic and cationic resins (Figure 7.7).[13]

Evaporation. Evaporation is a method of concentrating nonvolatile components in a solution or a dilute slurry by vaporizing the solvent. It is applied in nuclear plants to

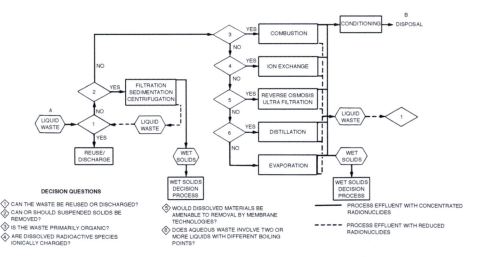

Figure 7.6 Technology selection chart for liquid LLW. From Low-Level Radioactive Waste Treatment Technology by EG&G Idaho, Inc., National LLW Management Program, DOE/LLW 13 Tc, U.S. Department of Energy, 1984. Reprinted by permission.

Table 7.6 Technologies often used for processing various types of liquid LLW

Type of technology or process	Regenerative solutions	Decontamination solutions	Oils	Other organics	Other inorganics	Scintillation vials and biological waste
Transfer technologies						
Filtration	X	X	X	—	—	X
Ion exchange	X	—	—	—	—	X
Ultrafiltration	X	—	—	—	—	X
Reverse osmosis	X	—	—	—	—	X
Concentration technologies						
Evaporation	X	X	—	—	—	X
Distillation	X	X	—	—	—	X
Crystallization	X	X	—	—	—	X
Flocculation	X	X	—	—	—	X
Precipitation	X	X	—	—	—	X
Sedimentation	X	X	—	—	—	X
Centrifugation	X	X	—	—	—	X
Integrated system	X	X	—	—	—	X
Transformation technologies						
Incineration	—	—	X	X	X	—
Calcination	—	—	X	X	X	—
Conditioning technologies						
Solidification	—	—	X	X	—	X
Absorption	—	—	X	X	X	X

Adapted from Low-Level Radioactive Waste Treatment Technology by EG&G Idaho, Inc., National Low-Level Radioactive Waste Management Program, DOE/LLW 13 Tc, 1984, U.S. DOE. Reprinted by permission.

concentrate aqueous wastes and to obtain relatively pure water for recycle or for acceptable discharge to the environs. PWRs and BWRs frequently use evaporators to process miscellaneous radioactive and chemical wastes (Figures 7.8 and 7.9).[13] Among the important elements in evaporator design are heat transfer, separation of evolved vapor from residual liquid, volume reduction, prevention of fouling of the heating surface, and conservation of energy. Figures 7.10 and 7.11 illustrate typical natural-circulation and forced-circulation evaporators, respectively, used in nuclear facilities. The advantages and disadvantages of these evaporator types are compared in Table 7.7 on page 228. Evaporators of these types are expected to yield concentrates containing 20–25 wt % solids. The solids content of the concentrates can be at least doubled by using a wiped-film evaporator (Figure 7.12) or an evaporator/crystallizer Figure 7.13.[13–15] The wiped-film evaporator is also called an agitated-, scraped-, thin-, or turbulent-film evaporator. The heating surface of a wiped-film evaporator is a single large-diameter cylindrical or tapered tube, and the liquid being concentrated is spread out into a thin, highly turbulent film by the blades of the rotor. Evaporator/crystallizers for the treatment of LLW are proposed as partial volume reduction (VR) systems that span the capabilities of current evaporator and dryer/calciner design; the VR factor (ratio of initial volume to volume after treatment) is expected to range between 2 and 5.[11,16]

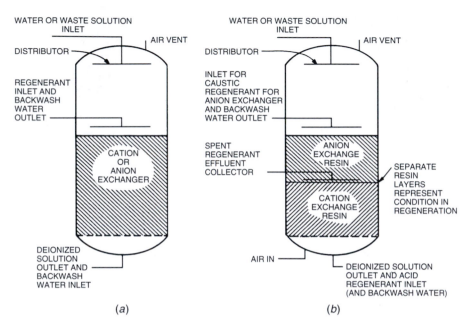

Figure 7.7 Schematic diagram of (a) separate-bed and (b) mixed-bed ion exchange systems. From LLW from Commercial Nuclear Reactors, Vol. 2. Treatment, Storage, Disposal and Transportation Techniques and Constraints by R. L. Jolley et al., ORNL/TM-9846/V2, Oak Ridge, Tenn., 1986. Reprinted by permission.

7.3.2 Technology Selection Procedure for Wet Solids

Figure 7.14 shows a technology selection chart for wet solid LLW. The basic system consists of steps for regeneration, incineration, pyrolysis, filtration, centrifugation, dewatering, drying, dehydration, or calcining (Table 7.8).[11]

Drying, dehydration, or dewatering. The drying process applies heat to remove liquid in a variety of ways, including in-drum drying, spray drying, and fluidized-bed drying; microwave drying is also being investigated.[11] When essentially all water is removed from wet wastes, the process is called dehydration. Bead-type ion exchange resin constitutes the major part of dewatered waste material. Dewatering technology uses pumping and/or gravitational flow to draw water from wet solids. In-container dewatering, which has been used at nuclear facilities for some time, involves disposable mechanical filter elements and a pump, all placed within a disposable container. This method has been used for the dewatering of resins (i.e., deep-bed ion exchanger resins) or the treatment of powdered resins.[11,13]

Filtration. Filtration separates solids from liquids by passing a suspension through a permeable medium. To force the fluid to flow through the filter medium, a pressure drop must be applied, which can be done by use of gravity, a vacuum, an applied pressure, or centrifugation. The use of filtration for nuclear facilities is described in reference 17 and

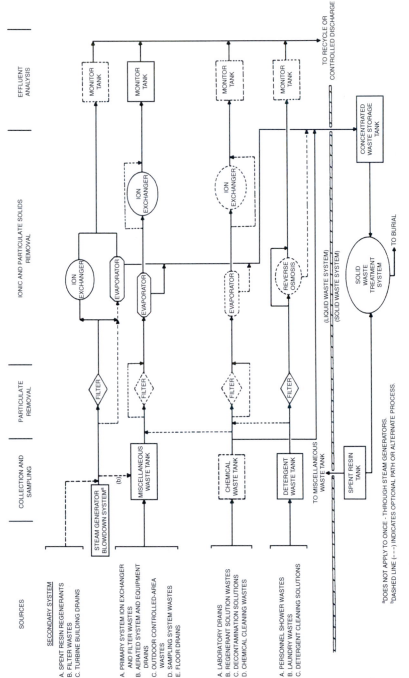

Figure 7.8 Flow diagram of a basic liquid radwaste processing system for PWR. From LLW from Commercial Nuclear Reactors, Vol. 2, Treatment, Storage, Disposal and Transportation Techniques and Constraints by R. L. Jolley et al., ORNL/TM-9846/V2, Oak Ridge, Tenn., 1986. Reprinted by permission.

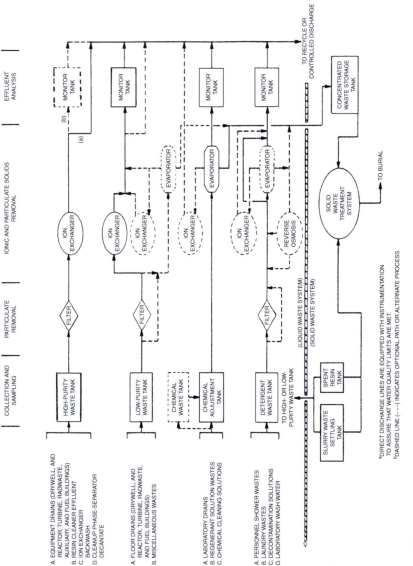

Figure 7.9 Flow diagram of a basic liquid radwaste processing system for a BWR. (a) Direct discharge lines are equipped with instrumentation to assure that water quality limits are met. (b) Dashed line indicates optional path or alternate process. From LLW from Commercial Nuclear Reactors, Vol. 2, Treatment, Storage, Disposal and Transportation Techniques and Constraints by R. L. Jolley et al., ORNL/TM-9846/V2, Oak Ridge, Tenn., 1986. Reprinted by permission.

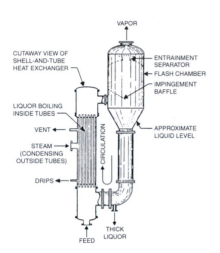

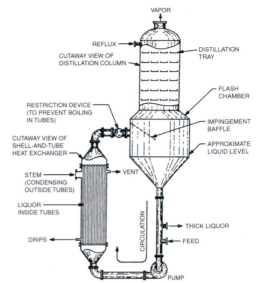

Figure 7.10 Natural circulation, rising-film, long-tube vertical evaporator with an external heater. From LLW from Commercial Nuclear Reactors, Vol. 2, Treatment, Storage, Disposal and Transportation Techniques and Constraints by R. L. Jolley et al., ORNL/TM-9846/V2, Oak Ridge, Tenn., 1986. Reprinted by permission.

Figure 7.11 Forced-circulation evaporator with an external, vertical single-pass heater and restriction device to prevent boiling in tubes. From LLW from Commercial Nuclear Reactors, Vol. 2, Treatment, Storage, Disposal and Transportation Techniques and Constraints by R. L. Jolley et al., ORNL/TM-9846/V2, Oak Ridge, Tenn., 1986. Reprinted by permission.

typical filter applications are given in Table 7.9.[11,13] Types of filters that have been used at nuclear facilities may be categorized in a number of ways; reference 17 suggests that they be classified in terms of their disposability (cartridge, screen, and bag) or reusability (Table 7.10).

Figure 7.15 shows a typical disposable type (cartridge) that is suitable for the removal of gross contamination from low-pressure, low-temperature systems, such as process streams of nuclear power plants. Multiple cartridges may be mounted in a single removable supporting structure so that the entire assembly can be replaced at one time. Cartridge filters usually consist of a fiber yarn wound around a perforated-metal core. Bag filters consist of nylon-mesh bags. Filters are changed when either the radioactivity level or the pressure differential reaches a preset value. Reusable filters may require a precoat (Figure 7.16), although some are used without one. Reusable filters are cleaned by backflushing, which pushes the filter cake from a filter as a sludge. The filter cake may also be recovered mechanically from a precoat filter. Precoat materials commonly used are diatomaceous earth, Solka floc, powdered resins, perlite, and asbestos. Several types of backflushable filters in use at nuclear facilities do not require precoats; examples are edge filters, porous metallic filters, porous ceramic filters, and stacked-disk filters (Figure 7.17).[11] The centrifugal-discharge filter is pressure-precoat filter; the precoat and filter cake supports are wire mesh screens mounted on horizontal

Table 7.7 Advantages and disadvantages of the evaporator types used in light-water reactor power plants

Evaporator type	Advantages	Disadvantages	Best applications	Frequent difficulties
Natural circulation	Low cost, large heating surface, in one body, low holdup, small floor space, good heat transfer coefficients at reasonable temperature differences (rising film), and good heat transfer coefficients at all temperature differences (falling film)	High headroom, generally unsuitable for salting and severely scaling liquids, poor heat transfer coefficients or rising-film version at low temperature differences, and recirculation usually required for falling-film version	Clear liquids, foaming liquids, corrosive solutions, large evaporation loads, high temperature differences for rising film, low temperature differences for falling film, and low-temperature operation for falling film	Sensitivity of rising-film units to changes in operating conditions and poor feed distribution to falling-film units
Forced circulation	High heat transfer coefficients; positive circulation; relative freedom from salting, scaling, and fouling	High cost, power required for circulating pump, and relatively high holdup or residence time	Crystalline product, corrosive solutions, and viscous solutions	Plugging of tube inlets by salt deposits detached from walls of equipment, poor circulation because of higher than expected head losses, salting because of boiling in tubes and corrosion—erosion
Submerged U-tube	Very low headroom, large vapor–liquid disengaging area, good heat transfer coefficients, and easy semiautomatic descaling	Unsuitable for salting liquids, high cost, and relatively high holdup or residence time	Limited headroom, small capacity, and severely scaling liquids	Slow response to changes in control settings and poor level control in vacuum units

From A State-of-the-Art Report on LLW Treatment by A. A. Kibbey and H. W. Godbee, ORNL/TM-7427, Oak Ridge National Laboratory, Oak Ridge, Tenn., 1980. Reprinted by permission.

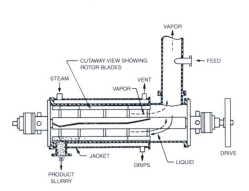

Figure 7.12 Wiped-film evaporator. From LLW from Commercial Nuclear Reactors, Vol. 2, Treatment, Storage, Disposal and Transportation Techniques and Constraints by R. L. Jolley et al., ORNL/TM-9846/V2, Oak Ridge, Tenn., 1986. Reprinted by permission.

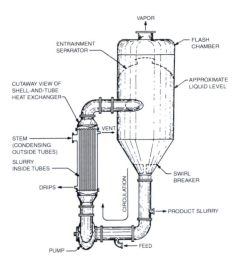

Figure 7.13 Evaporator/crystallizer. From LLW from Commercial Nuclear Reactors, Vol. 2, Treatment, Storage, Disposal and Transportation Techniques and Constraints by R. L. Jolley et al., ORNL/TM-9846/V2, Oak Ridge, Tenn., 1986. Reprinted by permission.

leaves attached to an axially mounted hollow vertical shaft (Figure 7.18). Reference 11 summarizes the potential advantages and disadvantages of filters for liquid waste (Table 7.11).

Traditionally, disposable cartridge-type filters have been used in PWRs and back-flushable tubular precoat filters in BWRs. Either type of filter may, however, be used for either type of reactor.[13] A superfine (SF) filter consisting of porous hollow fibers has been reported to show good filtering and backwash performance in plant testing.[18] The module rejects over 90% of particles in the size range 0.04 μm or larger and has an estimated life of over 3 years. Two commercial SF filter systems are operating, one at a BWR for treating low-conductivity water, the other at a PWR for clarification of refueling water.[13]

7.3.3 Technology Selection Procedure for Dry Solids

Figure 7.19 shows the technology selection procedure for DAW. The technologies often used for processing DAW are shown in Table 7.12. Decontamination is used for transfer technology with noncompactible and/or noncombustible wastes. Compaction, baling, and shredding are concentration technologies, while incineration is the major transformation technology. Solidification is the common conditioning technology for all forms of waste, gaseous (through transfer technologies), liquid, or solid. Decontamination processes include mechanical decontamination (e.g., high-pressure steam and water or sandblasting), chemical decontamination (solutions such as alkaline permanganate,

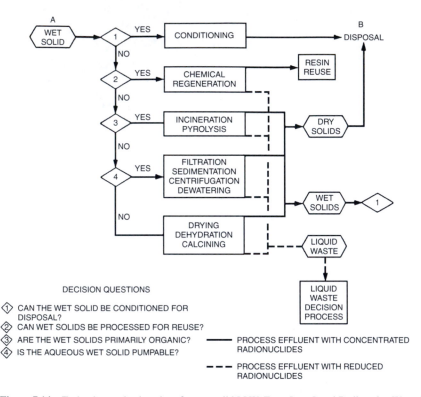

Figure 7.14 Technology selection chart for wet solid LLW. From Low-Level Radioactive Waste Treatment Technology by EG&G Idaho, Inc., National LLW Management Program, DOE/LLW 13 Tc, U.S. Department of Energy, 1984. Reprinted by permission.

mineral acids, or detergents), ultrasonics (vibrating finishing is a rapid and effective technique), and electrolytic decontamination (a smooth, polished surface is produced on metals and alloys and the object serves as the anode in an electrolytic cell). Small contaminated noncompactible and/or noncombustible waste items usually require no special treatment before being packaged for disposal. Large pieces of equipment sometimes require surface decontamination prior to size reduction by dismantling or torch cutting, unless surface decontamination can be effected to the point at which reuse or recycle is possible. Surface-contaminated scrap metals are sometimes reclaimed by smelting.[13]

The huge total volume of dry compactible and/or combustible waste suggests the need for suitable VR methods. Mechanical treatments designed to reduce LLW volume for storage and disposal include cutting, sawing and shearing, shredding and crushing, baling, compaction, and supercompaction. The options for VR and subsequent immobilization of DAW are shown in Figure 7.20.[13] Cutting, sawing, and shearing are used to prepare solid waste for packaging prior to transport or as a pretreatment for decontamination or incineration systems. Conventional cutting tools are often used in VR of metals and some plastics. The most widely used advanced technique is plasma-arc

Table 7.8 Technologies often used for processing various wet solids

Type of technology	Evaporator bottoms and miscellaneous sludges	Spent ion exchange resins	Filter sludges	Filter cartridges
Transfer technologies				
Filtration	X	—	—	—
Chemical regeneration	—	X	—	—
Concentration technologies				
Sedimentation	X	—	—	—
Centrifugation	X	—	—	—
Drying	X	—	X	—
Dewatering	X	X	X	—
Dehydration	X	X	X	—
Integrated systems	X	—	—	—
Transformation technologies				
Incineration	X[a]	X	X	—
Calcination	X	X	X	X
Conditioning technologies				
Solidification	X	X	X	X
High-integrity containers	—	X	X	X
Absorption	X	—	—	—

[a] Applicable to miscellaneous sludges only.

Adapted from Low-Level Radioactive Waste Treatment Technology by EG&G Idaho, Inc., National Low-Level Radioactive Waste Management Program, DOE LLW 13 Tc, 1984, U.S. DOE. Reprinted by permission.

cutting. It is small and easy to use, produces good cutting quality, and has a high cutting performance. The process can be used in air or under water, and cutting is possible in all positions.[19] The plasma-arc cutting technique has been used in the disassembly of vessels such as the Elk River Reactor and Sodium Reactor Experiment. By reducing the volume of control rods and fuel channels prior to burial, disposal costs have been reduced by 30%.[13,20] The other mechanical VR processes and incineration are described in the following sections.

Compaction, baling, and shredding. Compaction is a mechanical VR process in which material is compressed in disposal containers. Commercially available compacting devices are frequently used in radwaste treatment after making minor changes to accommodate hazardous airborne waste.[13,21] In general, a VR factor between 3 and 10 can be obtained by compaction, which has the advantage of being simple and inexpensive. In recent years, development has focused on improving the VR capacity of such units by preshredding, using anti-springback devices, and increasing the power of compaction (hence the name supercompaction). Presses used in compactors can have horizontal or vertical rams that generally apply pressures of 0.2–78 MPa (2–770 atm) to the waste, using a force of 0.04–13 MN (4.5–1500 tons). These cover both types of

Table 7.9 Radioactive liquid filters used in nuclear facilities[a]

Service	PWRs[b]	BWRs[b]
Fuel pool filters	Cartridge and etched disk (3 filters/reactor)	Precoat septum, precoat centrifugal, and cartridge etched disk (1 or 2 filters/reactor)
Boron recovery system filters	Cartridge and etched disk (5 filters/reactor)	NA
Solid-waste system filters	Cartridge and etched disk (1 filter/reactor)	None
Liquid-waste system filters	Cartridge and etched disk (1–4 filters/reactor)	Precoat septum, sand, precoat centrifugal, etched disk; flat bed (3 filters/reactor)
Reactor water cleanup filters	Cartridge and etched disk (5–7 filters/reactor)	Precoat septum and precoat cartridge (5–8 filters/reactor)
Powdered resin waste dewatering system filters	Precoat flatbed and precoat centrifugal (2 filters/reactor)	Precoat flat bed and precoat centrifugal (2 filters/reactor)

[a]PWR, Pressurized water reactor; BWR, boiling water reactor. Operating experience is limited for etched-disk, precoat clam shell, and precoat centrifugal filters.

[b]Used if a powdered resin settling tank system is not employed.

From LLW from Commercial Nuclear Reactors, Vol. 2, Treatment, Storage, Disposal and Transportation Techniques and Constraints by R. L. Jolley et al., ORNL/TM-9846/V2, Oak Ridge National Laboratory, Oak Ridge, Tenn., 1986. Reprinted by permission.

low-pressure systems (which typically apply forces up to about 100 tons) and high-pressure compactors or supercompactors (which apply forces of more than 100 tons).[13,21] The major parameters that affect the VR of the waste during compaction include the applied force, bulk density of the original waste, void space in the container, and springback of the material. Supercompactors can reduce the volume of noncombustible and traditionally noncompactible wastes with an expected VR of 2–4 for noncompactible waste.[13,22]

Table 7.10 Filters for liquids in nuclear facilities

Disposable	Reusable magnetic	Reusable with precoat
Pleated paper cartridge	Magnetite bed	Backflushable tubular bundle
Pleated wire screen	Electromagnetic	Centrifugal discharge
Wound cartridge		Clam shell
Woven mesh bag	Reusable without precoat	Flat bed
	Partially cleanable metallic	Pressure leaf
Reusable deep bed	Porous ceramic	Rotary vacuum
Crushed coal	Stacked etched disk	
Ground walnut shells		
Sand		

From The Use of Filtration to Treat Radioactive Liquid in LWR Power Plants by A. H. Kibby and H. W. Godbee, ORNL/NUREG-41 (NUREG/CR-0141), Oak Ridge National Laboratory, Oak Ridge, Tenn., 1978. Reprinted by permission.

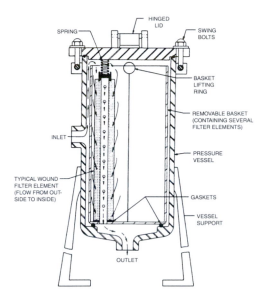

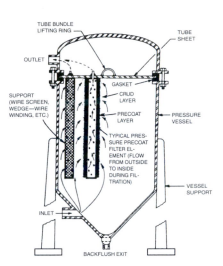

Figure 7.15 Typical disposable-cartridge filter illustrating liquid flow from outside to inside of element. From LLW from Commercial Nuclear Reactors, Vol. 2, Treatment, Storage, Disposal and Transportation Techniques and Constraints by R. L. Jolley et al., ORNL/TM-9846/V2, Oak Ridge, Tenn., 1986. Reprinted by permission.

Figure 7.16 Typical tubular-support pressure precoat filter. From LLW from Commercial Nuclear Reactors, Vol. 2, Treatment, Storage, Disposal and Transportation Techniques and Constraints by R. L. Jolley et al., ORNL/TM-9846/V2, Oak Ridge, Tenn., 1986. Reprinted by permission.

The most common low-pressure compactor is the 210-L drum compactor; waste is loaded into the drum and the power unit is activated to bring the platen down onto the material in the drum. Box compactors use one or two hydraulic pistons to operate a ram that compresses waste in steel boxes or wooden boxes with metal liners. Compaction of waste for disposal is relatively inexpensive. The largest single cost for disposal by compaction processes is the burial cost (approximately 57–81% of the total cost in 1983). By comparison, incineration (next section) is a capital-intensive process in which burial cost is only 5–10% of the total treatment cost.[13,23]

Incineration. There is much interest in incineration for volume reduction of DAW. It is a viable processing alternative for treating combustible DAW and is sufficiently developed for use in commercial power plants. Licensing of a radwaste incinerator may be a controlling factor in applying this technology at U.S. nuclear plants.[13,24] In converting organic material to gases and solid residue, incineration produces a less voluminous product than any other VR technique. Effective incineration completely eliminates organic hazards, destroys many toxic chemicals, and results in a chemically inert waste form that is compactible with recovery, immobilization, and disposal.[13,25] Several incinerator/immobilization systems are available for processing LLW and have performed satisfactorily for nuclear power plants worldwide.

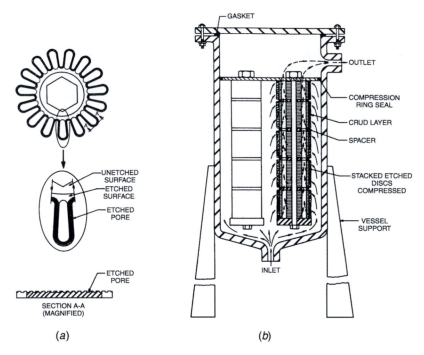

Figure 7.17 Views of an etched disk and schematic of a typical etched-disk filter. (a) Top and cross-sectional views of etched disk. (b) Assembled elements in filtration stage. From LLW from Commercial Nuclear Reactors, Vol. 2, Treatment, Storage, Disposal and Transportation Techniques and Constraints by R. L. Jolley et al., ORNL/TM-9846/V2, Oak Ridge, Tenn., 1986. Reprinted by permission.

Operational facilities and developmental prototypes have been used in the United States for the treatment of radwaste at Department of Energy and institutional facilities for a number of years.

The following types of incineration equipment are available: excess-air cyclone type (Mound facility; Figure 7.21, p. 238), controlled-air system (Los Alamos National Laboratory, LANL; Figure 7.22, p. 239), agitated hearth (Rocky Flats Plant, RFP; similar to Figure 7.23, p. 239), fluidized bed (RFP; Figure 7.24, p. 240), slagging pyrolysis process (INEL), rotary kiln (RFP; Figure 7.25, p. 240), and Penberthy molten glass system (Penberthy Electromelt International, Inc.; Figure 7.26, p. 241). The processes and operating variables of different types are compared in Table 7.13 (p. 242). Also included for comparison is the acid digestion process,[11] which is a chemical oxidation process that converts combustible organic waste to gaseous effluents and stable solid residue. One of the major concerns in this process is the treatment of the off gas and secondary wastes. Reference 11 compared the product characteristics, off-gas system, and secondary wastes associated with the different types of incinerators (Table 7.14) and the advantages and disadvantages of these types (Table 7.15).[16,26] The capital cost, operating cost, and energy cost are all highly dependent on the type of incinerator system used and the quantity and type of waste processed. These costs tend to be high in comparison with other treatment processes. Incinerator costs typically range between

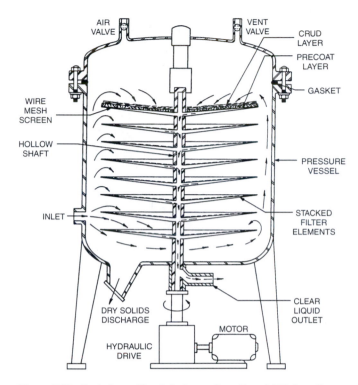

Figure 7.18 Typical centrifugal-discharge filter. From LLW from Commercial Nuclear Reactors, Vol. 2, Treatment, Storage, Disposal and Transportation Techniques and Constraints by R. L. Jolley et al., ORNL/TM-9846/V2, Oak Ridge, Tenn., 1986. Reprinted by permission.

$2 million and $8 million[27] and the total capital costs can be two to three times the equipment costs.[13]

Solidification/stabilization. Solidification is the process of converting LLW to a stabilized form to prevent degradation and release of radionuclides. The waste may be in liquid, slurry, sludge, or dry solid form and is solidified by mixing with an appropriate agent, or binder. A monolithic solid is produced either by chemical reaction with the waste, by forming microscopic cells that encapsulate the waste, or by coating and binding the individual waste particles together.[13] Thus, much of the solid waste and LLW is immobilized by solidification prior to disposal, and VR processes are often employed before solidification. Table 7.16 shows the types of waste that can be solidified.[13,28]

Solidification agents are shown in Table 7.17. Hydraulic cements are binders that harden by chemical interactions with water. Bitumen consists of high-molecular-weight hydrocarbons with both aliphatic and aromatic components and is a thermoplastic material. Vinyl esters are unsaturated polyesters used for the solidification of LLW, but the specific vinyl ester compositions are proprietary information. Selected solidification agents and waste form properties are compared in Table 7.18.[13,29–36] An

Table 7.11 Potential advantages and disadvantages of filters for liquids in LWR nuclear power plants

Type of filter	Advantages	Disadvantages
Disposable		
Wound cartridge	Compact; low solid waste volume; no backflush gas or liquid to treat; good solids removal	Remote and/or automatic changeout difficult because of nonuniformity and poor arrangement; changeout frequently done on radiation level rather than pressure drop; media migration may occur
Pleated paper cartridge	Compact; low solid waste volume; no backflush gas or liquid to treat; good solids removal	Remote and/or automatic changeout difficult because of nonuniformity and poor arrangement; changeout frequently done on radiation level rather than pressure drop; media migration may occur
Pleated wire screen	Can operate at elevated temperatures; good solids removal; little or no media migration	Fair mechanical strength when adequately supported; plugging may cause uneven flow and nonuniform cake buildup
Reusable without precoat		
Stacked etched disk	Short backflush time with thorough cleaning; expected to last for plant life; amenable to automatic and/or remote operation; low solid waste volume; compact; high mechanical strength	Low crud-holding capability; corrosion characteristics unknown; backwash waste to treat; low oil-holding capacity
Reusable with precoat		
Backflushable tubular bundle	Amenable to automatic and/or remote operation; powdered resin and/or diatomaceous earth precoat can be used; relatively compact	Precoat loss on loss of flow or fluctuation in pressure; excessive or uneven cake can cause strain and possible collapse of supporting screen; incomplete backflushing causes uneven precoat
Dry cake discharge		
Centrifugal discharge	High crud-holding capacity; can handle automatically and remotely all plant wastes with same filter; low maintenance requirements; no precoat loss caused by loss of flow, pressure, or power	Relatively high headroom; cake overloading can cause distortion; generates large sludge volume; some cake difficulty with Solka floc or resins alone
Flat bed	High crud-holding capacity; can handle automatically and remotely all plant wastes with same filter; no precoat loss caused by loss of flow, pressure, or power	Relatively large floor space and high headroom; cake overloading can cause belt wear; generates large sludge volume; some cake difficulty with resin alone; may require fairly high belt maintenance

From Low-Level Radioactive Waste Treatment Technology by EG&G Idaho, Inc., National Low-Level Radioactive Waste Management Program, DOE/LLW 13 Tc, 1984, U.S. DOE. Reprinted by permission.

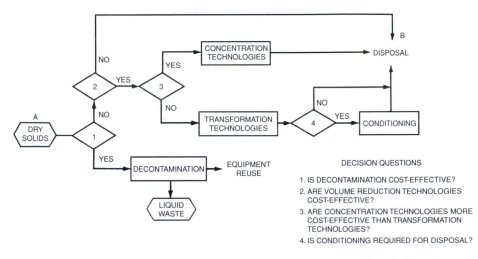

Figure 7.19 Technology selection chart for dry solid LLW. From Low-Level Radioactive Waste Treatment Technology by EG&G Idaho, Inc., National LLW Management Program, DOE/LLW 13 Tc, U.S. Department of Energy, 1984. Reprinted by permission.

in-line mixing technique in which the waste and solidification agent are mixed before being transferred to the disposable container is often used for waste solidification processes, as shown in Figure 7.27.

7.3.4 Volume Minimization Steps

Numerous VR methods have been described in previous sections. As summarized in reference 29, there are four major categories of VR processes for wet waste (dehydration,

Table 7.12 Technologies often used for processing various types of dry solid LLW

Type of technology	Trash	Contaminated equipment	Irradiated hardware
Transfer technologies			
Decontamination	—	X	—
Concentration technologies			
Compaction	X	—	—
Shredding or sectioning	X	X	X
Baling	X	—	—
Transformation technologies			
Incineration	X	—	—
Conditioning technologies			
High-integrity containers	—	—	X
Solidification (ash)	X	—	—

Adapted from Low-Level Radioactive Waste Treatment Technology by EG&G Idaho, Inc., National Low-Level Radioactive Waste Management Program, DOE/LLW 13 Tc, 1984, U.S. DOE. Reprinted by permission.

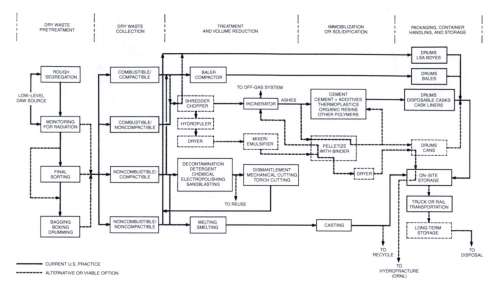

Figure 7.20 Flow diagram illustrating the management of DAW. (—) Current U.S. practice. (--) Alternative or viable options. From LLW from Commercial Nuclear Reactors, Vol. 2, Treatment, Storage, Disposal and Transportation Techniques and Constraints by R. L. Jolley et al., ORNL/TM-9846/V2, Oak Ridge, Tenn., 1986. Reprinted by permission.

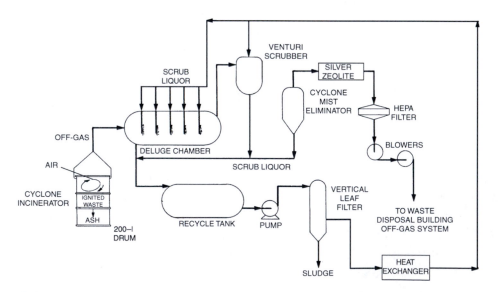

Figure 7.21 Excess-air cyclone-type incinerator. From LLW from Commercial Nuclear Reactors, Vol. 2, Treatment, Storage, Disposal and Transportation Techniques and Constraints by R. L. Jolley et al., ORNL/TM-9846/V2, Oak Ridge, Tenn., 1986. Reprinted by permission.

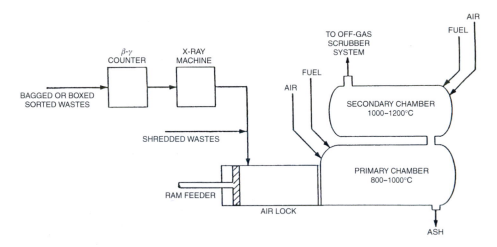

Figure 7.22 Los Alamos Scientific Laboratory controlled-air incinerator. From LLW from Commercial Nuclear Reactors, Vol. 2, Treatment, Storage, Disposal and Transportation Techniques and Constraints by R. L. Jolley et al., ORNL/TM-9846/V2, Oak Ridge, Tenn., 1986. Reprinted by permission.

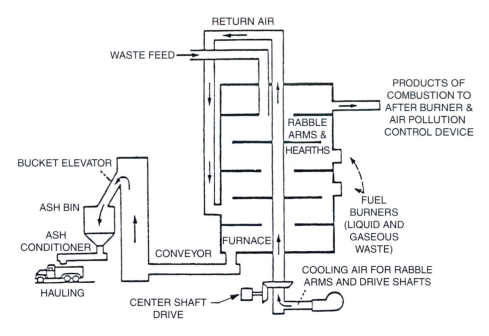

Figure 7.23 Multiple-hearth incineration system. From Generic Process Technologies Studies by Proctor & Redferm, Ltd., Weston Designers, Consultants, and Ontario Research Foundation, System Development Project, Ontario Waste Management Corporation, August 1982. Figure reproduced in LLW from Commercial Nuclear Reactors, Vol. 2, Treatment, Storage, Disposal and Transportation Techniques and Constraints by R. L. Jolley et al., ORNL/TM-9846/V2, Oak Ridge, Tenn., 1986. Reprinted by permission.

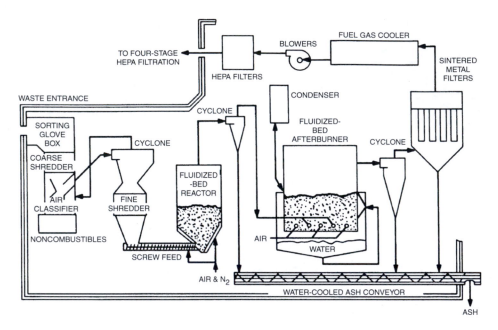

Figure 7.24 Rocky Flats Plant fluidized-bed incinerator. From LLW from Commercial Nuclear Reactors, Vol. 2, Treatment, Storage, Disposal and Transportation Techniques and Constraints by R. L. Jolley et al., ORNL/TM-9846/V2, Oak Ridge, Tenn., 1986. Reprinted by permission.

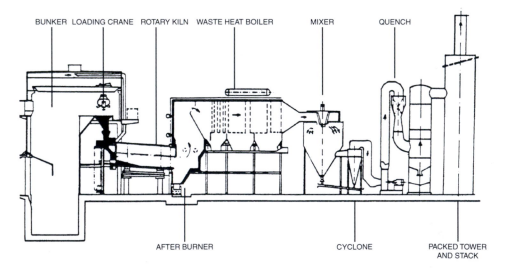

Figure 7.25 Example of modern hazardous waste rotary kiln incinerator with waste heat recovery and high-efficiency wet/dry flue-gas scrubber (Biebesheim, West Germany). From Generic Process Technologies Studies by Proctor & Redferm, Ltd., Weston Designers, Consultants, and Ontario Research Foundation, System Development Project, Ontario Waste Management Corporation, August 1982. Figure reproduced in LLW from Commercial Nuclear Reactors, Vol. 2, Treatment, Storage, Disposal and Transportation Techniques and Constraints by R. L. Jolley et al., ORNL/TM-9846/V2, Oak Ridge, Tenn., 1986. Reprinted by permission.

240

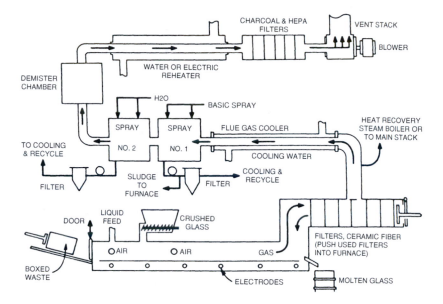

Figure 7.26 Sketch of the Penberthy molten-glass incinerator (electromelter) system proposed for treating low-level waste. From LLW from Commercial Nuclear Reactors, Vol. 2, Treatment, Storage, Disposal and Transportation Techniques and Constraints by R. L. Jolley et al., ORNL/TM-9846/V2, Oak Ridge, Tenn., U.S. Department of Energy, 1986. Reprinted by permission.

crystallization, incineration, and compaction) and two major VR processes for DAW (incineration and compaction). Faced with a potential shortage of disposal capacity and rapidly rising costs of available disposal facilities, the waste generators are opting for VR processing and packaging, not only to meet NRC requirements but also to reduce disposal costs and stay within reduced out-of-state limits imposed by host states for existing facilities under LLWPAA. In addition to the VR methods described in previous sections, an important step of minimizing waste generation and using a careful sorting/segregation process has been recognized. Because of lack of an adequate waste segregation process, substantial amounts of uncontaminated materials are being disposed of as LLW. An information exchange program has been established to inform generators of methods and practices for reducing the amount of waste they generate.[37] Numerous low-cost techniques for minimizing radwaste generation are being implemented at many power plants, and the annual waste generation has been reduced by up to 50%.[13,37] Figure 7.28 shows the total waste generation trends. The principal conclusions of an EPRI study were as follows:[37]

1. Plant factors such as size, system designs, and location have a limited effect on LLW generation at nuclear reactors.
2. Some plants generate less LLW than other similar plants mainly because of management attention and the overall attitude of workers and their awareness of the problems.

Table 7.13 Feed conditions and operating variables for the major incinerator types

Parameter	Acid digestion (HEDL)	Agitated hearth (RFP)	Controlled air (LANL)	Excess air (Mound)
Process description	Waste is chemically oxidized in H_2SO_4 containing a small percentage of HNO_3; acid is evaporated to produce residue product	Waste is batch fed into primary chamber; residue is raked by rotating rabble arm; final off-gas combustion in separate afterburning chamber	Waste is decomposed in starved-air primary chamber; volatiles and particulates are finally burned in secondary chamber	Waste is burned in steel storage drum in which air is injected in a spiral pattern to cool drum walls; continuous-feed option is available
Materials of construction	Digester and other items, glass, glass-lined, or Teflon-lined; off-gas equipment, standard for wet system	Incinerator is refractory lined; hearth is refractory lined; off-gas equipment, standard for wet system	Both chambers refractory lined with 5-mm mastic carbon steel outer shell; hearth is refractory; off-gas equipment is fiberglass-reinforced plastic, Hastelloy, and mastic	Incinerator is 316 SS; hearth is metal drum, off-gas equipment, standard for scrubbing corrosive acids
Solid feed capacity (kg/hr) Solid waste capabilities	10 Paper, rags, wood, rubber, PVC, ion-exchange resins, polyurethane; fluorinated wastes are corrosive to glass	70 Combustible solids with $<3 \times 10^{-3}$ mg Pu per gram; includes paper, PVC, polyurethane, PVC, and latex rubber cloth	45 Cellulosic compounds, polyethylene, PVC, latex rubber, and cartridge filters	27 Contaminated wastes containing paper, PVC, polyethylene, polypropylene, rubber, cloth, and tramp metal
Feed pretreatment— noncombustibles	Hand sorted to remove noncombustibles before shredding; also air classification for metal removal	Tolerates small noncombustibles, but large items must be removed before feeding	Large noncombustibles are hand sorted out; small noncombustibles are tolerated	In batch mode, noncombustibles do not interfere with burning; for continuous mode, hand sorting out of noncombustibles
Acid-producing solids— corrosion	Resistant to acids produced by PVC plastic, etc.; some acid vapors (HNO_3, HCL) escape off-gas scrubbing and are vented to atmosphere	Acid waste or PVC plastic may corrode metal rabble arms	Highly acidic materials may corrode steel incinerator shell and other head-end equipment	Acid gases at 800–1000°C, very corrosive to drum, upper combustion chamber, off-gas header, deluge tank, and transfer pipe between tank and venturi

Volatile solids	Minimal volatilization at 250°C; volatiles readily dissolved into acid system	Some volatilization occurs at 800–1000°C; lead deposition is possible in incinerator and refractory lining	Some volatilization at 800–1000°C; lead deposition occurs if lead-containing materials are burned	Some volatilization at 1100–1300°C; possible lead deposition
Operating temperatures	Primary: 250°C with H_2SO_4 and HNO_3 oxidation	Primary: 600–800°C, slightly reducing atmosphere; secondary: 1000°C	Primary: 500–800°C, starved air; Secondary: 1000–1500°C, oxygen enriched	Primary: 1100°C, excess air; Secondary: none

Parameter	Fluidized bed (RFP)	Rotary kiln (RFP)	Slagging pyrolysis (INEL)	Penberthy electromelt
Process description	Waste is fed to fluidized bed of Na_2CO_3 granules for partial combustion; volatiles are finally oxidized in fluidized bed of catalyst granules	Waste is charged into inclined, horizontal kiln that is rotating; final off-gas combustion occurs in a separate afterburning chamber	Stacked-kiln concept; waste loaded in top drying zone, descends into pyrolysis zone, and descends into melt zone to be discharged as molten slurry	Waste is burned and sorbed into a glass melt maintained by electromelt Joule heating; no second stage; product perfect for storage with no further treatment.
Materials of construction	Incinerator is made of all 316 SS; hearth distributor plate of 316 SS; off-gas equipment is made of non-acid-resistant materials	Incinerator, refractory brick-lined: off-gas equipment, typical of wet scrubbers	Incinerator interior, refractory tamped; off-gas equipment, standard wet and dry system	Incinerator, refractory lined; off-gas equipment, standard for wet system sorbing acidic gases
Solid feed capacity (kg/hr)	82	41	850	225
Solid waste capabilities	Paper, polyethylene, PVC, latex rubber, wood, leaded rubber, organic resins, cartridge filters, HEPA filters, and cloth	Designed for cellulose, polyethylene, PVC, rubber, organic resins, wood, and polyethylene	Designed to handle all kinds of combustibles, most types of noncombustibles, and contaminated soil	Combustible solids such as paper, wood, concrete, rubber, plastics, scrap glass, and metal
Feed pretreatment—noncombustibles	Large noncombustibles are hand sorted out before shredding; further removal is effected by air classification	Tolerates noncombustibles within size limitations of ram feeder and ash discharge port	Accepts all types of noncombustibles (soil) within size restraints of waste throat	Tolerates wide range of feeds; no data on noncombustibles; size limitation on feed

(continued)

Table 7.13 (continued)

Parameter	Fluidized bed (RFP)	Rotary kiln (RFP)	Slagging pyrolysis (INEL)	Penberthy electromelt
Acid-producing solids—corrosion	Acids produced from PVC or other source neutralized by bed of Na_2CO_3	Refractory lining unaffected by acids, some acid corrosion of steel outer tube or head-end feeding equipment	Acids may corrode metal components; in situ neutralization obtained by adding Na_2CO_3	Unknown
Volatile solids	Low temperature (550°C) minimizes lead deposition; Na_2CO_3 blocks lead, iron, and phosphorus; chloride remains in off gas	Refractory can be degraded by low-melting metals such as feed; some volatiles at 800–1000°C	High temperature (1650°C) volatilizes or melts most metals; lead in INEL waste could cause deposition and lining problems	Unknown
Operating temperatures	Primary: 525–625°C, 15–25% oxidation; secondary: catalytic at 550°C, oxygen enriched	Primary: 600–800°C, neutral to slightly reducing; secondary: 1000°C, oxygen enriched	Primary: 1500–1600°C, 30–60% excess air; Secondary: 1100–1200°C, 40% excess air	1200°C

From LLW from Commercial Nuclear Reactors, Vol. 2, Treatment, Storage, Disposal and Transportation Techniques and Constraints by R. L. Jolley et al., ORNL/TM-9846/V2, Oak Ridge National Laboratory, Oak Ridge, Tenn., 1986. Reprinted by permission.

Table 7.14 Product characteristics, off-gas system, and secondary wastes associated with the major incinerator types

Parameter	Acid digestion (HEDL)	Agitated hearth (RFP)	Controlled air (LANL)	Excess air (Mound)
Operation variables controlled	Temperature, HNO_3 addition and feed rate	Waste feed, fuel, and air rates, and time and temperature	Feed, fuel, and air rates; temperature is modulated for combustion	Feed, fuel, and air rates, and temperature
Product form	Dry salt cake, rich in sulfates and oxides, thermally stable, inert, and unstable for handling	Inert dry ash, not stable for storage	Dry, thermally and radioactively stable, inert, and unstable for storage	Dry ash and salt powder for sorption of HCl; not inert, not stable for storage
Off-gas system	Dilute acid scrubber, heater, HEPA filter, second scrubber, and final HEPA filter	Potassium hydroxide scrubber, venturi scrubber, gas–liquid separator, and HEPA filter	Water quenching, venturi scrubber, packed bed absorber, condenser, heater, roughing filter, and HEPA filters	Deluge chamber, filter
Secondary wastes	Scrubbed gases and alkaline scrub solution containing NaCl, $NaNo_3$, $NaNO_2$, and Na_2SO_4	Combustion gases and alkaline scrub solution	Combustion gases and neutralized, spent, NaCl scrubber solution	Combustion gases and spent, neutralized scrub solution
Status	Demonstrated for cold and radioactive wastes at 5 kg/hr; technology currently available	Demonstrated for cold wastes at 4 kg/hr; large-scale unit built for processing 70 kg/hr of LLW	Tested at LANL on cold and radioactive wastes; commercial units are available from several companies	Demonstrated in cold pilot plant for more than 5 kg/hr; technology available since 1978; currently marketed by several companies
Potential applications	Processes most solid wastes and high TRU activity wastes; good actinide recovery; best suited for low flow rates; because of liquid nature, has some advantages for remote operation	With minimal criticality problems, best suited for burning large volume LLW stream; because of anticipated high maintenance resulting from corrosion of metals and refractories, not a good candidate for remote operation	Good for large volumes of low-level TRU wastes; good nuclear safety features; subject to corrosion from PVC plastics; needs considerable testing before use in remote applications	The lack of criticality problems makes this unit a good candidate for burning LLW; especially adaptable to remote operation because of simplicity and low maintenance; good candidate for incineration of LLW at power plants

(continued)

Table 7.14 (continued)

Parameter	Fluidized bed (RFP)	Rotary kiln (RFP)	Slagging pyrolysis (INEL)	Penberthy electromelt
Operation variables controlled	Feed, nitrogen, and fuel rates, temperature, and Na_2CO_3 bed makeup	Feed, air, and fuel rates; temperature	Waste feed, sand, fuel, and air rates; temperature and pressure	Unknown
Product form	Inert dry oxide ash and dry salt; inert, nonstable for storage	Inert dry oxide, not stable for storage	Basaltic-type glassy slab, inert, and storage is dependent on leaching requirements	Glass melt containing ash and oxides; some small tramp metal can be entrapped
Off-gas system	Dry scrubbing—cyclone filter, sintered metal filter, and HEPA filters	Primary and secondary venturi scrubbers and HEPA filters	Wet and dry combination, preheater, heat exchanger, sand filter, caustic wet scrubber, preheater, HEPA filter, preheater, and stack	Not complete, but current plans include flue-gas cooler, basic and water scrubbers, demister, reheater, and charcoal and HEPA filters
Secondary wastes	Bed material (Na_2CO_3) and catalytic material ($Cr_2O_3 + Al_2O_3$) and acid-free combustion gas	Combustion gases and acidic and spent alkaline solutions including fly ash	Wet scrubber liquid and scrubbed gas	Wet scrubber liquids; sulfate-bearing wastes reduce VR factor because of sorption in liquid scrubber
Status	Demonstrated for cold wastes at 9 kg/hr; a radioactive waste demonstration designed for 80 kg/hr ready for full-scale operation in 1980; Aerojet Energy Conversion Co. is the only vendor in U.S.	Demonstrated at cold pilot plant, 2 kg/hr; plant being designed for 40 kg/hr of TRU wastes; not yet commercially available for radwaste	Mol Belgium (SCK, CEN) plant tested beta-gamma burning in 1978; not yet commercially available in the United States	Demonstration unit only

| Potential applications | Good for a wide variety of waste solids and liquids, for acid-producing wastes, and for burning liquids; can burn ion exchange resins; modular design and low maintenance make it a good candidate for remote operation | Good for wastes with recoverable TRU material and wastes with noncombustibles; if seal and bearing maintenance are small, this unit is a good candidate for remote operation | Designed for burning INEL retrievably stored wastes; good for burning municipal wastes; remote operation difficult because of equipment size and refractory maintenance | Wet and dry radioactive wastes, particularly glass; small tramp metal not a problem |

From LLW from Commercial Nuclear Reactors, Vol. 2, Treatment, Storage, Disposal and Transportation Techniques and Constraints by R. L. Jolley et al., ORNL/TM-9846/V2, Oak Ridge National Laboratory, Oak Ridge, Tenn., 1986. Reprinted by permission.

Table 7.15 Advantages and disadvantages of major incinerator types

Advantages	Disadvantages	Unique capabilities
Acid digestion, Hanford Engineering Development Laboratory		
Takes wide variety of wastes, but waste must be sorted and shredded; soluble residue for actinide recovery; low-temperature, single-stage operation; and processes high levels of radioactivity	Limited feed rate of 5.0 kg/hr maximum, difficult process control for acid feed, VR small without acid recycle, feed requires sorting and shredding, acid gases vented to atmosphere, and useful for only a limited range of organic liquids	Produces soluble inorganic sulfate and oxide residue for Pu recovery and has acid recycle
Agitated hearth, Rocky Flats Plant or Envirotech		
Mechanical agitation of waste during combustion produces efficient oxidation, minimal waste pretreatment, automatic ash removal, and nonrotating refractory has long life	Maintenance of mechanical equipment in the combustion chamber, accommodates only small amounts of activity, possible radioactivity buildup in refractory lining, and short life of seals	Positive agitation yields efficient combustion and has automatic ash discharge
Controlled air, Los Alamos National Laboratory		
Limited airflow in primary chamber reduces ash entrainment, built-in TRU assay and x-ray equipment, tolerates small noncombustibles, no shredding of feed, and commercially available	Possible corrosion of off-gas system by HCl, ash removal needed, possible migration of radioactivity in refractory lining, accommodates only low levels of fissile materials	Commercial equipment and minimal ash in off gas
Excess air, Mound Facility		
Low capital cost, no waste pretreatment with batch operation, low waste-handling requirements, and adaptable for in-plant operators	Subject to acid corrosion, high particulate loading in off-gas system, relatively high carbon content in ash, accommodates only low levels of fissile material, and reported VR factors do not include subsequently produced waste	Allows for incineration of waste in storage drum without sorting or pretreatment
Fluidized bed, Rocky Flats Plant		
In situ neutralization of acids, low-temperature combustion eliminates refractories, agitation of waste by fluidization during combustion, dry off-gas system, continuous ash removal, low-temperature fired ash for actinide recovery, good for high levels of activity, and half the size of conventional incinerators	Sorting and shredding required for feed, feed should be free of metals and other combustibles to eliminate unnecessary loading of fluidized bed, expensive catalyst needed for off-gas burning, and some insoluble catalyst in ash	Neutralization of acids in fluidized bed of Na_2CO_3, has dry off-gas system, and has no refractories

Table 7.15 (continued)

Advantages	Disadvantages	Unique capabilities
	Rotary kiln, Rocky Flats Plant	
Continuous discharge of ash, minimizing criticality problem; can burn melted or liquid materials; industrially proven success; positive automatic ash removal; minimal waste pretreatment; tumbling action enhances combustion; and processes high levels of fissile material	Rotary seal maintenance, short refractory life possible, incomplete graphite combustion possible, and possible radioactivity migration and buildup in the refractory linings	Positive agitation of wastes for complete combustion and has automatic ash discharge system
	Slagging pyrolysis, Idaho National Engineering Laboratory	
Can process unsegregated wastes with high percentage of noncombustibles, minimal waste sizing or pretreatment, product is a stabilized residue, slag is continuously discharged, in situ neutralization of acids if Na_2CO_3 is used, and is commercially available	Large volume of waste in unit could cause nuclear safety concern, slag residue is unacceptable for recovery of actinides, weight reduction problems because of required additives for slag formation, high capital and operating costs	Slag-type residues are produced requiring no further fixation and system accepts noncombustibles
	Penberthy electromelt	
Product has excellent storage properties with no further treatment, tolerates wide range of feeds including liquids, no second stage required, and melt is easily removed	Refractory lining required and off gases carrying radioactivity will require solidification of scrubber solution resulting in minimal VR advantage for these wastes	Excellent product characteristics and handles wide variety of feeds

From LLW from Commercial Nuclear Reactors, Vol. 2, Treatment, Storage, Disposal and Transportation Techniques and Constraints by R. L. Jolley et al., ORNL/TM-9846/V2, Oak Ridge National Laboratory, Oak Ridge, Tenn., 1986. Reprinted by permission.

3. Performance of the same radwaste process equipment may vary from plant to plant as a result of the attitude toward and attention given to radwaste equipment and problems.
4. Radwaste systems should have sufficient redundancy to permit necessary preventive maintenance.

According to reference 37, such waste minimization techniques can be classified as general/administration, wet waste, and dry waste. Selected techniques that have been applied at six or more nuclear plants are mentioned here.[13,37]

General/administrative:
1. Use of dedicated compacting and decontamination crews
2. Establishment of consolidated and dedicated radwaste organization
3. Restriction of clean material to contaminated area
4. Limited access to the contaminated area

Table 7.16 Types of wastes that can be solidified

Liquids (including slurries)	Dry solids (contaminated trash excluded)
Evaporator concentrates (viscous slurries)	Incineratory ash (by type of feed)
Borates (5–50 wt %)	Dry active waste only
Sulfates (8–50 wt %)	Ion exchange resins
Mixed borates and sulfates (5–50 wt %)	Filter sludges
Reverse osmosis concentrates (3–10 wt %)	Mixtures of the above
Miscellaneous decontamination liquids	Dryer residues
Contaminated oils	Sodium sulfate[a]
	Sodium borate/boric acid[a]
Wet solids	Future possibilities
Ion exchange resins (bead)	Dried resin beads
Ion exchange resins (powdered)	Dried powdered resins
Sludges	Dried filter sludges
Diatomaceous earth	
Cellulose fibers	
Mixed cellulose fibers and powdered resins	
Non-precoat filter	
Resin cleaning	

[a] Sodium salts are typical; other metal salts may be produced by different processing methods.

From LLW from Commercial Nuclear Reactors, Vol. 2, Treatment, Storage, Disposal and Transportation Techniques and Constraints by R. L. Jolley et al., ORNL/TM-9846/V2, Oak Ridge National Laboratory, Oak Ridge, Tenn., 1986. Reprinted by permission.

Wet waste:
1. Portable demineralizers
2. Metallic backflushable filters
3. Improvement of bed life by adjusting resin ratios
4. Use of high-integrity containers

Table 7.17 Solidification agents

Hydraulic cements
  Portland (original agent)
  Gypsum
  Pozzolanic (mixture of Portland cement and ASTM class F fly ash)
  Aluminous (high-alumina cement composed primarily of monocalcium aluminate for aqueous tritiated water)
  Masonry cement (mixture of Portland cement and slaked lime)

Bitumen (asphalt)
  High-molecular-weight hydrocarbons containing both aliphatic and aromatic components (a thermoplastic material)

Unsaturated polyester polymers
  Vinyl esters—composition proprietary

Other agents
  Urea-formaldehyde (discontinued due to failure to meet evolving performance criteria with occasional drainable free liquids)
  Polymer-modified gypsum cement (blended as a powder with liquid waste)
  Glass (similar to solidification for HLW)

Table 7.18 Comparison of solidification agents and waste form properties

Waste form/binder property	Portland cement[a]	Asphalt	Unsaturated polyester	Urea-formaldehyde
Waste form				
Product density (kg/m^3)	1500–2000	1000–1500	1100–1300	1000–1300
Water binding strength	High	NA	Moderate–high	Moderate
Free-standing water	Occasionally	Never	Seldom	Often
Compressive strength (MPa)	0.1–25	—	8–20	0.4–3
Mechanical stability	High	Moderate	Moderate–high	Low[b]
Flammability	None	Moderate	Low–moderate	Low
Leachability	Moderate	Low–moderate	Moderate	High
Corrosivity to mild steel	Protective	Noncorrosive	Noncorrosive	Corrosive
Binder				
Shelf life	Several years	Months[c]	~6 mo	4 mo–1 yr
Storage temperature (°C)	−40 to 50	50[c]	20	10–20
Chemical tolerance[d]				
Boric acid concentrate	Poor–fair	Good	Good[e]	Good
Sodium sulfate concentrate	Fair–good	Fair	Good	Reduced efficiency
Alkaline waste solution	Good	Good	Good	Reduced efficiency
Detergent waste solution	Poor	Fair	Fair	Poor
Organic liquids	Poor	Fair	Poor	Poor
Ion exchange resins	Fair	Fair	Good	Good
Sludges	Good	Good	Good	Good
Volumetric efficiency[f]	0.5–0.9	>2	0.6–0.7	0.6–1.0

[a] Includes Portland cement and cement with additives.

[b] Loses water and strength on exposure to air.

[c] Stored heated for ease of use. Can be stored at −40 to 40°C for an indefinite period.

[d] For urea-formaldehyde, does not consider imposed free liquid criteria.

[e] May require pretreatment.

[f] Ratio of the as-generated waste volume to the solidified waste volume.

From LLW from Commercial Nuclear Reactors, Vol. 2, Treatment, Storage, Disposal and Transportation Techniques and Constraints by R. L. Jolley et al., ORNL/TM-9846/V2, Oak Ridge National Laboratory, Oak Ridge, Tenn., 1986. Reprinted by permission.

5. Adequate feed stream characterization
6. Leak detection and repair

Dry waste:
1. Segregation of contaminated trash drums
2. Removal and recycle of reusable items
3. Use of metal low-specific-activity (LSA) boxes
4. Chemical and ultrasonic decontamination of tools, small equipment, etc.
5. Storage of contaminated equipments for reuse in centralized location
6. Reusable anticontamination clothing

The benefit of purchasing VR equipment for a utility can be evaluated through use of a VR cost–analysis computer code, VRTECH, which uses an extensive data base.[38]

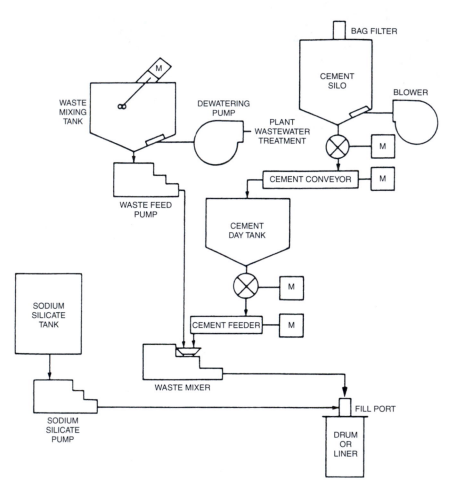

Figure 7.27 Simplified process flow diagram for an in-line mixing system for solidifying radioactive waste with cement–sodium silicate. From LLW from Commercial Nuclear Reactors, Vol. 2, Treatment, Storage, Disposal and Transportation Techniques and Constraints by R. L. Jolley et al., ORNL/TM-9846/V2, Oak Ridge, Tenn., 1986. Reprinted by permission.

Reference 38 indicated that radwaste generation rates and future burial price increases are the key factors in assessing the economic value of VR.

7.3.5 Mobile Process Systems

Mobile process systems and services have evolved in response to operating difficulties incurred with installed liquid waste treatment evaporators and solidification systems.[39] Various service companies have provided mobile or portable equipment to process nuclear plant waste streams. Systems often used are filtration, ion exchange, compaction, and solidification processes. Figures 7.29–7.31 show examples of such nuclear services for LLW management equipment. A mobile solidification system is shown in

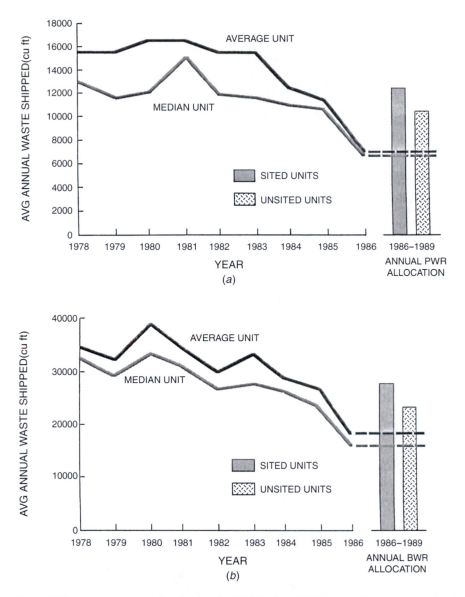

Figure 7.28 Total waste generation trends: (a) all BWRs; (b) all PWRs. From Radwaste Generation Survey Update, Vols. 1 and 2, by G. S. Daloisis and C. P. Deltete, EPRI NP-5526 Final Report, Electric Power Research Institute, Palo Alto, Calif., 1988. Copyright 1988. Electric Power Research Institute. Reprinted by permission.

Figure 7.29,[40] a mobile compaction system in Figure 7.30,[41] and a proposed mobile incineration system in Figure 7.31.[42] The economics of using mobile services are affected by several variables, including service company fees, availability of solidification processes, the nature of the feed streams, and the requirement for reuse or discharge of the processed water.

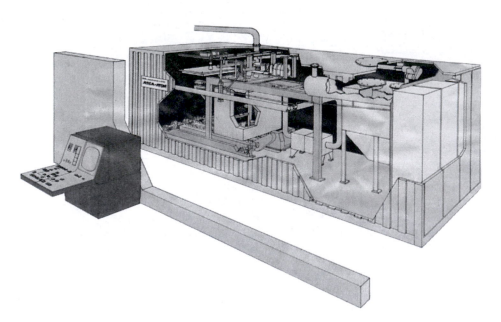

Figure 7.29 Mobile solidification system (MOSS)—ASEA-ATOM Company. MOSS is designed to immobilize radioactive waste generated by BWR and PWP plants as well as by other nuclear facilities. From MOSS-Mobile Solidification System by American Nuclear Society, *Nucl. News*, March 1985. Copyright 1985 American Nuclear Society. Reprinted by permission.

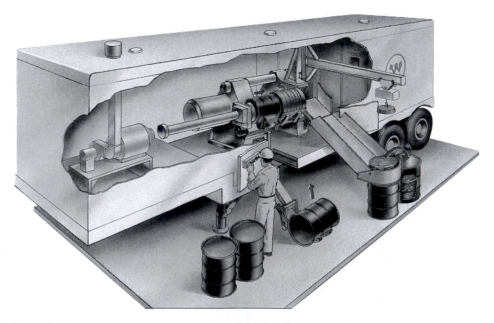

Figure 7.30 Mobile compaction system—Westinghouse Hittman compact 1. From Comprehensive LLW Management—The Westinghouse Hittman Approach by Westinghouse Electric Corporation, Westinghouse Hittman Nuclear Inc., Columbia, Md., 1984. Copyright 1984 by Westinghouse Electric Corporation. Reprinted by permission.

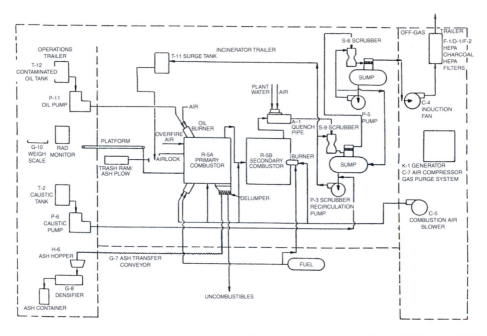

Figure 7.31 Mobile volume reduction system—AECC mobile incinerator. From Mobile VR System AECC, Topical Report AECC-4-NP-A, Rev. 1, Prepared for NRC, Aerojet Energy Conversion Company, 1986. Reprinted by permission.

7.3.6 Survey and Assessment of Treatment Technologies

Reference 13 presented a matrix of waste streams versus treatment technologies, based on technical feasibility; it did not consider location-specific parameters such as licensing requirements, public perception, and management philosophy, which may constrain the selection of technologies. Tables 7.19–7.21 reproduce major utility streams from reference 13, with the available technologies being divided into "extensively applied" (usually preferred) and "sparingly applied" categories and the potentially applicable technologies placed in an "advanced or developmental" category. The last category may be considered in the decision process to identify promising developments that may be cost-effective in the near future.

A survey was made in the United States in June 1985 of treatment technologies that were currently being used, that had been discontinued, and that were under consideration. The survey results included information about problem areas, areas needing research and development, and the use of mobile treatment.[43] The questionnaire was sent to 76 nuclear power plants, including 55 in operation and 21 under construction, and 41 of the plant operators responded. Table 7.22 shows the assessment of the treatment methodologies from reference 13 in six major categories: dewatering, thermal/physicochemical, sorting/segregation, decontamination, mechanical (VR), and solidification. The most frequently used technologies in mid-1985 were mechanical VR, dewatering, decontamination, and solidification processes. Compaction was the principal mechanical VR treatment used for solid wastes. Evaporation

Table 7.19 Treatment technologies for LLW streams—wet waste streams

	Treatment methods		
Utility streams	Extensively applied	Sparingly applied	Advanced or developmental
Spent resins	Batch drying, cement solidification, sorbent treatment, dewatering, encapsulation/ containerization	Centrifugation, asphalt solidification, organic polymer solidification	Drying, drying/ pyrolyzing, drying/ incineration (e.g., glass furnace), incineration, plasma reactors, acid digestion, solidification (glass, slagging)
Spent filter cartridges	Batch drying, sorbent treatment, dewatering, encapsulation/ containerization	Substitution	Solidification (glass, organic polymer, slagging)
Pump and lube oils— contamination	Filtration, dewatering	Coagulation/flocculation, distillation, incineration, solidification (asphalt, cements)	Evaporation, freeze– thawing, ion exchange (inorganic), stripping, solvent extraction, incineration, plasma reactors, acid digestion, solidification, superfine hollow fiber filter
Scintillation fluids	Distillation, sorbent treatment	Evaporation, substitution	Evaporation, incineration, supercritical water oxidation, solidification, bioabsorption
Filter sludges	Solidification (cements), sorbent treatments, dewatering, encapsulation, containerization	Evaporation, batch drying, centrifugation, coagulation/ flocculation	Freeze–thawing, wet- air oxidation, oxidation– reduction, biological, solidification
Regenerant wastes (Na_2SO_4)	Evaporation, evaporation/ crystallization, filtration, distillation	Evaporation, evaporation/ drying, batch/ continuous drying, solidification	—
Radwater evaporation concentrate	Filtration, solidification, sorbent treatment, encapsulation/ containerization	Centrifugation, coagulation/ flocculation, solidification (asphalt, organic polymers)	Wet-air oxidation, oxidation– reduction, biological, solidification, superfine hollow fiber filters

Table 7.19 (continued)

	Treatment methods		
Utility streams	Extensively applied	Sparingly applied	Advanced or developmental
Spent fuel pool runoff	Filtration, ion exchange (organic/inorganic)	—	Reverse osmosis/ hyperfiltration, ZrO_2 membrane ultrafiltration, supercritical water oxidation, biological, superfine hollow fiber filter
Floor and equipment drains	Evaporation, filtration	Reverse osmosis/hyperfiltration, coagulation/flocculation	ZrO_2 membranes, ultrafiltration, electrodialysis, wet-air oxidation, ultraviolet/ozone oxidation, biological, sorbent treatment, superfine hollow fiber filter
Chemical cleanup wastes	Filtration, evaporation	Evaporation (thin film), coagulation/ flocculation	Stripping, solvent extraction, incineration, plasma reactors, oxidation–reduction
Detergent solutions (including laundry personal cleanup, etc.)	Filtration	Evaporation, reverse osmosis/hyperfiltration, solidification (cements)	ZrO_2 membranes, ultrafiltration, electrodialysis, biological, bioadsorption, substitution, superfine hollow fiber filter

From LLW from Commercial Nuclear Reactors, Vol. 2, Treatment, Storage, Disposal and Transportation Techniques and Constraints by R. L. Jolley et al., ORNL/TM-9846/V2, Oak Ridge National Laboratory. Oak Ridge, Tenn., 1986. Reprinted by permission.

and filtration were the principal dewatering methods for relatively highly concentrated aqueous wastes, and drying was the principal method for wet residues. Mechanical, ultrasonic, and chemical methods were most commonly used for decontamination; cement was the principal solidification agent. Treatment methodologies being considered for the future involved wider use of ultrafiltration and evaporation, incineration, electrolytic decontamination, supercompaction, shredding and grinding, and solidification with cement and asphalt. Mobile facilities were being used or

Table 7.20 Treatment technologies for LLW streams—dry waste streams

Utility streams	Treatment methods		
	Extensively applied	Sparingly applied	Advanced or developmental
Vent air filters	Solidification (cements), encapsulation/ containerization	Dismantlement, compaction	Cutting/sawing, shredding and grinding, baling, crushing, supercompaction, solidification, decontamination
Reactor components	Dismantlements, cutting/sawing/ shearing, encapsulation/ containerization	Shredding/grinding, decontamination (chemical, electrolytic, vibratory)	Electrolytic processes, smelting/melting, supercompaction
Miscellaneous metals	Sorting–segregation, cutting/sawing/ shearing, shredding/grinding, encapsulation/ containerization	Solidification (cements) decontamination (chemical, electrolytic, mechanical)	Electrolytic processes, smelting/melting, solidification (slagging)
Wood (large items)	Encapsulation/ containerization	Incineration, sorting–segregation, cutting/sawing/ shearing, shredding/grinding, crushing	Incineration, wet-air oxidation, supercompaction
Trash: cloth and paper, plastics, PVC, rubber, glass	Sorting–segregation, shredding/grinding, baling, crushing, compaction, encapsulation/ containerization	Incineration, sorting–segregation, shredding/grinding, compaction	Incineration, plasma reactors, supercompaction, (decontamination chemical, electrolytic, vibratory, mechanical)

From LLW from Commercial Nuclear Reactors, Vol. 2, Treatment, Storage, Disposal and Transportation Techniques and Constraints by R. L. Jolley et al., ORNL/TM-9846/V2, Oak Ridge National Laboratory, Oak Ridge, Tenn., 1986. Reprinted by permission.

under consideration principally for filtration, ion exchange, compaction, and solidification with cement.

7.4 LOW-LEVEL WASTE PACKAGING AND TRANSPORTATION

Packaging and transportation of LLW for disposal are governed by a complex set of federal, state, and local laws and waste acceptance criteria established by the disposal facilities. State and local laws and waste acceptance criteria vary across the United States. Therefore, this section will focus on the two federal regulation: (1) 49CFR173 from the Department of Transportation[44] and (2) 10CFR71 from the Nuclear Regulatory Commission.[45]

Table 7.21 Treatment technologies for LLW streams—process and other waste streams

Utility streams	Treatment methods		
	Extensively applied	Sparingly applied	Advanced or developmental
Primary coolant	Filtration, ion exchange (organic)	—	Reverse osmosis/hyperfiltration, ZrO_2 membranes, ultrafiltration, ion exchange (inorganic), dewatering
Steam generator condensate, steam generator blowdown	Filtration, ion exchange (organic)	—	Reverse osmosis/ hyperfiltration, ZrO_2 membranes, ultrafiltration, ion exchange (inorganic), dewatering
Boron recycle feed	Evaporation, filtration, ion exchange (organic)	—	Evaporation (thin film), ion exchange (inorganic), bioadsorption, dewatering
Industrial LLRW	Evaporation, batch-drying, centrifugation, filtration, ion exchange (inorganic) distillation, incineration, oxidation–reduction, sorting–segregation, dismantlement cutting/sawing/ shearing, shredding/grinding, baling, crushing, compaction supercompaction	Evaporation (thin film), reverse osmosis/ hyperfiltration, precipitation, coagulation/ flocculation, stripping, solvent extraction, smelting/melting, solidification (asphalt, cements), substitution	Evaporation/crystallization, evaporation/extrusion, drying, freeze–thawing, incineration, plasma reactors, oxidation–reduction, biological, bioadsorption, superfine hollow fiber filter

From LLW from Commercial Nuclear Reactors, Vol. 2, Treatment, Storage, Disposal and Transportation Techniques and Constraints by R. L. Jolley et al., ORNL/TM-9846/V2, Oak Ridge National Laboratory, Oak Ridge, Tenn., 1986. Reprinted by permission.

Both of these regulations define three types of *low specific activity* (LSA) material, which is generally considered to be LLW. The definitions are as follows:

LSA-I:

1. Ores containing only naturally occurring radionuclides (e.g., uranium, thorium) and uranium or thorium concentrates of such ores.

2. Solid unirradiated natural uranium or depleted uranium or natural thorium or their solid or liquid compounds or mixtures.

Table 7.22 Assessment of treatment methodologies

Treatment	Currently used	Being considered	Discontinued
Dewatering	(172)	(26)	(19)
Evaporation	38	6	5
Drying	49[a]	1	6
Centrifugation	5	1	4
Filtration	38[a]	6[a]	0
Ultrafiltration	4	8	0
Reverse osmosis and other	38[a]	4	4
Thermal/physicochemical	(56)	(29)	(3)
Ion exchange	41[a]	2	0
Distillation	11[a]	3	3
Incineration	2	24	0
Electrolytic and other	2	0	0
Sorting/segregation	(84)	(16)	(0)
Decontamination	(142)	(27)	(13)
Mechanical	45	4	0
Electrolytic	11	10	4
Ultrasonic	38	6	2
Chemical and other	48	7	7
Mechanical treatment (VR)	(181)	(72)	(0)
Cutting, sawing, etc.	36	3	0
Shredding, grinding	16	29	0
Compaction	74	6	0
Supercompaction	3	30	0
Dismantlement, baling, etc.	52	3	0
Solidification	(121)	(48)	(11)
Cement	83[a]	24[a]	10
Asphalt	9	14	0
Organic polymers	4	5	1
Sorbents and other	25	5	0

[a] Including mobile facilities.

From LLW from Commercial Nuclear Reactors, Vol. 2, Treatment, Storage, Disposal and Transportation Techniques and Constraints by R. L. Jolley et al., ORNL/TM-9846/V2, Oak Ridge National Laboratory, Oak Ridge, Tenn., 1986. Reprinted by permission.

3. Class 7 (radioactive) material, other than fissile material, with an unlimited value of A_2, the number of Curies specified for each radionuclide in a table included within the regulation.

4. Mill tailings, contaminated earth, concrete, rubble, other debris, and activated material in which the Class 7 (radioactive) material is essentially uniformly distributed and the average specific activity does not exceed $10^{-6} A_2/g$.

LSA-II:

1. Water with tritium concentration up to 0.8 TBq/L (20.0 Ci/L).

2. Material in which the Class 7 (radioactive) material is distributed throughout and the average specific activity does not exceed $10^{-4} A_2/g$ for solids and gases and $10^{-5} A_2/g$ for liquids.

LSA-III. Solids (e.g., consolidated wastes, activated materials) that meet the requirements of Section 173.468 and for which the following are satisfied:

1. The Class 7 (radioactive) material is distributed throughout a solid or a collection of solid objects, or is essentially uniformly distributed in a solid compact binding agent (such as concrete, bitumen, ceramic, etc.).

2. The Class 7 (radioactive) material is relatively insoluble, or it is intrinsically contained in a relatively insoluble material, so that, even under loss of packaging, the loss of Class 7 (radioactive) material per package by leaching when placed in water for 7 days would not exceed $0.1A_2$.

3. The average specific activity of the solid does not exceed $2 \times 10^{-3}A_2$/g.

Surface contaminated objects (SCOs) are also generally transported and disposed of as LLW. 10CFR71.4 defines an SCO to be "a solid object that is not itself classed as radioactive material, but which has radioactive material distributed on any of its surfaces." The regulation specifies limits on surface contamination for SCOs.

Packages used for transporting LSA waste and SCOs must meet general design requirements specified in 49CFR173.410 and general standards specified in 10CFR71.43. These requirements include package size, types of seals on openings, construction materials, and ability to withstand normal transportation conditions, such as vibration, without releasing radioactive material. 10CFR71.47, which governs external radiation doses, states that the radiation level will "not exceed 2 mSv/h (200 mrem/h) at any point on the external surface of the package." If a package exceeds this limit, it must be transported on a vehicle that carries only packages from the shipper of the radioactive material, and people loading and unloading the packages must have written instructions. Limits on radiation doses external to the transport vehicle are specified.

7.5 OPERATIONAL EXPERIENCE WITH VOLUME REDUCTION SYSTEMS

Several historical operational experiences with volume reduction (VR) systems have been reported in the literature.[46] The Palisades power station VR system, transportable VR and solidification system, and British Nuclear Fuels plc (BNFL) VR system are discussed below.

7.5.1 Palisades VR System[47]

The first operating VR system in the United States, according to the Consumers Power Company's Palisades Power Station, consists of an extruder evaporator that evaporates water from liquid wastes while simultaneously encapsulating the residual solids in an asphalt binder. Startup testing of the VR system showed that boric acid, bead resin, powdered resin, and cartridge filters could be solidified. Radioactive concentrates were first processed by the VR system on January 30, 1984. Neither contaminated bead resin nor cartridge filters were processed during the first year. The use of powdered resin at Palisades has been discontinued and contaminated powdered resins have therefore not been processed. From January 31, 1984 to January 30, 1985, about 41,000 gal of boric acid was processed, resulting in 150 asphalt drums. The previous cement process (30 gal of waste per drum) would have produced 1370 drums, so a VR ratio of 9 was achieved. A net VR ratio of 12.2 has been observed for individual drums. Consumables per drum, consisting of sodium hydroxide for boric acid neutralization, the asphalt binder, and the container, cost about $94. The average weight of the drums produced was 532 lb, with an average loading of 232 lb of sodium borate salts per drum. The surface dose rate of these drums ranged from a high of 1 R/hr to a low of 12 mR/hr, with an average of 94 mR/hr. Only Class A waste was solidified during that time, and the first shipment of asphalt waste to a disposal site was in May 1985. A drum fill requires about 10 hr and operators use a walk-pass method to run the VR system, requiring about 2.5 person-hr per drum. Some equipment modifications were made based on operating experience, including drum conveyor change, condensate boiler feed pump replacement, ventilation prefilter change, and steam dome spray nozzle modifications.

7.5.2 Transportable VR and Solidification System (TVR-III)[48,49]

The first liquid waste TVR-III system began processing radmaterial in August 1986 at the Palo Verde Nuclear Generating Station and completed the first campaign of processing wastes in January 1987. The transportable unit was mounted on a double low-bay trailer 3 m (10 ft) wide by 14 m (46 ft) long. The complete stand-alone system has enclosed weather-protected modules; is equipped with spill containment and drainage, filtration, radiation monitoring, shielding, and HVAC control; and includes ALARA considerations. TVR-III is an one-step VR and bitumen solidification concept (Figure 7.32). The end product characteristics are shown in Table 7.23.

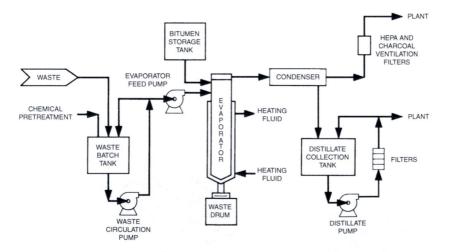

Figure 7.32 Basic flow diagram TVR-III system. From Use of VR and solidification services at Arizona Nuclear Power Projects Palo Verde Nuclear Generating Station by T. P. Hillmer and R. D. Doyle, in *Waste Management '87*, R. G. Post, ed., University of Arizona Press, Tucson, Ariz., 1987. Copyright 1987 by University of Arizona. Reprinted by permission.

Table 7.23 End product characteristics of TRV-III

Unit weight	58,430 kg (128,000 lb)
Number of drums produced	52
Gross drum weight	250 kg (550 lb)
Solid content	50–60%
Free water	0
Fill	
One pass	88–97%
Two passes	94–100%
Surface dose rate	60–180 mR/hr
Cost comparison	
Cement service	
Unit cost	$5.11/L ($19.33/gal)
Annual cost	$1,450,600
Asphalt service	
Unit cost	$4.80/L ($18.16/gal)
Annual cost	$1,362,600
Annual savings with asphalt service	$88,000

7.5.3 BNFL Sellafield Reprocessing Plant VR System

Since 1971, BNFL has been responsible for operating the Sellafield site (formally Windscale Plant) and providing fuel cycle services.[50] Consequently, BNFL has been responsible for the treatment of any wastes generated as a result of such services. Some liquid effluents are discharged to sea after treatment. To minimize the amount of radioactivity discharged to the environment, the company is introducing a new plant that will reduce annual Sellafield discharges to about 0.74 TBq (20 Ci) alpha and 300 TBq (8000 Ci) beta. The reduction is being achieved by three principal processes:

1. A Site Ion Exchanger Plant (SIXEP) with pressurized columns, which was brought into operation in 1985 to treat water from the fuel storage ponds, mainly to remove Cs and Sr (Figure 7.33).
2. Evaporation and decay storage of liquid wastes from solvent extraction by the salt evaporator, also commissioned in 1985 (Figure 7.34).
3. An Enhanced Actinide Removal Plant (EARP), was scheduled to be operational by 1992, to remove alpha activity (actinides) and some beta activity from the LLW effluent (about 250 m^3/day or 8830 ft^3/day) not treated by SIXEP and currently discharged to sea (Figure 7.35).

Experienced and anticipated results are as follows:

1. SIXEP: During 20 months of operation, a decontamination factor (DF) of about 900 was achieved for Cs at a flow of 3000 m^3 (1.06 × 10^5 ft^3) of purge per day, costing $180 million.
2. Salt evaporator: A new dedicated evaporator is used with a caustic scrubber to remove radioactive iodine at a flow of about 150m^3/day (5300 ft^3/day), costing $22 million.

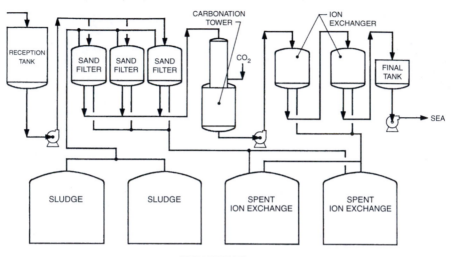

SEA DISCHARGE TREATMENT PLANT

Figure 7.33 Simplified process flow diagram for SIXEP. From BNFL's objectives and achievements in the reduction of radioactive discharges from reprocessing plant by C. S. Mogg and W. Heafield, in *Waste Management '87*, vol. 3, p. 613, R. G. Post, ed., University of Arizona Press, Tucson, Ariz., 1987. Copyright 1987 by University of Arizona. Reprinted by permission.

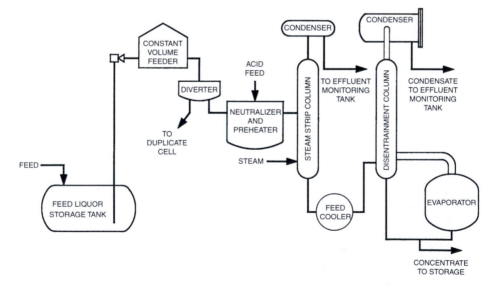

Figure 7.34 Simplified process flow diagram for salt evaporator. From BNFL's objectives and achievements in the reduction of radioactive discharges from reprocessing plant by C. S. Mogg and W. Heafield, in *Waste Management '87*, vol. 3, p. 613, R. G. Post, ed., University of Arizona Press, Tucson, Ariz., 1987. Copyright 1987 by University of Arizona. Reprinted by permission.

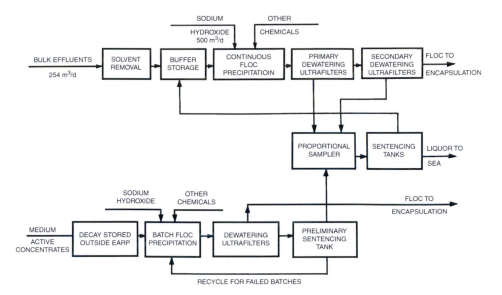

Figure 7.35 EARP simplified block diagram. From BNFL's objectives and achievements in the reduction of radioactive discharges from reprocessing plant by C. S. Mogg and W. Heafield, in *Waste Management '87*, vol. 3, p. 613, R. G. Post, ed., University of Arizona Press, Tucson, Ariz., 1987. Copyright 1987 by University of Arizona. Reprinted by permission.

3. EARP: Flocculation is caused by the addition of NaOH to the iron-bearing acidic streams to increase the pH. Almost all alpha activity coprecipitates with the floc, leaving a virtually inactive aqueous phase. EARP has the capacity for 1000 m³/year of MAW (medium active waste),* 2400 m³/year (8.48×10^4 ft³/year) of salt evaporator concentrates, and 250 m³/day (8830 ft³/day) of LLW, costing $300 million.

7.6 SHALLOW LAND DISPOSAL

Throughout the world, shallow land burial and ocean disposal are the two methods practiced for the disposal of LLW. The latter method has been discontinued by most countries, although the development of potential subseabed disposal has been actively pursued in the European communities. These two approaches represent two different philosophies. The land burial approach favors isolation and VR and sequestering of the waste radioactivity until decay has essentially eliminated any hazard—the concentrate-and-contain philosophy. The ocean disposal approach relies on the enormous volume and continuous motion of the ocean to prevent undesirable radionuclide concentrations in the ocean environment if the wastes escape from containers—the dilute-and-disperse philosophy.[13] A third alternative, extraterrestrial

---

*MAW is a waste classification used in Europe to refer to material with radioactivity $>10^{-4}$ μCi/mL but <10 μCi/mL.

disposal, has been considered but has not yet been shown to be feasible. Therefore, shallow land burial is the method currently used to dispose of LLW. To provide greater assurance that radionuclides will not migrate significantly, the concept of greater confinement is under consideration, using deep trenches, engineered structures, and shallow repositories.[13,51]

7.6.1 Environmental Safety of Shallow Land Disposal

The following safety objectives of near-surface disposal facilities are specified in 10CFR61:

1. Protection of the general public from release of radioactivity.
2. Protection of the individuals from inadvertent intrusion.
3. Protection of individuals during operations.
4. Ensure stability of the site after closure.

The most likely scenario for waste release from a land disposal involves water intrusion. Leaching and movement of wastes into groundwater may occur with subsequent transport of radionuclides and chemicals to off-site groundwater or surfacewater bodies. For above-grade disposal facilities, runoff directly into surface waters of leached radionuclides is possible.[13] The concentration of radionuclides that may be released from the disposal site to the environment in groundwater, air, soil, plants, or animals must not result in a yearly dose for any individual exceeding an equivalent of 25 mrem to the whole body, 75 mrem to the thyroid, and 25 mrem to any other organs. The environmental safety of the waste disposal system is predicated on the stability of the waste and the disposal site so that once the waste is emplaced and covered, access of water to it is minimized. The classification scheme for radwaste (Tables 7.4A and 7.4B) is designed to distinguish among LLW in the following manner: (1) wastes that do not have to meet stability requirements, (2) wastes for which a stable form (i.e., maintaining gross physical properties and identity) is required to protect the disposal site integrity, and (3) wastes that require protection against inadvertent intruders. These three classes are known as class A (segregated wastes), class B (stable wastes), and class C (intruder wastes). Section 7.2.3 listed the concentration limits for these classes. Table 7.24 presents the NRC regulatory scheme for LLW disposal (10CFR61).[13] In addition to the three classes mentioned above, the table shows the exempt wastes, which are below class A concentrations or "below regulatory concern,"[52] and prohibited wastes, which are above class C concentrations, or class C+. These two extremes are not explicitly mentioned in 10CFR61 but are included for completeness. All three waste classes must meet minimum requirements that include the following (10CFR61.56):

1. Waste form and packaging must meet DOE and NRC transportation requirements.
2. Waste cannot be packaged in cardboard or fiberboard containers.
3. Packages containing liquids must contain sufficient absorbent material to absorb twice the volume of liquid.

Table 7.24 The NRC regulatory scheme in 10CFR61 for LLW disposal

Waste classification	Exempt wastes[a]	Class A segregated wastes	Class B stable wastes	Class C intruder wastes	Prohibited wastes[a]
Time period before reaching acceptable concentration levels	Already at acceptable levels	During site occupation	<100 yr	100–500 yr	>500 yr
Protective measures	Not required	Decay	Institutional controls	Intrusion barriers	Disposal by means other than near-surface disposal
Waste form requirements	None	Meets minimum requirements	Meet both minimum and stability requirements	Meet special requirements for class C wastes	Specific approved form
Minimum disposal depth	None	None	None	5 m or use of engineered barriers	>5 m
Container stability	Not required	Not required	300 yr	300 yr	
Radionuclide characteristics	Concentrations less than in Col. 1, Table 7.4B	Maximum concentrations less than in Col. 1, Table 7.4B	Concentrations greater than in Col. 1 but less than in Col. 2, Table 7.4B	Concentrations greater than in Col. 2 but less than in Col. 3, Table 7.4B	Concentrations at or exceeding Col. 3, Table 7.4B

[a]These classes are not explicitly mentioned in 10CFR61 but are included for completeness.
From LLW from Commercial Nuclear Reactors, Vol. 2, Treatment, Storage, Disposal and Transportation Techniques and Constraints by R. L. Jolley et al., ORNL/TM-9846/V2, Oak Ridge National Laboratory, Oak Ridge, Tenn., 1986. Reprinted by permission.

4. Waste must not be readily capable of deterioration, undergo explosive decomposition at normal temperatures and pressures, or explosively react with water.
5. Wastes generally must not contain or be capable of generating quantities of toxic gases, vapors, or fumes harmful to persons handling or transporting wastes.
6. Pyrophoric materials shall be treated, prepared, or packaged to be nonflammable.
7. Waste in gaseous form must not exceed 1.5 times atmospheric pressure and must have total activity less than 100 Ci per container.
8. Wastes containing biological pathogenic or infectious material must be treated to reduce to maximum extent practicable the potential hazard.

Requirements that apply to waste generators and waste processors, related to the preparation of more complete manifests and descriptions of the waste (10CFR20) and the need for segregating and classifying waste, are shown in Table 7.25.[13]

7.6.2 Overview of Commercial Disposal Sites

As shown in Table 7.1, four of the six commercial disposal sites in the United States have been closed. A brief description of each of the six commercial sites as well as the private Envirocare site is given below.

Beatty, Nevada, site.[52,53] The site consists of shallow trenches of varying dimensions ranging from 91 to 244 m (300–800 ft) in length, 1.2 to 106 m (4–350 ft) in width, and

Table 7.25 Procedures for preparing LLW for near-surface disposal

Requirement	Information required	Required of
Shipment manifest	Identity of persons generating and persons transporting wastes Type, volume, and class of wastes, radionuclide identity and concentration; total radioactivity, chemical form, and solidification agent	Waste generator, waste collector, and waste processor (treats or repackages wastes)
Certification	Certification that wastes are properly classified, described, packaged, marked, and labeled and are in proper condition for transport under NRC and DOT regulations	Waste generator
Waste preparation	Meet classification requirements (Section 61.55) and waste characteristic requirements (Section 61.56) of 10CFR61	Waste generator and waste processor
Labeling	Label each package to identify the waste type: class A, class B, or class C	Waste generator and waste processor
Other requirements	Conduct quality assurance program and management audits, and investigate missing shipments	Waste generator and waste processor

From LLW from Commercial Nuclear Reactors, Vol. 2, Treatment, Storage, Disposal and Transportation Techniques and Constraints by R. L. Jolley et al., ORNL/TM-9846/V2, Oak Ridge National Laboratory, Oak Ridge, Tenn., 1986. Reprinted by permission.

1.8 to 15.2 m (6–50 ft) in depth. A regional groundwater table lies at a depth of 79 and 100 m (260 and 330 ft) below the surface in the alluvial soils. U.S. Ecology, Inc., monitors the concentration of radioactivity in groundwater in on- and off-site wells and in air, soils, and vegetation on a quarterly basis and has not found evidence of migration of radioactivity from the burial trenches through any environmental pathways. The Nevada Division of Health is the current regulatory agency. Between March 1976 and December 1979, a series of events involving improper handling and disposal of LLW resulted in the site being closed for intervals due to temporary suspensions of the operator's license. These closings were to protect the public by preventing unsafe shipments of LLW on Nevada's highways and were not indicative of unsafe practices by the disposal site operators.

Barnwell, South Carolina, site.[53,54] The site consists of slit trenches and conventional shallow trenches. Two slit trenches, 76 and 152 m (250 and 500 ft) long, 0.9 m (3 ft) in width, and 6.1 m (20 ft) deep, are used for the disposal of special (class C) waste having high surface radiation levels. Conventional shallow land burial trenches varying from 61 m (200 ft) long, 15 m (50 ft) wide, and 4.6 m (15 ft) deep to 305 m (1000 ft) long, 30.5 m (100 ft) wide, and 6.4 m (21 ft) deep have been used for most of the LLW. To facilitate the collection and removal of leachate, the trenches are constructed with a floor that slopes to one side. A French drain runs along the side of the trench and is sloped about 0.3%. Water collection sumps and standpipes for removal of leachate are placed at 152-m (500-ft) intervals along the drain. Lateral infiltration of moisture into the trenches from the surrounding soil is reduced by replacing any sand lenses or layers in the trench walls with compacted clay. After backfilling with sand, a layer of soil (minimum 0.9 m or 3 ft) is placed over the trench, followed by a layer of compacted clay (minimum 0.6 m or 2 ft) and topsoil with cover crop. The regional groundwater table lies at depths ranging from 9.1 to 18.3 m (30 to 60 ft), while the principal source of potable water in the area is from a depth in excess of 107 m (350 ft). With a mean annual precipitation of 1.2 m (47 in.), surface water runoff occurs only after unusually heavy rainfall. The site operator, Chem-Nuclear System, Inc. (CNSI), monitors the concentration of radioactivity in on-site and off-site wells at quarter-yearly and annual intervals. A comprehensive study by the U.S. Geological Survey (USGS) indicates that migration of tritium is occurring within 3 m (10 ft) of some older burial trenches.[55] There have not been significant problems specifically related to either site operations or on-site waste management, but waste handling difficulties have arisen related to improperly packaged waste received or to other violations.[56] During 1979, the number and impact of shipping violations were serious enough to cause governors of three states with operating burial sites to demand corrective action by the federal government. At the Barnwell site, from 1971 through 1982 over 3.9×10^5 m³ (14×10^6 ft³) of waste was disposed of. When the volume climbed to an annual rate of 68,000 m³ (240,000 ft³) in October 1979, the governor of South Carolina imposed a phased-in volume limitation of 3400 m³ (120,000 ft³) per year, effective in October 1981. Through that year, approximately 40% of the licensed acreage remained available for future use.

Richland, Washington, site.[53,54] This is the only commercial site located on federal land and consists of 100 acres of land leased by U.S. Ecology, Inc. It was licensed and opened as a commercial venture in 1965 and except for a period in 1979 has operated since that time. The burial trenches vary dimensionally, with typical sizes ranging from 91 to 104 m (300–340 ft) in length, 7.6 to 42.7 m (25–140 ft) in width, and 6.1 to 13.7 m (20–45 ft) in depth. The trend is toward trenches 107 m (350 ft) long and 13.7 m (45 ft) deep. The waste containers are placed in the trench, leaving a minimum distance of 2.4 m (8 ft) from the original ground surface. After backfilling with the excavated soil, a soil cover is formed into a mound at least 1.5 m (5 ft) thick at the centerline and 0.9 m (3 ft) thick near the trench edge. A 15-cm (6-in.) layer of riprap (gravel and cobble) is placed on the mound to protect the cover soil against wind erosion and intrusion by burrowing animals. Water infiltration into the trenches has not been a problem, and the bottoms of the trenches are hundreds of feet above the water table. The depth of the water table in this region ranges from 59 to 107 m (195–350 ft). The climate at the site is mild and quite dry, with an average annual precipitation of about 15 cm (6 in.) and an annual potential evaporation rate of 1.4 m (55 in.). The facility has experienced no problems related to site operations or waste containment, but, in common with the Barnwell and Beatty sites, it did experience difficulties from violations of transportation and packaging regulations. These problems led in 1979 to generally unified actions to tighten generator and transporter compliance procedures along with actions to upgrade waste acceptance standards at the three sites.

Maxey Flats, Kentucky, site.[53,57] The site consists of 46 closed trenches, one open trench with a temporary roof, a number of caissons, and several special pits. The trenches range from 45.7 to 183 m (150–600 ft) in length, 3 to 22.9 m (10–75 ft) in width, and 2.7 to 9.1 m (9–30 ft) in depth. The flow of each trench slopes at 1° toward a sump constructed at the low end to permit water collection and removal. The caissons (generally 4.6 m or 15 ft deep by several feet in diameter) were used to dispose of high-activity gamma sources, while special trenches were used for large volumes of high-activity waste (spent resins). The trenches lie entirely within a green shale interlain with siltstone and sandstone, which contains perched groundwater in the soil zone at a depth of 0.9–1.8 m (3–6 ft). There is a continuous groundwater table at a depth of 9.1–15.2 m (30–50 ft), but no regional aquifer is in the area. When filled with waste, the trench was covered with a minimum of 0.9 m (3 ft) of a clayey soil in compacted layers, a mounded cap was developed over each trench to assist water runoff, and a layer of topsoil was added to support a vegetative cover. A significant amount of water accumulated in the burial trenches, which required a water management program that started in 1973 and continues to this day. One possible explanation for the water problem is that the backfill material of the trench is less dense than the surrounding materials, causing water to be perched in the trench.[58] This technical factor, along with other contributing factors, led to the closure of the site in 1977, with a total waste deposition of 1.36×10^5 m³ (4.8×10^6 ft³). Since then, site conditions have been analyzed and a remedial plan developed, along with investigations to determine potential migration pathways and extent of dispersion. Efforts to stabilize the site prior to decommissioning are under way.[59]

West Valley, New York, site.[52,56] There are 14 burial trenches in north and south areas, typically 10.7 m (35 ft) wide, 6.1 m (20 ft) deep, and 183–244 m (600–800 ft) long. The surface water in the vicinity of the area consists of Frank's Creek on the east and an unnamed tributary on the north and west sides, although the existence of aquifer has not been proved. This is part of the Nuclear Service Center, which was established and operated by Nuclear Fuel Services (NFS) until March 1975. At that time operations were suspended by NFS and the site remained closed after an overflow of contaminated water from two of the trenches in the north area was detected. As there are no records describing site preparation activities for these trenches, there is some question as to the removal and disposition of the surface soils and the approach used for capping the trenches. In March 1975, water that had infiltrated into two trenches with only 1.2 m (4 ft) of cover seeped through the covers at a rate of approximately 1 gal/day. NFS stopped burial operations immediately and terminated commercial operations at the site. The water accumulation in the trenches had been monitored during the first 2–3 years and reached an essentially stable level except in few cases. It was thought to have resulted from the impermeability of the silty till soil in conjunction with the normal heavy local precipitation. Water seeping through the cover was pumped to bring the trench water content to the lowest practical level. Another problem that arose during operation was the indication that there was localized surface and slope erosion induced by surface water flow. Remedial measures at the site since then have included reworking the covers to achieve greater compaction and eliminate cracks and depressions and occasional pumping of the trench water. The site remains closed, with necessary monitoring and maintenance being continued by a small crew. A total waste volume of 6.5×10^4 m^3 (2.3×10^6 ft^3) was deposited at the site. In addition to the state-licensed site for commercial LLW, there is a separate, NRC-licensed burial area primarily for HLW from the one-time reprocessing plant.

Sheffield, Illinois, site.[53,57] The site has 21 separate trenches typically 152 m (500 ft) long, 15–18 m (50–60 ft) wide, and 6–7.6 m (20–25 ft) deep, with a minimum of 3 m (10 ft) at the surface between trenches. Waste packages were placed in trenches to within 0.6 m (2 ft) of the original ground level. A minimum of 1 m (3 ft) of compacted clay was used to form a cap and cover the trenches. The regional groundwater aquifer is about 90 m (300 ft) below the site, but there is a saturated zone in the glacial drift 4.6–20 m (15–65 ft) below grade. The USGS estimated the velocity of groundwater to be between 1 and 2 m (3 and 7 ft) per year, but studies showed differently. In December 1977, tritium was found in samples from monitoring wells and appeared to be migrating from the first trench. Between December 1978 and March 1979 the USGS obtained information indicating that sand and other coarse-grained deposits far more extensive than originally thought appeared to underlie most of the site continuously (sand lens). In 1976, as the site capacity was too small, the operator (Nuclear Engineering Company Inc., NECO; now U.S. Ecology, Inc.) applied to the NRC for license renewal and expansion, but it was ruled that no waste could be placed in the new trench, which was constructed in the originally licensed 20-acre tract. This led to a de facto closure of the site. U.S. Ecology withdrew its application in April 1978 and closed the site to burial of radmaterials because the expansion application was left pending and there was no space in other trenches.

Envirocare. The Envirocare facility near Clive, Utah, is a private disposal facility that was first licensed in 1988 by the state of Utah to dispose of naturally occurring radioactive materials. In 1991, its license was amended to allow Envirocare to accept some Class A LLW with limits on the types and concentrations of radionuclides as well as on chemical and physical properties of the waste. Envirocare also has a license issued by the NRC to accept uranium and thorium mill tailings. The facility was designed to accept 247 million ft^3 of LLW, and through 1998, it had disposed of 619,000 ft^3 of operating waste and nearly 10 million ft^3 of wastes resulting from cleanup of nuclear facilities.[9] Envirocare accepts wastes not only from commercial generators, but also from DOE, the Department of Defense, and the Environmental Protection Agency. While the other six LLW disposal facilities in the United States have buried waste in trenches, at Envirocare, the waste is put in above-ground cells which are capped when they are full.

Conclusions. The operating experience at the commercial disposal sites has provided an excellent basis for understanding the problems associated with shallow land burial. The combined effects of poor hydrologic isolation of the waste and physical stability of the disposal unit in relation to surface water infiltration and subsequent migration of radionuclides were major problems.[60] Disposal unit stability can be ensured by providing stable waste packages and waste forms, compacting backfill material, filling the void space between packages, and installing self-supporting trench caps with suitable materials. Hydrologic isolation can be achieved through a combination of proper site selection, subsurface drainage controls, internal trench drainage systems, and immobilization of the waste.

7.6.3 Disposal Development in the European Community

All major European countries have worked on projects for the disposal of low- and medium-level radwaste by shallow land burial (e.g., France and the U.K.), disposal in mines (West Germany), or disposal in rock caverns (Sweden). Some shallow land burial sites have been operating for many years (e.g., the Centre de la Manche in France and Drigg in the U.K.); however, the capacity of these sites is limited. A fairly generic system was developed at Riso, Denmark, which consists of a regular hexagon-shaped concrete container with concrete bunkers or cylindrical construction of conventional technology and soil cover of various thicknesses.[51] As a part of the engineered barrier, molten bitumen is used to seal the interface between individual hexagonal units. No release will occur if it is possible to keep water out of contact with the waste for a sufficiently long time, and only very slow releases can be expected if the transport out of the repository is exclusively by diffusion after saturation has taken place. Thus the design approach may aim at the goal that major defects in barriers will be slow to develop and that significant water flow driven by hydraulic gradients, solution density difference, or gas pressures will be improbable.[51] Adequate safety of shallow land burial of LLW may also be attained without using a complex system of barriers if the facilities are protected against percolating

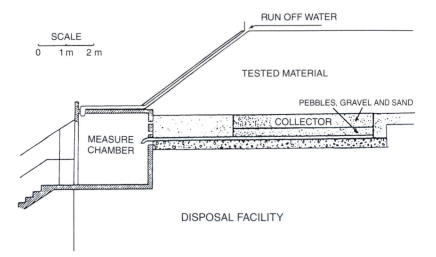

Figure 7.36 Experiment with cover material for trenches. From Technology for the Improvement of Shallow Land Burial by K. Broderson and R. Andre Jehan, in *Radwaste Management and Disposal, Proc. 2nd European Community Conference*, edited by R. Simon, Luxembourg, April 1985.

rainwater by some type of watertight covering. ANDRA, the organization responsible for LLW disposal in France, has initiated a program to develop and test such coverings, using a system of 6 × 4 m (20 × 13 ft) rectangular collector plates of stainless steel with edges raised to a height of 0.5–1 m (1.6–3.2 ft), which is nearly filled with high-permeability materials. The cover material is spread in a thickness of 3–4 m (10–13 ft) above the permeable material in the collector (Figure 7.36). Physicochemical measurements include conductivity, pH, temperature, turbidity, dissolved oxygen, and major anions and cations, and results demonstrate that watertight covers can be made if proper materials are selected.[51]

7.7 COMPUTER CODES

PAGAN 1.1. Code system for performance assessment ground-water analysis for low-level nuclear waste. This code, developed at Sandia National Laboratories in the early 1990s, is used by the U.S. NRC when evaluating license applications for low-level waste disposal facilities. It can be used to model radionuclide transport in the vadose zone and aquifers. It is written in FORTRAN 77 and runs on a PC.

PRESTO-II. Code system for low-level waste environmental transport and risk assessment. This code, developed at Oak Ridge National Laboratory in the mid-1980s, can be used to model transport of radionuclides from a low-level waste burial trench, through the environment, and to a nearby population. Exposure of the population and health effects are also modeled. The code is written in FORTRAN 66.

7.8 DISCUSSION QUESTIONS AND PROBLEMS

1. What are the two types of DOE activities that generate LLW?

2. What are the five categories of commercial LLW generators? Name at least one way LLW is produced by each category of generator.

3. What are the different classes of LLW? How is each class defined?

4. Which sites in the United States are currently accepting LLW?

5. To which compact does your state belong? What other states are members? Where do LLW generators in your state send their waste?

6. Describe a typical LLW burial facility.

7. What are the advantages and disadvantages of creating regional LLW disposal facilities?

8. Define the three classes of treatment processes discussed in this chapter. Briefly explain each one.

9. What steps are being taken to reduce the volume of LLW created and/or disposed? Why is it important to generators to reduce the volume of waste buried?

REFERENCES

1. U.S. Department of Energy, Integrated Data Base Report-1993: U.S. Spent Nuclear Fuel and Radioactive Waste Inventories, Projections, and Characteristics, DOE/RW-0006, Rev. 10, U.S. Department of Energy, Washington, D.C., December 1994.
2. Idaho National Engineering and Environmental Laboratory, Manifest Information Management System. Available at *http://mims.inel.gov* [December 22, 2000].
3. Fentiman, A. W., M. E. Jorat, and R. J. Veley, Factors That Affect the Cost of Low-Level Radioactive Waste Disposal, RER-66, OSU Extension Fact Sheet, 1993.
4. U. S. Nuclear Regulatory Commission, Report on Waste Disposal Charges: Changes in Decommissioning Waste Disposal Costs at Low-Level Waste Burial Facilities, NUREG-1307, Rev. 9, 2000.
5. Shapiro, F. C., *Radwaste*, Random House, New York, 1981.
6. EG&G Idaho, Inc., Directions in Low-Level Radioactive Waste Management—Planning State Policy on LLRW, Report DOE/LLW-6Ta, National LLRW Management Program, Idaho Falls, Idaho, 1982.
7. U.S. Congress, 96th, Low-Level Radioactive Waste Policy Act, Public Law 96-573, 1980.
8. U.S Congress, 99th, Low-Level Waste Policy Amendments Act (LLWPAA), Public Law 99-240, 1986.
9. U. S. General Accounting Office, Low-Level Radioactive Wastes: States Are Not Developing Disposal Facilities, GAO/RCED-99-238, September 1999.
10. EG&G Idaho, Inc., Radwaste Management: A Summary of State Laws and Administration, Report DOE/LLW-18T, Rev. 5, National LLRW Management Program, Idaho Falls, Idaho, 1985.
11. EG&G Idaho, Inc., Low-Level Radioactive Waste Treatment Technology, National LLRW Management Program, Report DOE/LLW 13 Tc, Idaho Falls, Idaho, 1984.
12. Coplan, B. V., and P. J. Mayo, Liquid waste management, in *Radioactive Waste Technology*, A. A. Moghissi et al., eds., p. 223, ASME, New York, 1986.
13. Jolley, R. L. et al., LLW from Commercial Nuclear Reactors. Vol. 2, Treatment, Storage, Disposal and Transportation Techniques and Constraints, ORNL/TM-9846/V2, Oak Ridge National Laboratory, Oak Ridge, Tenn., 1986.
14. Gilbert/Commonwealth, State-of-the-Art Review of Radwaste Volume Reduction Techniques for Commercial Nuclear Power Plants, ORNL/SUB-79/13873/2, Oak Ridge National Laboratory, 1980.
15. Kibbey, A. A., and H. W. Godbee, A State-of-the-Art Report on LLW Treatment, ORNL/TM 7427, Oak Ridge National Laboratory, 1980.

16. Trigilio, G., VR Technologies in Low Level Radioactive Waste Management, NUREG/CR-2206, Nuclear Regulatory Commission, Washington, D. C., 1981.

17. Kibby, A. H., and H. W. Godbee, The Use of Filtration to Treat Radioactive Liquid in LWR Power Plants, NUREG/CR-0141, ORNL/NUREG-41, Oak Ridge National Laboratory, 1978.

18. Koshiba, Y., T. Nakatani, H. Kuribayashi, and N. Kurosaka, Operational experience of non-precoated type filters (super fine filters) in PWR & BWR nuclear power plants, *Waste Management '85*, R. G. Post, ed., University of Arizona Press, Tucson, Ariz., 1985.

19. LaGuardia, T. S., Decommissioning of nuclear facilities, in *Radioactive Waste Technology*, A. A. Moghissi et al., eds., p. 510, ASME, New York, 1986.

20. Briesmeister, A., J. Harper, B. Reich, and J. L. Warren, Los Alamos transuranic waste size reduction facility, in *The Treatment and Handling of Radwaste*, A. G. Blasewitz et al., eds., p. 87, Battelle Press, Richland, Wash., 1983.

21. IAEA, Conditioning of Low and Intermediate-Level Radwaste, Technical Report Ser. 222, International Atomic Energy Agency, Vienna, 1983.

22. Sathrum, C. H., and D. L. Stember, Volume reduction of DAW by a mobile supercompactor system, in *Waste Management '85*, R. G. Post, ed., University of Arizona Press, Tucson, Ariz., 1985.

23. Stock Technical Bulletin, Stock Equipment Co., Chagrin Falls, Ohio, 1985.

24. EPRI, Radwaste Incinerator Experience, EPRI-NP-3250, Electric Power Research Institute, Palo Alto, Calif., 1983.

25. Godbee, H. W., and E. L. Compere, DOE radwaste incineration technology, *Nucl. Safety*, vol. 27, no. 1, p. 56, 1981.

26. Ziegler, D. L., G. D. Lehmkuhl, and L. T. Meile, Nuclear Waste Incineration Technology Study, Transuranic Waste Management Program, RI Rocky Flats Plant Report RFP-3250, Golden, Colo., 1981.

27. Huebner, R. E., Economic consideration in management of LLW, in *Radioactive Waste Technology*, A. A. Moghissi et al., eds., p. 687, ASME, New York, 1986.

28. Tucker, R. F., Jr., R. A. Nelson, and C. C. Miller, LLW Solidification, EPRI-NP-2900, Electric Power Research Institute, Palo Alto, Calif., 1983.

29. Tucker, R. F., Jr., and R. A. Nelson, Solidification, packaging, and volume reduction, in *Radioactive Waste Technology*, A. A. Moghissi et al., eds., p. 362, ASME, New York, 1986.

30. U.S. ERDA, Alternatives for Managing Wastes from Reactors and Post-Fission Operations in the LWR Fuel Cycle, ERDA 76-43, vol. 2, Energy Research and Development Administration [now DOE], Washington, D.C., 1976.

31. Neilson, R. M., and L. R. Dole, Chemical considerations for the immobilization of LLW, in *Radioactive Waste Technology*, A. A. Moghissi et al., eds., pp. 339, 347, ASME, New York, 1986.

32. Columbo, P., and R. M. Neilson, Jr., Properties of Radioactive Wastes and Containers, BNL-50957, Brookhaven National Laboratory, Upton, N.Y., 1975.

33. IAEA, Bituminization of Asphalt Waste, Technical Report Ser. 116, International Atomic Energy Agency, Vienna, 1970.

34. Werner and Pfleiderer Corp., Radwaste Volume Reduction and Solidification System Topical Report, WPC-VRS-1, Werner and Pfleiderer Corporation, Waldwick, N.J., 1976.

35. Burns, R. H., Solidification of low and intermediate-level wastes, *Atomic Energy Rev.*, vol. 9, no. 3, p. 547, 1971.

36. Dow Chemical, The Dow System for Solidification of LLW from Nuclear Power Plants, Topical Report, Dow Chemical Company, Midland, Mich., 1978.

37. Gilbert Associates, Inc., Identification of Radwaste Sources and Reduction Techniques, EPRI NP-3370 Final Report, vol. 2, Palo Alto, Calif., 1984; Daloisis, G. S., and C. P. Deltete, Radwaste Generation Survey Update, vols. 1 and 2, NP-5526 Final Report, Electric Power Research Institute, Palo Alto, Calif., 1988.

38. EPRI, Long-Term Low Level Radioactive Waste VR Strategies, EPRI-NP-3763, vol. 1, Electric Power Research Institute, Palo Alto, Calif., 1984.

39. EG&G Idaho, Inc., Feasibility and Conceptual Design for a Mobile Incineration System for Combustible LLW, EGG-2217, Idaho Falls, Idaho, 1983.

40. American Nuclear Society, *Nucl. News*, March 1985.

41. Westinghouse Electric Corp., Comprehensive LLW Management—The Westinghouse Hittman Approach, Westinghouse Hittman Nuclear Incorporated, Columbia, Md., 1984.

42. AECC, Mobile VR System, Topical Report AECC-4-NP-A, Rev. 1, prepared for NRC, Aerojet Energy Conversion Company, 1986.
43. Jolly, R. L., and B. R. Rodgers, A survey of LLW treatment methods and problem areas associated with commercial nuclear power plants, in *Waste Management '87*, vol. 3, p. 675, R. G. Post, ed., University of Arizona Press, Tucson, Ariz., 1987.
44. U.S. Department of Transportation, 49CFR173, Shippers—General Requirements for Shipments and Packagings, Subpart I, Class 7 (Radioactive) Materials, 1999.
45. U.S. Nuclear Regulatory Commission, 10CFR71, Packaging and Transportation of Radioactive Material, 1999.
46. Tang, Y. S., Review of LLW management, in *Population Exposure from the Nuclear Fuel Cycle*, E. L. Alpen et al., eds., p. 229, ORNL, Oak Ridge, Tenn., September 1987.
47. Neal, T. P., C. C. Miller, and M. D. Naughton, Operational experience of the Palisades VR system—The first 12 months, in *Waste Management '85*, R. G. Post, ed., University of Arizona Press, Tucson, Ariz., 1985.
48. Hillmer, T. P., and R. D. Doyle, Use of VR and solidification services at Arizona Nuclear Power Projects Palo Verde Nuclear Generating Station, in *Waste Management '87*, R. G. Post, ed., University of Arizona Press, Tucson, Ariz., 1987.
49. Day, J. E., R. D. Doyle, S. B. Rastberger, and E. Tchemitcheff, Transportable VR bitumen solidification system from design to operation, in *Spectrum '86, Proc. ANS International Topical Meeting—Waste Management and Decontamination and Decommissioning*, p. 288, J. M. Pope et al., eds., Niagara Falls, N. Y., September 1986.
50. Mogg, C. S., and W. Heafield, BNFL's objectives and achievements in the reduction of radioactive discharges from reprocessing plant, in *Waste Management '87*, vol. 3, p. 613, R. G. Post, ed., University of Arizona Press, Tucson, Ariz., 1987.
51. Broderson, K., and R. Andre Jehan, Technology for the Improvement of Shallow Land Burial, in *Radwaste Management and Disposal, Proc. 2nd European Community Conference*, R. Simon, ed., Luxembourg, April 1985.
52. Robinson, P. J., et al., Parametric studies of the impact of conventional disposal of waste below regulatory concern—Using the IMPACTS computer code with the EPRI IMPACTS-PLUS interface, in *Waste Management '87*, vol. 3, p. 249, R. G. Post, ed., University of Arizona Press, Tucson, Ariz., 1987.
53. Bonda, G. The disposal of radmaterial, in *Radioactive Waste Technology*, A. A. Moghissi et al., eds., p. 463, ASME, New York, 1986.
54. U.S. DOE, Low Level Radwaste Disposal: Currently Operating Commercial Facilities, DOE/LLW-9, 1983.
55. Cahill, J. M., Hydrology of the LLW Burial Site and Vicinity near Barnwell, South Carolina, U.S. Geological Survey Open-File Report 82-863, 1982.
56. SC-DHEC, Department Regulation 61-63, Radmaterial Title A, Part I—General Provisions, Part III—Standards for Protection Against Radiation and Part IV—Notices, Instructions, and Report to Workers, Inspectors, and Department Regulation 61-83, Transportation of Radwaste into or within S.C., Department of Health and Environment Control, Columbia, S.C., 1985.
57. U.S. DOE, Low Level Radwaste Disposal: Commercial Facilities No Longer Operating, DOE/LLW-6TF, 1982.
58. Levin, G. B., and L. J. Mezgor, Alternative Techniques for Shallow Land Burial of LLW, ORNL/NFW-84/18, DOE LLW Management Program, Oak Ridge, Tenn., 1984.
59. PNL, Research Program at Maxey Flats and Consideration of Other Shallow Land Burial Site, NUREG/CR-1832 or PNL-3510, 1981.
60. Onishi, Y., R. J. Herne, E. M. Arnold, C. F. Cowan, and F. L. Thompson, Critical Review: Radionuclide Transport, Sediment Transport, and Water Quality Mathematical Modeling, and Radionuclide Adsorption/Desorption Mechanisms, NUREG/CR-1322 or PNL-2901, 1981.

EIGHT

URANIUM ORE MILL TAILINGS MANAGEMENT

8.1 INTRODUCTION

Uranium mining is the starting point for the nuclear fuel cycle. Uranium is widely distributed in the earth's crust with an average abundance of about 2 g/ton. In general, uranium ore deposits considered suitable for mining contain 0.03–0.5% uranium by weight. Whether it is economical to mine a particular ore deposit depends on several factors including the richness of the ore, the market price for uranium, the mining technique used, and the associated health and environmental costs. Two methods have been commonly used for mining uranium ore: underground mines and open pit mines. Both techniques have been used historically to mine other ores or coal, and both are well understood. Another uranium mining method called *in situ* or *solution mining* has been tried on an experimental basis. Underground mining of uranium has associated with it all of the hazards of underground mining of other natural resources as well as elevated concentrations of radon gas, radon being a radioactive decay product in the uranium chain. Open pit mining requires moving large amounts of overburden to reach the ore body and then replacing the soil, an expensive process that can leave scars if the site is not properly reclaimed. Most of the uranium mined in the United States has come from open pit mines. The in situ mining involves pumping a solvent through the ore body in the ground and removing the uranium in solution. This method avoids the dangers associated with underground mining and the environmental damage done by open pit mining. However, it does not recover as much of the uranium present as the other two methods, and there are concerns about contaminating groundwater with the solvent.

If the uranium ore is removed from the ground, it is sent to a mill for processing to recover the uranium. For every ton of uranium ore that is milled in the United States, not more than 2 kg (or 5 lb) of uranium is extracted, leaving the rest to be discharged as finely ground, sandy tailings. The tailings contain other naturally radioactive substances (e.g., radium), which are responsible for more than three-fourths of the radioactivity that was originally in the ore (for the decay series of uranium, see Table 2.2, Section 2.2.3). From the mill, the tailings go as a slurry into a tailings pond. Through drying, they form a large, spreading delta around the pond, and in this way huge tailings piles have been created. About 90,000 m^3 ($3 \times 10^6 ft^3$) of tailings is created in producing the yellowcake, U_3O_8, needed for each gigawatt-year of electric power generation (corresponding to the annual power output of a 1250-MW plant operating at 80% capacity factor). These tailings are typically left near the uranium mill. Should they ever have to be moved, they would fill ten 1-mile-long trains of hopper cars.[1] The amount of tailings generated in fueling a large nuclear power plant is almost one-third greater than the total amount of fly ash, bottom ash, and scrubber sludge left from the operation of a coal-fired electric plant of the same size.[1,2]

In the United States some 121 million m^3 ($4.3 \times 10^9 ft^3$) of tailings had accumulated by the end of 1983, distributed among piles at 24 active or recently active sites and at more than a score of sites shut down some years ago. The largest pile is at the Kerr-McGee mill near Grants, New Mexico. It covers 250 acres and rises to a height of about 100 ft.[1] Failures of containment at mill tailings impoundments have occurred.[3] For instance, in July 1979, when a tailings dam gave way at the mill near Churchrock, New Mexico, the escape of some 100 million gallons of tailings solution left 60 miles or more of the Rio Puerco contaminated along its course through Navajo lands in New Mexico and Arizona. The actual hazard was found to be slight, although the Navajos suffered much inconvenience and anxiety, particularly with regard to stock watering. An especially troublesome and costly problem associated with abandoned mill tailings piles has been the tendency of unwitting or irresponsible individuals to use them as sources of landfill in construction of homes, commercial buildings, and even schools. This happened in numerous communities near uranium mining and milling sites in the West.

8.1.1 Content of the Wastes

Uranium mill tailings are the material left after uranium has been removed from the ore. As noted earlier, only about 5 lb of uranium is recovered from 1 ton of ore, leaving approximately 1995 lb of tailings. To remove the uranium from ore, the ore is crushed and leached with acid. The uranium dissolves in the acid, and the solution is drained from the solids. The remaining solids are then washed in clean water, and the resulting slurry is pumped to a tailings pond where the water evaporates or seeps into the ground. The dry tailings are about 70–80% by weight sand-sized particles and the other 20–30% fine, claylike particles called slimes.[4] The daughter products of uranium remain in the tailings and represent about three-fourths of the radioactivity in the ore.

8.1.2 Nature of Hazards

Radiological hazards of radon (^{222}Rn). Many of the radioactive daughters of uranium, while of little direct consequence environmentally, pose a radiologic health risk to humans that in most cases is small compared with other risks routinely experienced. Because of the long half-lives of the parents, these potential impacts continue into the distant future. The behavior of the wastes must be considered over times that are unusually long compared with human experience. Radon-222, an inert radioactive gas with a 3.8-day half-life, and the decay product of radium-226 continually escape to the atmosphere from surface soils. The escape rate from uranium mill tailings is usually higher than that from normal soils, depending on the physical properties of the tailings and any cover over the tailings.[5] The radiologic risk to the public from ^{222}Rn results from the irradiation of lungs by its short-lived airborne alpha-emitting daughters. At and very close to the tailings, the concentration of their nuclides in air can be measurably greater than normal background levels. The risk to an individual living close to tailings areas can be estimated.[6] For instance, the individual lifetime risk of lung cancer associated with living continuously about 1 km (0.6 miles) downwind from a tailings pile releasing 500 TBq of radon per year is 0.2%. This is about the same as the individual lifetime risk of the lung cancer arising from exposure to levels of radon indoors in an average U.S. or European house. Beyond a few kilometers from a tailings area any increase in radon concentration owing to releases from that area cannot be distinguished from background levels.

Aquatic dispersal of soluble tailings material. Radionuclides such as ^{230}Th, ^{226}Ra, ^{210}Pb, and ^{210}Po may be leached from waste or tailings piles; the ratio depends on the hydrogeochemistry of the tailings and surrounding rock or unconsolidated sediments, on the integrity of any retaining structure, and on the permeability of any capping material. The exposure to radiation from these leached radionuclides may be a result of transfer along aquatic food chains and through contamination of drinking water. Aquatic exposure pathways are particularly important in wet climates. The concentration of these radionuclides may be measurable only close to a tailings site, while farther away such contributions to normal levels in diet and the environment are only calculable.

Dispersal of insoluble waste rock and tailings material. Radionuclide in waste rock particles or tailings dust may be dispersed by wind and waste erosion if the waste piles are not contained by dams and cappings or if such containment has been breached as a result of natural erosion or human intrusion. The rates of erosion of cap materials for tailings depend on the engineering design of the capping and on geomorphologic processes. The estimates would become speculative if tailings were uncovered after having undergone hundreds or thousands of years of internal geochemical activity.[7] As mentioned earlier, tailings can also be removed by humans for misuse. Prediction of exposures in the future also requires prediction of the behavior of the tailings and associated structures.

8.2 HISTORY AND CURRENT MANAGEMENT OF TAILINGS

8.2.1 History

Uranium mines and mills operating between the mid-1940s and the mid-1960s supplied uranium to the U.S. government for use in its nuclear weapons program. In later years, the mines and mills also supplied uranium for commercial nuclear power plants. Early regulations governing radioactive materials did not address mill tailings. Tailings were the rock and soil left *after* the uranium had been removed and were not considered to be dangerous. In fact, some contractors used the tailings as fill material to level the land before constructing homes, schools, and other buildings. In 1966 high levels of radon discovered in buildings in Grand Junction, Colorado, were determined to have come from uranium mill tailings beneath the buildings. In 1972, the U.S. Congress passed the Grand Junction Remedial Action Program (Public Law 92-314), which provided funds to remediate about 1000 structures with high radon levels. Tailings were removed or solidified to keep the radon from moving through them. Ventilation systems were installed in many of the buildings to pump out the radon-laden air.

After the Grand Junction discovery, studies were conducted to determine hazards associated with mill tailings. In 1978, Congress passed the Uranium Mill Tailings Radiation Control Act (UMTRCA; Public Law 95-604), which was designed to minimize those hazards. Title I of the act made the Department of Energy responsible for cleaning up uranium mill tailings sites that were already inactive when the law was passed. Most of those tailings had been generated while processing uranium for use by the federal government. Title II made the owners/operators of privately owned mills still active at the time the law was passed responsible for cleaning up their own sites. Most of those mills processed uranium primarily for use in commercial nuclear reactor fuel, although some of the uranium was for the U.S. Department of Energy.

Under the Uranium Mill Tailings Radiation Control Act, the Environmental Protection Agency (EPA) was required to establish cleanup standards for the mill tailings sites and did so in 40CFR192. Since the privately owned mills were required to hold licenses from the Nuclear Regulatory Commission (NRC), the NRC was to ensure that they met the cleanup standards. In addition, the NRC was also to review remedial action plans at the Title I sites and license any mill tailings disposal cells. The NRC's regulations related to uranium mill tailings can be found in 10CFR40.

The cost of cleaning up inactive (Title I) sites is to be shared by DOE and the government of the state in which the site is located. DOE will pay 90% of the cost, and the state government is to pay 10%.[8] If the inactive site is on Indian tribal lands, DOE will pay 100% of the cost. The cost of cleaning up Title II sites was to be borne primarily by the mills' owners. If some portion of the tailings at a site was generated when milling uranium sold to DOE, then the federal government is to reimburse the owners for the cost of cleaning up those mill tailings. See 10CFR765 for details.

8.2.2 Current Management

Twenty-four sites are covered by Title I. Over 5000 properties near the sites have been found to be contaminated by mill tailings and will also be cleaned up by DOE. The Title I mill tailing site locations and remediation status are shown in Figure 8.1. Current plans call for the mill tailings to be stabilized in place or stabilized on site at 11 of these sites. At the other 13, the mill tailings will be removed from site and disposed of at a remote location, probably one also owned by the DOE. Long-term surveillance and maintenance will be required for all of the permanent disposal sites.

Twenty-six sites are covered by Title II. The locations of these sites are shown in Figure 8.2. After the owners have cleaned these sites, they will be turned over to the

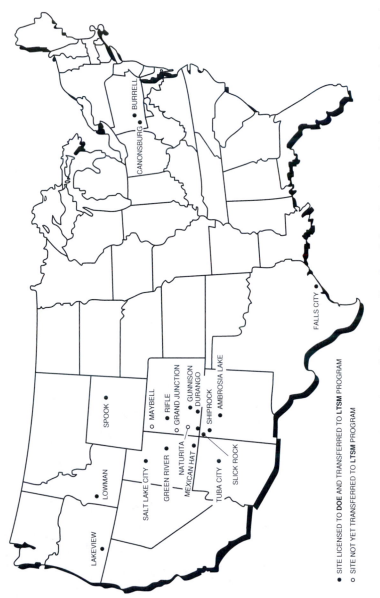

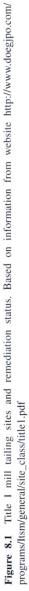

• SITE LICENSED TO **DOE** AND TRANSFERRED TO **LTSM** PROGRAM

○ SITE NOT YET TRANSFERRED TO **LTSM** PROGRAM

Figure 8.1 Title 1 mill tailing sites and remediation status. Based on information from website http://www.doegipo.com/programs/ltsm/general/site_class/title1.pdf

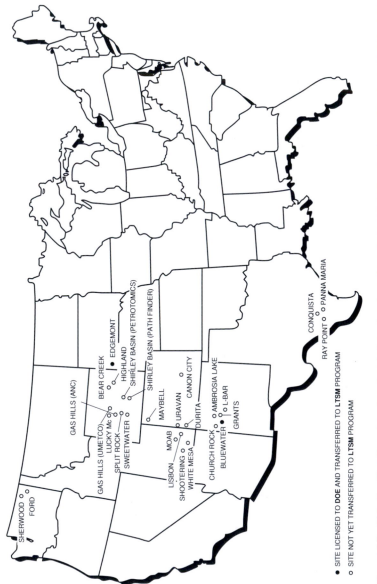

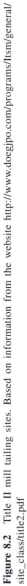

Figure 8.2 Title II mill tailing sites. Based on information from the website http://www.doegipo.com/programs/ltsm/general/site_class/title2.pdf

federal government for long-term monitoring and maintenance. The NRC is to work with the owners to ensure that they provide adequate funds to pay for the long-term care. Federal funds are not be used to pay for the long-term care.[9]

Both surface and groundwater remediation are required by the Uranium Mill Tailings Radiation Control Act of 1978. Surface remediation generally involves covering the tailings with a sloped cap of compacted clay and then adding rocks or vegetation on top of the clay. The cover substantially reduces the release of radon, controls erosion, minimizes leaching of the tailings by minimizing infiltration of rain water, and discourages people or animals from digging in the tailings. The caps will be used on piles of mill tailings that are left in place or on those which are moved to a disposal cell in a new location. Burying mill tailings is another way to minimize their impact on human health or the environment. However, that is a much more expensive option and is not likely to be used extensively. Surface remediation at all Title I sites, which have been inactive for more than two decades, was nearly complete by the 1998 target date.

Once the mill tailings are stabilized and material from the tailings ceases to leach into the groundwater, that water can be cleaned. Groundwater contamination will be measured and monitored. Results will be used to conduct a risk assessment, select cleanup methods, and prioritize remediation efforts. Two options are considered for groundwater remediation: (1) letting the water clean itself over time, called natural flushing, and (2) actively treating the water by pumping it out of the ground, removing the contaminants, and returning it to the ground. Groundwater remediation at the Title I sites has not yet begun at most sites and is to be completed by 2014.

8.3 CASE STUDIES OF MANAGEMENT AND DISPOSAL OF MILL TAILINGS AND WASTES

8.3.1 Reference Sites and Management Scenarios

To illustrate the engineering factors that have to be considered, the following hypothetical reference sites are defined for three distinctive regions with environments (i.e, geology, hydrology, climate, demography) reasonably characteristic of a real, possibly composite region:

1. A reference site with a tropical monsoon climate—a ring dike with a waste rock cover.
2. A reference site with a northern temperate climate—a valley dam impoundment with a vegetable cover.
3. Two reference sites in a semiarid desert region: (a) an unconfined deposit with a small starter dam and (b) a deposit below grade in a specially dug pit.

The management options selected for the study, of necessity, are either presently in use or those for which research has been carried out and reasonable estimates of costs are available. Three kinds of management strategies aimed at reducing radiation doses from tailings were used:

1. Increasing the isolation of the tailings from the biosphere (e.g., below-grade options).
2. Reducing the activity of radionuclides associated with the tailings by changing the mining method or by removing radionuclides for disposal during processing (e.g., radium, thorium removal).
3. Allowing the radionuclides to disperse in a predictable way (e.g., waterborne contaminants to be lost to deep lake or ocean sediments, where they are buried by natural sedimentation processes).

8.3.2 Results for the Reference Site in the Tropical Region

With management options consisting of various types of erosion-resistant engineered soil/clay/rock covers (Figure 8.3), the possible optimum management strategy involves the burial of the waste rock. This option arises because the waste rock at this site was assumed to contain low-grade uranium mineralization, and it is optimum only when collective doses are assessed into the far future (10,000 years) but not far enough for the collective dose commitments to have become dominated by deflated (wind-dispersed) tailings. In addition, a final stage of operation of the mill at breakeven grade provided some small reduction in collective dose, so this option, for which the incremental costs

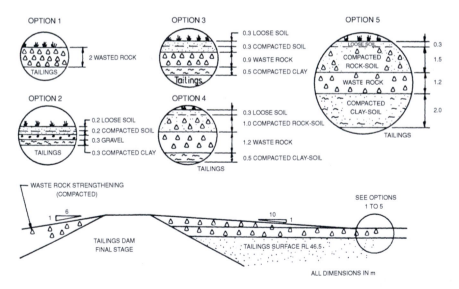

Figure 8.3 Management scenarios for the reference site in the tropical region with a monsoonal climate. Options: (1) base case, rock cover 2 m; (2) soil/gravel cover 1 m thick; (3) soil/rock cover 2 m thick; (4) soil/rock cover 3 m thick; (5) soil/rock cover 5 m thick. From Long-Term Radiological Aspects of Management of Waste from Uranium Mining and Milling by NEA, OECD, Report of the Committee on Radiation Protection and Public Health and the Radioactive Waste Management Committee, September 1984. Copyright 1984 by NEA, OECD. Reprinted by permission.

were assumed to be zero, is obviously preferred over the base case. The incremental cost-effectiveness for this option with respect to the base case with breakeven operation for the last year was $11,500 per person-Sv. The possible options with various thicknesses of soil/rock cover could incur incremental costs of $100,000 per person-Sv. With the integrating time of 10,000 years for the collective dose, the selection of optimum would therefore vary with the value assigned to unit reduction in collective dose a in the range $10,000–100,000 per person-Sv. The uncertainties in the estimates of collective dose could only be guessed; it is conceivable that the dose estimates were an order of magnitude high. This factor is similar to the relative range of values of incremental cost-effectiveness, which renders the comparison less decisive except for the extreme options. The maximum dose rates to the most highly exposed individual were predicted to be from 20 to 620 μSv/year, depending on the options, and to occur several thousand years in the future.

8.3.3 Results for the Reference Site in the Northern Temperate Region

With management options consisting of improved engineering work to render the dam more impermeable, provision of a vegetable cover, and removal of radium and thorium from the tailings (Figure 8.4) for all integrating times, a dam with low permeability was the most cost-effective option with respect to the base case. The estimated costs varied from $230 to $3600 per person-Sv for integrating times in the range of 100–10,000 years. It should be noted that the occupational doses and the costs associated with handling and storing of the removed radium and thorium were not considered; the actual incremental cost-effectiveness would thus be higher than indicated. Because of the large factor between the values of incremental cost-effectiveness for these options, the choice of optimum was fairly insensitive to integrating time and the choice of a. It was also insensitive to the modeling of radon distribution globally since aquatic pathways were also important for this reference site. The estimates of maximum doses to the most highly exposed individuals were very sensitive to modeling parameters in the aquatic pathways, which were not well known. They ranged from 100 to 4400 μSv/year, depending on the option, and were predicted to occur within the first few hundred years.

8.3.4 Results for the Reference Site in the Semiarid Region

Two base cases were considered: an uncovered above-grade pile with a small starter dam, and a covered below-grade pit. Management options for both cases consisted of various soil/clay/rock covers (Figure 8.5). Also, for the below-grade pit, the options of having a pit available for backfilling with tailings and of having a specially dug one were considered. For an integrating time of 100 years a sequence of management options can be identified with increasing impermeability to radon and increasing cost, with values of incremental cost-effectiveness from $15,000 per person-Sv up. If longer integrating times were chosen, the above-ground erodible covers deteriorated sufficiently that the reductions in collective dose rate were no longer attained. Only those options with gravel cappings and suitably engineered side slopes were then

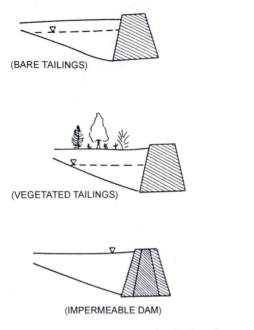

(BARE TAILINGS)

(VEGETATED TAILINGS)

(IMPERMEABLE DAM)

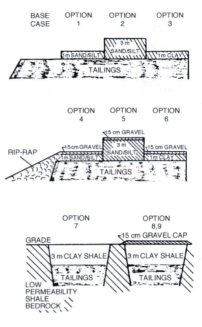

Figure 8.4 Management scenarios for the reference site in the northern temperate area. Base case, bare dry tailings behind a permeable dam. Options: (1) vegetated dry tailings behind a permeable dam; (2) fully saturated tailings behind a low-permeability dam; (3) bare dry tailings, from which 90% of the leachable radium and thorium has been removed, retained behind a permeable dam; (4) vegetated dry tailings, from which 90% of the leachable radium and thorium has been removed, retained behind a permeable dam; (5) fully saturated tailings, from which 90% of the leachable radium and thorium has been removed, retained behind a low-permeability dam. From Long-Term Radiological Aspects of Management of Waste from Uranium Mining and Milling by NEA, OECD, Report of the Committee on Radiation Protection and Public Health and the Radioactive Waste Management Committee, September 1984. Copyright 1984 by NEA, OECD. Reprinted by permission.

Figure 8.5 Management scenarios for the reference site in the semiarid desert area. Base case, above grade, no cover. Options: (1) above grade, 1 m sand/silt cover; (2) above grade, 3 m sand/silt cover; (3) above grade, 1 m clay cover; (4) above grade, 1 m sand/silt, riprap 15 cm gravel cap; (5) above grade, 3 m sand/silt, riprap 15 cm gravel cap; (6) above grade, 3 m clay, 15 cm riprap gravel cap; (7) below grade, 3 m clay/shale; (8) below grade, 3 m clay/shale, 15 cm; (9) below grade in existing pit, 3 m clay/shale, 15 cm. From Long-Term Radiological Aspects of Management of Waste from Uranium Mining and Milling by NEA, OECD, Report of the Committee on Radiation Protection and Public Health and the Radioactive Waste Management Committee. September 1984. Copyright 1984 by NEA, OECD. Reprinted by permission.

selected as a possible optimum. For an integrating time of 10,000 years, disposal into an available pit would be selected in preference to disposal into a specially dug pit. The former would clearly be a very site-specific option. The maximum dose rates to the most highly exposed individuals from the options with erodible covers increased with time over the period modeled, ranging up to 1500 μSv/yr. The need to ensure that little erosion takes place may be an important constraint on the optimization for this kind of tailings site.

8.3.5 Uncertainties of the Analysis

Uncertainties in the predicted values of collective doses are the major limitation on the practical application of the formal quantitative approach to optimization of radiation protection, and in many cases they obscure any differences between the effectiveness of various management options in reducing collective doses. The choice of the management option for managing tailings can be very sensitive to the choice of time for which the collective dose commitment is calculated, to values placed on detriment from radiation, and to the value judgments that might be made with respect to the different distributions of dose in the exposed population.

Estimates of the costs of protection are subject to many uncertainties, which are not uncommon in economic assessments. Estimates of dosimetric quantities involve modeling of the release and transport of contaminants from tailings and various uncertainties: errors in the models used and errors in values of quantities, or parameters, used in the model. The importance of the various uncertainties for the outcome of the optimization depends to some degree on whether a change in exposure route is involved. As long as the principal transport mechanisms are the same, errors associated with various steps in the transport chain tend to have a proportional influence on the doses for all alternatives and should not influence the relative ranking of the options. When different transport mechanisms are involved, as in the case comparing doses from isotopes having different chemical properties and thus different transport behavior, the effects of the uncertainties are much more difficult to predict. Sometimes direct validation of a model and estimation of uncertainties are possible by comparison with new experimental data. Otherwise, indirect methods have to be used to estimate uncertainties.

8.4 COMPUTER CODES

MILDOS-AREA. Calculation of radiation dose from uranium recovery operations for large-area sources. Available at http://www-rsicc.ornl.gov/nrc/allcodes/codes/ccc/ccc6/ccc%2D608.html. "MILDOS-AREA estimates the radiological impacts of airborne emissions from uranium mining and milling facilities or any other large-area source involving emissions of radioisotopes of the uranium-238 series. Wind frequency data are provided by the user. The transport model includes the mechanisms of dry deposition of particulates, resuspension, radioactive decay and progeny ingrowth, and plume reflection. Deposition buildup and ingrowth of radioactive progeny are considered in estimating surface concentrations, which are modified by radioactive transformation, weathering, and other environmental processes. MILDOS-AREA allows the user to vary the emission rates of the sources as a step-function of time. Impacts to humans through such pathways as inhalation, external exposure, and ingestion are estimated based on calculated annual average air concentrations of nuclides. Individual, total individual, annual population, and environmental dose commitments are calculated with conversion factors derived from recommendations of the International Commission on Radiological Protection (ICRP) and Oak Ridge National Laboratory. Age-specific dose factors are calculated."

"A validation study of MILDOS-AREA was conducted using measured Rn-222 concentration and flux data from the Monticello, Utah uranium mill tailings impoundment. The results of this study demonstrated that use of MILDOS-AREA can result in generally good agreement between model-generated and measured Rn-222 concentrations."

AREAC. Radiological emission analysis code system. Available at http://www-rsicc.ornl.gov/nrc/allcodes/codes/ccc/ccc4/ccc%2D438.html. "AREAC was designed to calculate potential radiological impact of atmospheric releases of radionuclides from area sources. It represents an initial attempt at developing a quantitative model for analyzing the potential radiological impact of airborne, constant, continuous releases of gaseous radionuclides from area sources (principally inactive uranium tailings piles). It can calculate radionuclide concentrations and individual inhalation doses at up to six specific receptor locations and at up to 192 general locations around an area source. Population doses can also be calculated."

"AREAC is useful for more accurately assessing close-in doses from large area sources, such as uranium mill tailings piles. Results are more accurate than those calculated with existing air pathway models."

8.5 DISCUSSION QUESTIONS AND PROBLEMS

1. Name two common methods used for mining uranium ore and an experimental method. Briefly describe these methods.

2. What radioactive gas is found in elevated concentrations near uranium ore mill sites?

3. What is the ratio of extracted uranium to uranium ore removed from the ground? What is done with the excess?

4. Describe three hazards associated with mill tailings.

5. How many atoms of radon per liter are in the EPA radon limit of 4 pCi/L?

6. If tailings are 0.1% radium by weight, how many atoms of radon would be produced in 1 kg of tailings in 1 day?

7. What is the difference between Title I and Title II sites under UMTRCA?

8. What are the advantages of putting caps on mill tailings?

9. What contaminants might be found in ground water near mill tailings? How is the ground water contaminated and how is it cleaned up?

10. How would surface water be contaminated?

REFERENCES

1. Carter, L. J., *Nuclear Imperative and Public Trust—Dealing with Radwaste*, Resources for the Future, Washington, D.C., 1987, p. 12.
2. U.S. Health Physics Society, *Proceedings of the LLW Management Conference*, EPA/520/3-79-002, U.S. EPA, Washington, D.C., 1979.
3. Walter, W. H., Overland Erosion of Uranium-Mill Tailings Impoundments: Physical Processes and Computational Methods, NUREG/CR-3027 (PNL-4523), 1983; available from National Technical Information Service, Springfield, Va.

4. U.S. Department of Energy, Integrated Data Base Report-1996: U.S. Spent Nuclear Fuel and Radioactive Waste Inventories, Projections, and Characteristics, DOE/RW-0006, Rev. 13, December 1997.

5. Thomas, V. W., K. K. Nielson, and M. L. Mauch, Radon and Aerosol Release from Open Pit Uranium Mining, NUREG/CR-2407 (PNL-4071), 1982; available from National Technical Information Service Springfield, Va.

6. U.S. EPA, Final Environmental Impact Statement for Remedial Action Standards for Inactive Uranium Processing Sites, Report EPA 520/4082-013-1, Washington, D.C., 1982.

7. NEA, OECD, Long-Term Radiological Aspects of Management of Waste from Uranium Mining and Milling, Report of the Committee on Radiation Protection and Public Health and the Radioactive Waste Management Committee, September 1984.

8. U.S. Department of Energy, 1996 Baseline Environmental Management Report. Available at http://www.em.doe.gov/bemr96/index.html.

ADDITIONAL READINGS

U.S. Department of Energy, Linking Legacies, Connecting the Cold War Nuclear Weapons Production Processes to Their Environmental Consequences, DOE/EM-0319, January 1997.

U.S. General Accounting Office, Uranium Mill Tailings—Cleanup Continues, but Future Costs Are Uncertain, GAO/RCED-96-37, December 1995.

U.S. General Accounting Office, Uranium Mill Tailings—Status and Future Costs of Cleanup, Testimony before the Subcommittee on Energy and Power, Committee on Commerce, U.S. House of Representatives, February 28, 1996.s

U.S. Nuclear Regulatory Commission, 10CFR40, Domestic Licensing of Source Material, 1991.

CHAPTER NINE

MIXED WASTES

9.1 INTRODUCTION

Mixed wastes contain both radioactive and hazardous wastes. Therefore, a dual regulatory framework exists for mixed wastes, with the U.S. Environmental Protection Agency (EPA) or authorized states regulating the hazardous wastes using the Resource Conservation and Recovery Act (RCRA) and the U.S. Nuclear Regulatory Commission (NRC), NRC agreement states, or the U.S. Department of Energy (DOE) regulating the radioactive waste using the Atomic Energy Act (AEA). NRC generally regulates commercial and non-DOE federal facilities. DOE is currently self-regulating and its orders apply to DOE sites and contractors. The requirements of RCRA and AEA are generally consistent and compatible. However RCRA takes precedence over AEA in the event provisions or requirements of the two are inconsistent.

Almost all of the commercially generated (non-DOE) mixed waste is composed of low-level radioactive waste (LLRW) and hazardous waste and is called low-level mixed waste (LLMW) or sometimes, mixed low-level waste (MLLW). Commercially generated LLMW is produced in all 50 states at industrial, hospital, and nuclear power plant facilities. Radioactive and hazardous materials are used in a number of processes such as medical diagnostic testing and research, pharmaceutical and biotechnology development, and pesticide research, as well as nuclear power plant operations. Under the 1984 Amendments to RCRA, Land Disposal Restrictions (LDR) regulations prohibit disposal of most mixed waste including LLMW until it meets specific treatment standards. Most of the commercial mixed waste can be treated to meet the LDRs by commercially available technologies. Commercial mixed waste volumes are very small (approximately 2%) compared to the volume of mixed waste being generated or stored

Table 9.1 Summary of estimated total MLLW inventories and FY1996 generation

	Volume (m^3)	
Category	Total inventory	FY1996 generation[a]
DOE sites		
RCRA and RCRA PCB MLLW	71,710[b]	608
Non-RCRA PCB MLLW	4,530[b]	73
DOE MLLW total	76,240	681
Major commercial sites[c]	2,116	3,949
Other commercial sites[d]	31,014	0

[a]Except where indicated.

[b]Based on ref. 2. The currentness of these data for the various DOE sites ranges from September 1995 to July 1997.

[c]Reported for calendar year 1990.

[d]Wastes from commercial- and government-sponsored (DOE, EPA, DOD) activities that are disposed of at other commercially operated disposal facilities.

by the Department of Energy (DOE). There are three main types of mixed waste being produced or stored at DOE facilities, low-level, high-level, and transuranic. Table 9.1[1] provides a summary of the estimated total inventories for the commercial- and DOE-generated mixed wastes.

9.2 HISTORICAL BACKGROUND

The Resource Conservation and Recovery Act was promulgated in 1976 and required the EPA to regulate hazardous wastes from cradle to grave. However, RCRA specifically excludes source, special nuclear, and by-product materials as defined by the Atomic Energy Act of 1954. In 1984 the court mandated that RCRA would apply to DOE's hazardous wastes. It was not until October 6, 1992 that Congress passed the Federal Facility Compliance Act, which required DOE and other federal facilities to comply with EPA and state hazardous waste regulations and sanctions. Since that time DOE, NRC, and EPA have jointly worked to clarify the rules and regulations for handling, packaging, and disposal of mixed wastes.

9.3 SOURCES, CLASSIFICATION, AND INVENTORIES

Mixed wastes are generated in every state in the United States, at all federal facilities working with nuclear materials, and at many hospitals, universities, and private facilities. The largest portion by far is produced at DOE sites. The wastes produced at DOE sites include MLLW, HLW, and transuranic waste (TRU). Tables 9.2 and 9.3[1] provide the combined total volume inventories (at all DOE sites) by waste form.

Table 9.2 Total volume (m³) of inventory and generation of DOE RCRA and RCRA PCB MLLW, by physical form[a]

		Volume (m³)				
				FY projections		
MPC name	MPC code	Current inventory	Actual 1996	1997	1998–2006	2007–2030
Liquids	L0000	148.34	12.15	19.27	545.96	107.10
Aqueous liquids/ slurries	L1000	4,903.52	4.67	255.18	6,705.85	1,232.45
Organic liquids	L2000	1,311.53	69.36	140.06	1,440.56	2,935.09
Solids	S0000	490.52	40.93	62.78	1,842.79	361.18
Homogeneous solids	S3000	21.51	0.86	9.77	54.98	59.89
Inorganic homogeneous solids	S3100	49,991.40	89.98	148.60	11,279.31	7,999.89
Organic homogeneous solids	S3200	501.68	0.33	1.87	13.49	25.48
Soil/gravel	S4000	1,459.34	21.43	63.36	7,737.51	2,638.25
Debris waste	S5000	784.43	24.32	0.25	2.29	6.10
Inorganic debris	S5100	1,353.16	105.93	218.95	16,671.29	36,146.50
Organic debris	S5300	2,257.89	67.10	96.96	3,238.62	2,834.41
Heterogeneous debris	S5400	6,146.02	110.56	325.31	13,672.19	8,702.63
Unknown/other matrix	U9999	331.56	0.79	1.30	35.46	6.96
Lab packs	X6000	527.13	13.57	39.82	2,558.53	923.98
Special waste	X7000	5.70	1.83	—	—	—
Elemental mercury	X7100	8.11	0.51	0.49	1.39	1.07
Elemental hazardous metals	X7200	870.41	33.52	56.52	2,713.79	1,126.41
Beryllium dust	X7300	5.25	—	—	—	—
Batteries	X7400	26.05	0.42	1.29	3.95	27.04
Reactive metals	X7500	394.63	0.00	0.20	1.80	1.20
Explosives/propellants	X7600	15.79	9.37	7.19	30.94	40.07
Compressed gases/aerosols	X7700	9.89	0.00	—	0.14	0.03
Immobilized forms	Z1000	145.75	0.00	14.00	97.53	1,067.63
Decontaminated solids	Z2000	0.00	—	—	293.98	134.64
Total		71,709.82	607.61	1,463.17	68,942.36	66,377.99

[a] Based on Ref. 2.

9.4 REGULATIONS AND STANDARDS FOR MIXED WASTES

Table 9.4 gives a chronological list of regulatory activities with regard to mixed wastes.

9.5 WASTE MINIMIZATION METHODOLOGIES

There are many technologies being developed by industry and within the National Laboratory system. However, there are only a few that have been adequately demonstrated and approved by NRC and EPA. The Chemical Technology Division at Oak Ridge National Labs has been working on developing processes for separating the radioactive

Table 9.3 Total volume (m^3) of inventory and generation of DOE non-RCRA PCB MLLW, by physical form[a]

MPC name	MPC code	Volume (m^3) Current inventory	Actual 1996	FY projections 1997	1998–2006	2007–2030
Liquids	L0000	20.75	—	—	—	—
Aqueous liquids/slurries	L1000	5.73	—	—	—	—
Organic liquids	L2000	59.99	0.00	0.00	0.00	0.00
Solids	S0000	316.48	70.16	0.79	2.03	3.97
Inorganic homogeneous solids	S3100	20.12	0.60	0.97	5.51	2.31
Organic homogeneous solids	S3200	1.37	—	—	—	—
Soil/gravel	S4000	2,667.38	—	—	—	—
Debris waste	S5000	8.06	0.41	11.11	69.85	34.80
Inorganic debris	S5100	1,102.63	1.00	1.80	1.00	0.50
Organic debris	S5300	82.54	0.56	1.44	7.46	19.90
Heterogeneous debris	S5400	228.01	0.07	0.07	0.63	1.68
Unknown/other matrix	U9999	5.28	0.00	0.40	0.10	0.10
Lab packs	X6000	11.55	—	—	—	—
Total		4,529.89	72.80	16.58	86.59	63.26

[a]Based on Ref. 2.

components from the nonradioactive wastes. These same engineers are also working on soil washing techniques and other technologies for in situ soil decontamination. These technologies are not suficiently developed yet to be approved by the regulators.

The following technologies have been approved by the regulators for use in treating mixed wastes.

Pulsed plasma supercritical organics treatment This technology is available from Solar Wind Environmental, located in Mountain View, California (information available at http://www.techknow.org). This system uses an underwater plasma discharge, which creates an electrical, shaped pulse in the water. The result is a violent supercritical electrochemical reaction releasing hundreds of millions of electrons into the water. An intense continuous reaction occurs from the pulse corona flow. Organic molecules are split into their basic parts. For example, air and water separate into oxygen peroxy and hydroxyl radicals.

Matrix enhanced treatment system This system is available from Earthworks Environmental Inc. (information available from http://www.techknow.org). The matrix enhanced treatment system is an ex situ treatment methodology to treat organic and inorganic soil contamination in a single throughput of the unit. The treatment unit is mobile and it is logistically very easy to mobilize and treat soils at the source. Treatment costs are claimed to be less than 65% of the cost for landfilling while providing a permanent solution.

Table 9.4 Regulations and standards for mixed wastes

1976	The Resource Conservation and Recovery Act (RCRA) is promulgated giving EPA the authority to regulate hazardous waste from cradle to grave
1981	NRC recognizes joint regulation for mixed waste rulemaking
1984	In LEAF vs. Hodell, the Court mandated that RCRA be applied to DOE hazardous waste,
1984	Hazardous Solid Waste Amendments to RCRA increase the stringency of hazardous waste requirements
1986	EPA publishes notice clarifying RCRA jurisdiction
1986	Land disposal restrictions for California are promulgated by EPA
1987	NRC and EPA publish draft guidance on the definition and identification of LLMW.
1987	NRC and EPA publish joint guidance on siting guidelines
1987	EPA promulgates LDR standards for solvents and dioxins
1987	DOE clarifies its position on by-product materials
1987	DOE Order 5400.3, Hazardous and Radioactive Mixed Waste Programs, establishes policy guidelines and minimum requirements
1988	DOE Order 5820-2A, Management of Defense LLW, establishes policies for handling LLW
1988	EPA publishes a notice which clarifies requirements for facilities that treat, store, or dispose of radioactive mixed wastes
1988	EPA publishes hazardous waste injection restrictions including mixed wastes
1990	A conditional 10-year no-migration variance for limited amounts of untreated waste for the purpose of testing and experimentation is granted to the DOE WIPP facility
1992	EPA issues a policy of giving a reduced priority to civil enforcement of the RCRA storage prohibition
1992	Congress passes the Federal Facilities Compliance Act, which defines mixed waste and requires DOE and other federal facilities to comply with RCRA
1994	EPA extends its policy of giving a reduced priority to civil enforcement of the RCRA storage prohibition for 2 years
1995	EPA/NRC publishes Draft Guidance on the Storage of LLMW
1996	EPA announces a limited extension of its policy of prohibition of storage of mixed waste until April 20, 1998
1997	NRC and EPA jointly publish final guidance on the testing requirements for mixed wastes

Reduction of SO_x/NO_x and particulates This system is available from Specialty Chemical Consultants, Inc., located in New Berlin, Wisconsin (information available from http://www.techknow.org). The treatment system utilizes combustion catalyst additives for heavy/mazut or bunker C oil-fired utilities for reduction of unburned carbon particulates, sulfur oxides, and nitrogen oxides. It improves the thermal and heat transfer efficiency of boiler operation and saves on fuel consumption. It also reduces sulfur corrosion and low-temperature corrosion in the system. This technology reduces ash load factors to 60% of that achieved by magnesium hydroxide/oxide slurry-based products currently used by utilities.

9.6 WASTE PACKAGING AND DISPOSAL

The National Laboratories along with other organizations are continuing to develop methodologies to encapsulate and package mixed wastes for ultimate disposal. Brookhaven National Laboratories (BNL) has done a lot of work in this area (information available from http://www.dne.bnl.gov). Examples of their efforts follow.

BNL has developed a low-density polyethylene encapsulation process for low-level radioactive, hazardous, and mixed wastes that provides greater long-term stability than products from conventional solidification technologies. Polyethylene is an inert, low-permeability, thermoplastic material that is highly resistant to chemical attack, microbial degradation, and radiation damage. This process can be used for either microencapsulation or macroencapsulation. The microencapsulation process involves heating the polyethylene above its melting point and combining it with the dry waste to form a homogeneous mixture that is allowed to cool into a monolithic solid waste form in which small particles are interspersed within the polymer matrix. Macroencapsulation involves pouring molten polyethylene into a waste container in which large pieces of waste have been suspended or supported. Upon cooling, the polyethylene forms a solid layer sorrounding the waste.

Sulfur polymer cement, developed by the U.S. Bureau of Mines in 1972, is a thermoplastic material that is easily melted to a low-viscosity liquid at 120°C. BNL has developed an encapsulation process that can be used to solidify fly ash in sulfur polymer cement. The process involves using a dual-action mixer to simultaneously heat the waste and binder with several additives to form a homogeneous mixture which simply cools into a solid monolith waste form without the need for chemical reactions. As much as 2.5 times more incinerator fly ash can be solidified in sulfur polymer cement than in hydraulic cement and with improved compressive and tensile properties.

The Environmental and Waste Management Group at BNL has been developing and characterizing innovative thermosetting polymers for waste management use for many decades. BNL has developed and characterized thermosetting polymers for encapsulation of hazardous, mixed, and radioactive waste and for container materials. These materials cover broad ranges of chemical and physical durability, performance, viscosity, and cost. The polymers selected are innovative materials with desirable properties in both their fluid and solid states. This makes them suitable for applications where impermeability, chemical resistivity, high strength, and long-term durability are required.

BNL has proposed using new low-temperature glasses and glass-ceramics based on advanced phosphate formulations for the treatment of low-level and mixed wastes. The high temperature (1200–1500°C) required for vitrification using borosilicate glass is a major drawback because volatization of certain isotopes and heavy metals can occur. Alternative glass compositions have been prepared with melt temperatures between 450°C and 900°C and with improved durability.

9.7 COMPUTER CODES

There are several computer codes used to address issues associated with mixed wastes. Some of these codes are listed here, but this is not intended to be a complete list. The following codes were obtained from a software catalog that is maintained by the U.S. Department of Energy, Office of Scientific and Technical Information, Energy Science and Technology Software Center, P.O. Box 11020, Oak Ridge, Tenn. 37831. Several other computer codes are provided elsewhere in this book.

CAN. Canister model systems analysis. This package provides a computer simulation of a system for packaging nuclear waste in canisters. The canister model calculates overall programatic cost, number of canisters, and waste inventories.

COMRADEX4. Accidental release radiologic dose. This code was developed to evaluate potential radiologic doses in the near environment of radioactive releases, especially postulated accident releases.

PATH. Gamma dose calculations and shielding analysis. This code is a highly flexible shielding code utilizing the common point-kernel integration technique.

GTIPS. Greater than Class C (GTCC) tracking, inventory and projection system. This code was developed as part of the National Low Level Waste Management Program's effort to characterize GTCC wastes.

RANCHMD. Radionuclide migration geologic media. This code is a one-dimensional transport code for transport of radionuclide chains through layered geologic media, taking into account longitudinal dispersion, convection, and retention.

WAPPA. Waste package performance assessment. This code is intended to serve as a tool for evaluating both the relative and the absolute performance of waste package design concepts.

9.8 DISCUSSION QUESTION

1. Which regulatory agency rules are applicable in the event of disagreement between the requirements of RCRA and AEA?

REFERENCES

1. Integrated Data Base Report-1996, U.S. Spent Nuclear Fuel and Radioactive Waste Inventories, Projections, and Characteristics, DOE/RW-0006, Rev. 13, December 1997.
2. U.S. Department of Energy, Office of Environmental Restoration, Office of Waste Management, Technical Information Collection Data Base. Updated through October 30, 1997.

CHAPTER TEN

ENVIRONMENTAL RESTORATION

10.1 INTRODUCTION

Dozens of sites around the nation are contaminated with radioactive material or a mixture of radioactive and hazardous waste. Scientists and engineers from many disciplines will be involved in cleaning up these sites over the next several decades. The purpose of this chapter is to provide the reader with an understanding of the types of contaminants found at these sites, the technologies currently available to remove or at least contain the contaminants, and the rules and regulations that govern site restoration. We will focus on 16 major sites in the U.S. Nuclear Weapons Complex. These sites have some of the highest levels of radioactive contamination in the United States, and the Department of Energy, which is responsible for the sites, has an active program to clean them up. That program includes development and testing of technologies that will be available for restoration of other sites.

10.2 DESCRIPTION OF THE NUCLEAR WEAPONS COMPLEX

The Nuclear Weapons Complex (NWC) has been in operation for over half a century. The first facilities were built to support development of the atomic bomb. At that time, the nation's resources were focused on winning World War II. Contamination of the environment resulting from nuclear weapons development and testing was not a major concern, for several reasons. First, the consequences of losing the war were far more immediate and more serious than the consequences of contaminating land in what were then isolated areas of the United States. Second, scientists had limited knowledge of the nature of radioactive material and its effects on the environment. The first controlled fission did not occur until December 2, 1942, and the times demanded that the new technology be used before all of its ramifications were known. In addition, instruments to measure radiation were crude. The relatively low levels of radiation that are now of concern did not register on the early detection devices, so the extent of contamination was not readily apparent. As the Cold War developed after World War II, large numbers of nuclear weapons were built, requiring significant amounts of radioactive material and producing large volumes of waste. Security throughout the Nuclear Weapons Complex was very tight, and only those with a "need to know" were given access to weapons sites. Thus, there was no monitoring of releases of contaminants to the environment by anyone other than weapons complex employees.

Sixteen major sites in the Nuclear Weapons Complex are shown on the map in Figure 10.1. Livermore, Los Alamos, and Sandia were the primary sites for weapons research and development. Although Los Alamos' primary activity was to design, develop, and test nuclear weapons, the lab also produced nuclear weapons components and small quantities of plutonium. Livermore Laboratory was opened in 1952 to share the large amount of work in the nuclear field with Los Alamos. Work at Livermore included planning nuclear weapons experiments and designing thermonuclear weapons.[1] Sandia National Laboratory was established in 1949 to design nonnuclear components for nuclear weapons. In later years, Sandia scientists conducted projects related to security of nuclear sites and transportation and disposal of radioactive waste.

Four of the sites supplied materials for weapons. At the Hanford site in Richland, Washington, nine reactors were operated, and five chemical separation facilities were

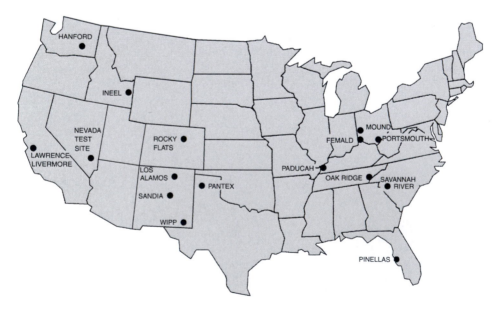

Figure 10.1 Location of 16 major laboratories in the U.S. Department of Energy's Nuclear Weapons Complex.

built to separate plutonium from the spent reactor fuel. As of 1987, all of the reactors at the Hanford site were closed. The Savannah River Site near Aiken, South Carolina, included five reactors and two chemical separation plants to produce plutonium. In addition, tritium for hydrogen bombs was processed at the Savannah River Site. Reactors at this site are also closed. The Idaho National Engineering Laboratory (now the Idaho National Engineering and Environmental Laboratory), about 60 miles west of Idaho Falls, Idaho, reprocessed spent fuel from reactors that powered naval vessels and recovered enriched uranium for use in weapons production. The Feed Materials Production Center (now the Fernald Environmental Management Project) in Fernald, Ohio, fabricated uranium metal targets to be used in reactors that produced plutonium. In 1989 all production at the Fernald site ended, and the site was designated as a test site for environmental cleanup technologies.[2]

Oak Ridge National Laboratory in Oak Ridge, Tennessee, was the first facility to separate plutonium for weapons. It operated on a small scale, supporting only the very first weapons. Since then Oak Ridge National Laboratory has been a leader in peaceful uses of atomic energy, such as studies of the health effects of radiation, evolution of both power and research reactors, and creation of radiation-resistant materials. The research at Oak Ridge National Laboratory now includes not only nuclear topics but life science, computer science, and other areas. Studies of energy technology in areas such as magnetic fusion, conservation, and renewable energy sources are also important.[3]

The gaseous diffusion plants in Paducah, Kentucky, and near Portsmouth, Ohio, enriched uranium for early atomic weapons. Later, when plutonium was used for atomic weapons, the gaseous diffusion plants produced high enriched uranium for use in reactors that power the Navy's ships. In recent years, the two plants have produced low enriched uranium for commercial nuclear power plants. Since the mid-1990s, the plants have been operated by a private corporation, the United States Enrichment Corporation.

Many facilities in the Nuclear Weapons Complex fabricated parts for nuclear weapons. The Pinellas Plant in Pinellas, Florida, produced nonnuclear components of weapons. The components were primarily metal and electrical. Triggers for the weapons were produced at the Mound Plant near Dayton, Ohio. The Rocky Flats Plant in Rocky Flats, Colorado, produced triggers as well, but it also fabricated other uranium and beryllium weapons components in addition to recovering plutonium from weapons parts and production scrap.

During the Cold War nuclear weapons were assembled at the Pantex Plant near Amarillo, Texas. Also at Pantex, selected weapons which had been removed from storage were dismantled and their components were tested to ensure that they were functioning properly. In recent years some weapons have been permanently dismantled at Pantex.

Two of the sixteen major sites being considered in this section are now used for nuclear waste disposal. The Waste Isolation Pilot Plant (see Chapter 6) is the permanent disposal facility for transuranic waste. The Nevada Test Site (NTS) is now used for the disposal of low-level radioactive waste from some other DOE facilities. However, in the 1950s and 1960s, the NTS was used for underground testing of nuclear weapons. Table 10.1 provides a concise description of the activities of the 16 Nuclear Weapons Complex sites discussed above.

Table 10.1 Operations and wastes at 16 major laboratories in the DOE's Nuclear Weapons Complex

Name	Weapons process categories	Operations involving weapons production	Remnants
Fernald Environmental Management Project	Fuel and target fabrication; mining, milling, and refining	1950s–1989 Opened as the Feed Materials Production Center Uranium ore converted into uranium metal Uranium metal fabricated into target elements for reactors that produced plutonium and tritium	Contaminated environmental media, release sites, waste, materials in inventory, surplus facilities
Hanford	Reactor operations; chemical separations; research, development, and testing; fuel and target fabrication; component fabrication	1942–present Government-owned nuclear weapons production site Fabricated reactor fuel Operated five chemical separation facilities Operated nine reactors Produced plutonium components for nuclear weapons Now researches applications of nuclear energy	Contaminated environmental media, release sites, waste, materials in inventory, surplus facilities
Idaho National Engineering Laboratory	Chemical separations; research, development, and testing	1949–present Opened as the National Reactor Testing Station 1953–1992, reprocessed spent fuel from naval reactors to recover enriched uranium for reuse in nuclear weapons production Stores TRU and LLW from Rocky Flats Plant Performed minor nuclear weapons R&D work	Contaminated environmental media, release sites, waste, materials in inventory, surplus facilities
Lawrence Livermore National Laboratory—Main Site	Research, development, and testing	Opened as a flight training base and engine overhaul facility In 1950, began nuclear weapons research	Contaminated environmental media, release sites, waste, materials in inventory, surplus facilities
Lawrence Livermore National Laboratory—Site 300	Research, development, and testing	Remote high-explosives testing area Areas for high-explosive component testing Instrument firing tables Particle accelerator	Contaminated environmental media, release sites, waste, materials in inventory, surplus facilities
Los Alamos National Laboratory	Component fabrication; research, development, and testing	1943–present Opened especially to design, develop, and test nuclear weapons Also produced small quantities of plutonium metal and nuclear weapons components Later included academic and industrial research	Contaminated environmental media, release sites, waste, materials in inventory, surplus facilities

Site	Mission	Activities	Legacy
Mound Plant	Component fabrication	1946–1995 Government-owned site Developed and fabricated nuclear and nonnuclear components for the weapons program, including polonium–beryllium initiators Built detonators, cable assemblies, and other nonnuclear products Began in 1969 to retrieve and recycle tritium from dismantled nuclear weapons Nonweapons activities included the production of plutonium-238 thermoelectric generators for spacecraft	Contaminated environmental media, waste, materials in inventory, surplus facilities
Nevada Test Site	Research, testing, and development	1950–present Historically used for atmospheric and underground testing of nuclear explosives at a full-scale level Currently used for LLW disposal.	Contaminated environmental media, release sites, materials in inventory, waste, surplus facilities
Oak Ridge National Laboratory	Chemical separations; research, development, and testing; reactor operations	1942–present First to produce and separate gram quantities of plutonium Now principally supports nonweapons programs, radioisotope production, and research in a variety of fields Has supplied isotopes for the nuclear weapons program	Contaminated environmental media, release sites, waste, materials in inventory, surplus facilities
Paducah Gaseous Diffusion Plant	Uranium enrichment	1954–present Opened solely for the enrichment of uranium for weapons production Eventually supplied enriched uranium for naval and commercial reactor fuel Produced UF6 feed until 1960s Still enriching uranium for commercial customers, primarily nuclear power utilities	Contaminated environmental media, release sites, waste, materials in inventory, surplus facilities
Pantex Plant	—	1952–present Converted to a high-explosives component fabrication and weapons assembly plant in 1951 Currently disassembles weapons and stores fissile material	Contaminated environmental media, release sites, waste, materials in inventory, surplus facilities

(continued)

Table 10.1 (continued)

Name	Weapons process categories	Operations involving weapons production	Remnants
Pinellas Plant	Component fabrication	1957–1994 Produced nonnuclear weapons	Waste, materials in inventory, surplus facilities
Portsmouth Gaseous Diffusion Plant	Uranium enrichment	1956–present Began as a producer of high enriched U (HEU) for weapons Later, used to produce HEU for naval reactors In 1994, part of the site was leased to the government-owned United States Enrichment Corporation Now enriches uranium for commercial customers	Contaminated environmental media, release sites, waste,
Rocky Flats Environmental Technology Site	Component fabrication	1951–1989 Produced triggers in nuclear weapons as well as other uranium, beryllium, and steel weapons components Recovered plutonium from returned weapons parts, production scrap, and residues	Contaminated environmental media, release sites, waste, materials in inventory, surplus facilities
Sandia National Laboratory/New Mexico	Weapons operations; research, development, and testing	1949–present Formed to design nonnuclear components of nuclear weapons Housed a weapon assembly line 1946–1957	Contaminated environmental media, release sites, waste, materials in inventory, surplus facilities
Savannah River Site	Research, development, and testing; fuel and target fabrication; heavy water enrichment; chemical separations; reactor operations	1950–present Produces, purifies, and processes plutonium, tritium, and other radioisotopes for nuclear weapons programs and other purposes Fabricates fuel Operates five reactors and two chemical separation plants Conducts research and development Produces heavy water and processed tritium Produces plutonium-238 for use in thermoelectric generators	Contaminated environmental media, release sites, waste, materials in inventory, surplus facilities
Waste Isolation Pilot Plant	—	—	—

Compiled from tables in Linking Legacies: Connecting the Cold War Nuclear Weapons Production Processes to Their Environmental Consequences, U.S. Department of Energy, January 1997.

10.3 WASTE INVENTORIES AND CONTAMINATION AT NUCLEAR WEAPONS COMPLEX SITES

Large amounts and many types of radioactive wastes have been generated at Nuclear Weapons Complex sites. Other chapters of this book focus on specific types of waste and include information on those wastes from the Nuclear Weapons Complex along with methods for treating and disposing of them. However, it is useful to have a summary of wastes currently stored at DOE sites. That summary can be found in Table 10.2. Of particular importance are the high-level wastes being stored, primarily at Hanford, the Savannah River Site, the Idaho National Engineering and Environmental Laboratory, and the West Valley Demonstration Plant (see Table 10.3). Much of this waste is stored in underground tanks, many of which have leaked in the past, leading to site contamination. In addition, future on-site transportation and treatment of the wastes presents opportunities for further contamination. In the past, some low-level and transuranic wastes have been buried or otherwise disposed of at the Nuclear Weapons Complex sites (see Table 10.4). Some of the wastes are migrating from their storage areas and contaminating additional land. At some time in the future, there may be a demand to remove those wastes or to treat them in situ.

Nuclear Weapons Complex sites are contaminated with both hazardous and radioactive materials. In some locations, the radioactive and hazardous materials are found together, presenting a particularly difficult cleanup problem.

Sources of the contamination can be traced back to the work done at the various NWC sites. For example, chemical separation facilities at Hanford and Savannah River were used to separate plutonium from spent fuel taken from reactors on the same sites. The separation required dissolution of the spent fuel in a strong acid. Highly radioactive fission products remained in that acid, and the solution was pumped to underground storage tanks. Similar facilities at the Idaho National Engineering and Environmental Laboratory were used to reprocess spent fuel from the navy's reactors and resulted in similar wastes.

Some of the sites, such as the Kansas City Plant and Mound Laboratories, produced nonnuclear components for weapons. These components were primarily metal and electronic. In order to ensure good electrical connections, the metals were carefully cleaned and sometimes electroplated. A wide variety of effective, and hazardous, solvents and degreasers were used at these sites and are now commonly found in soils, drainpipes, and buildings at the DOE sites.

Work on characterization of site contamination began in the late 1980s. A report on contaminants found at the major NWC sites was published by the Office of Technology Assessment in 1991.[4] This report listed the radioactive and hazardous wastes found in air, soil, surface water, ground water, and sediment at each site. Table 10.5, which lists the contaminants identified at Oak Ridge National Laboratory, is an example of the type of information found in the report. The National Defense Authorization Act for Fiscal Year 1995 required the Department of Energy to report on waste streams resulting from nuclear weapons production. The report was prepared in response to that law and was published in January 1997.[2] It thoroughly catalogs contaminants found at NWC sites.

Table 10.2 Volume and activity of radioactive wastes stored or disposed of at Nuclear Weapons Complex sites[a]

Sites	Nuclear weapons volume (m³)	Nuclear weapons radioactivity (Ci)	Nonweapons volume (m³)	Nonweapons radioactivity (Ci)
Falls City	2,900,000	870	1,500,000	460
Grand Junction Mill Tailing Site	2,300,000	2,500	1,200,000	1,300
Old Rifle and New Rifle	2,000,000	1,700	1,100,000	890
Ambrosia Lake	1,900,000	1,600	1,000,000	880
Maybell	1,700,000	310	930,000	160
Mexican Hat	1,400,000	990	746,000	530
Salt Lake City	1,400,000	1,100	720,000	610
Monticello Remedial Action Project	1,300,000	1,300	690,000	710
Durango	1,300,000	1,300	670,000	680
Riverton	900,000	300	480,000	160
Hanford Site	850,000	330,000,000	83,000	28,000,000
Savannah River Site	820,000	500,000,000	10,000	42,000,000
Shiprock	800,000	580	420,000	310
Fernald	490,000	8,100	0	0
Nevada Test Site	480,000	9,800,000	0	0
Monument Valley	470,000	35	250,000	20
Lakeview	460,000	82	250,000	43
Tuba City	390,000	350	210,000	190
Gunnison	360,000	170	190,000	90
Slick Rock Union Carbide and North Continent	320,000	58	120,000	21
Naturita	270,000	20	150,000	10
LANL	260,000	1,800,000	0	0
Niagara Falls Storage Site	200,000	2,200	0	0
Weldon Spring Site Remedial Action Project	190,000	NA	0	0
Green River	190,000	22	100,000	12
Y-12 Plant	170,000	11,000	0	0
Spook	160,000	104	84,000	55
INEL	140,000	56,000,000	150,000	11,000,000
Canonsburg	110,000	360	60,000	190
K-25 Site	100,000	69	48,000	34
Bowman	64,000	3	34,000	2
Lowman	64,000	16	34,000	8
Middlesex Sampling Plant	51,000	NA	0	0
Portsmouth	36,000	64	23,000	42
Belfield	29,000	3	15,000	1
Latty Avenue Properties	24,000	NA	0	0
Rocky Flats	20,000	86,000	0	0

Table 10.2 (continued)

Sites	Nuclear weapons volume (m³)	Nuclear weapons radioactivity (Ci)	Nonweapons volume (m³)	Nonweapons radioactivity (Ci)
Paducah	16,000	77	10,000	50
Edgemont Vicinity Properties	15,000	NA	8,000	NA
LLNL	10,000	19,000	0	0
Mound	9,200	1,400,000	0	0
ORNL	7,400	130,000	240,000	4,300,000
Sandia National Lab/NM	3,300	9,300	0	0
Reactive Metals Incoporated	2,900	30	0	0
Grand Junction Projects Office	780	NA	370	NA
Pantex Plant	480	12	0	0
Pinellas Plant	66	30,000	0	0
Kansas City Plant	33	1	0	0
Sandia National Lab/California	27	13	0	0
Nonweapons Sites	0	0	98,000	26,000,000
Total	24,000,000	900,000,000	12,000,000	110,000,000

[a]NA, Not available.

Source: Linking Legacies: Connecting the Cold War Nuclear Weapons Production Processes to Their Environmental Consequences, U.S. Department of Energy, January 1997.

While contaminated media include soil, surface water, sediment, groundwater, and air, the vast majority of the contamination is found in soil and groundwater. The total volume of contaminated solid media at all of the NWC sites is approximately 79 million m³. Of this amount, 95% is soil. Seventy percent of the soil is contaminated with radionuclides, 14% with hazardous waste, and 16% with a mixture of radioactive and hazardous waste. The total volume of contaminated water at the NWC sites is estimated

Table 10.3 High-level radioactive wastes stored at DOE sites

	Nuclear weapons volume (m³)	Nuclear weapons radioactivity (Ci)	Nonweapons volume (m³)	Nonweapons radioactivity (Ci)
Hanford	220,000	320 million	19,000	27 million
Savannah River Site	120,000	490 million	10,000	42 million
Idaho National Engineering Laboratory	11,000	52 million	0	0
West Valley Demonstration Project	0	0	2,100	25 million

Compiled from information in Ref. 2.

Table 10.4 Buried transuranic and low-level wastes at Nuclear Weapons Complex sites

	Transuranic wastes			
Site	Nuclear weapons volume (m³)	Nuclear weapons radioactivity (Ci)	Nonweapons volume (m³)	Nonweapons radioactivity (Ci)
Hanford	55,000	150,000	8,800	24,000
INEL	53,000	230,000	4,500	20,000
LANL	14,000	5,600	0	0
Savannah River Site	4,900	31,000	0	0
ORNL	5	7	170	233
Sandia	1	1	0	0
Nonweapons sites	0	0	1,350	652,000

	Low-level wastes	
Site	Nuclear weapons volume (m³)	Nonweapons volume (m³)
Savannah River Site	680,000	0
Hanford	560,000	53,000
Nevada Test Site	480,000	0
LANL	220,000	0
INEL	37,000	110,000
ORNL	6,800	220,000
Fernald	340,000	0
Y-12 Plant (TN)	150,000	0
K-25 Site (TN)	54,000	27,000
LLNL	9,100	0
Portsmouth	7,300	4,800
Paducah	4,600	3,000
Sandia	3,200	0
Pantex Plant	130	0
Nonweapons ocean disposal	0	19,000

Source: Ref. 2.

to be 1800 million m³ (475 billion gal), of which 99% is ground water. Fifty-seven percent of that groundwater is contaminated with radionuclides, 14% is contaminated with hazardous wastes, and 29% is contaminated with a mixture of radioactive and hazardous wastes.[2]

10.4 LAWS AND DEPARTMENT OF ENERGY ORDERS THAT APPLY TO WEAPONS COMPLEX SITES

Prior to the 1980s, facilities operated by the Department of Energy and its predecessors were not subject to regulation by other agencies. Since then, the Nuclear Weapons Complex sites have been required to comply with some laws and regulations enforced

Table 10.5 Contaminants identified at Oak Ridge National Laboratory

Contaminant	Air	Soil	Surface water	Groundwater	Sediment
Radionuclides	Questionable	Am-241	Am-241	Sb-125	Am-241
		Cs-137	Cs-137	Cs-137	Cs-137
		Co-60	Co-60	Co-60	Co-60
		Cm-244	Cm-244	Eu	Cm-244
		Pu-238	Gross beta	Gross alpha	Eu
		Pu-239	Sr	Gross beta	Pu-238
		Ra-228	H-3	Pu	Pu-239
		Sr-90		Ru-106	Sr-90
		U-232		Tc-99	U-232
		U-233		Th-232	U-233
		U-234		H-3	U-234
		U-235		U-232	U-235
		U-238		U-233	U-238
				U-234	
				U-235	
				U-238	
Metals	Pb	Hg	Cl	As	Cr
Inorganic compounds	Questionable			Ba	Pb
				Cd	Hg
				Cr	
				Pb	
				Hg	
Volatile organic compounds (VOCs)	Questionable			Acetone	Undefined VOCs
				Benzene	
				Carbon tetrachloride	
				Chloroform	
				1,1-Dichloroethylene	
				trans-1,2-Dichloroethylene	
				Dimethyl phthalate	
				Ethylbenzene	
				Methylene chloride	
				Naphthalene	
				1,1,2,2-Tetrachloroethane	
				Trichloroethylene	
				Xylene	
Miscellaneous		Stored petroleum products	Fecal coliform	Endrin	PCBs
			Total suspended solids	Stored petroleum products	

Source: Ref. 4.

by other federal agencies concerned with human health and the environment, such as the Environmental Protection Agency (EPA) and the Nuclear Regulatory Commission (NRC). In addition, several states have agreements with either the EPA or the NRC or both that allow the states to enforce most regulations normally enforced by the federal agency. These "agreement states" have enacted legislation making state regulations at least as stringent as the federal regulations, designated a state enforcement agency, hired and trained staff, and demonstrated to the appropriate federal agency that they have the ability to enforce the regulations. Thus, environmental cleanup projects at many NWC sites are governed by a mixture of federal and state regulations in addition to a set of internal DOE orders. This section lists and briefly discusses some of the laws and DOE orders that commonly apply to NWC sites.

Three federal laws that affect many NWC sites are the Resource Conservation and Recovery Act of 1976 (RCRA), the Comprehensive Environmental Response, Compensation, and Liability Act of 1980 (CERCLA), and the National Environmental Policy Act of 1969 (NEPA). RCRA gives the EPA responsibility for regulating the generation, transportation, treatment, storage, and disposal of hazardous wastes. Many of the NWC sites generate both hazardous wastes and mixtures of hazardous and low-level radioactive wastes, both of which are covered under RCRA. The Hazardous and Solid Waste Amendments of 1984 (HSWA) expanded RCRA's scope. Under HSWA, burial of untreated hazardous wastes was prohibited, and the EPA was required to establish treatment and disposal standards for the wastes. Furthermore, when a facility seeks a permit under RCRA to dispose of wastes, the EPA can require the operator of the facility to take corrective action to eliminate releases from any other solid waste disposal unit on the site, regardless of when the waste was buried. This portion of the HSWA has been applied to many NWC sites.

CERCLA, more commonly known as the Superfund Act, was passed to provide for the cleanup of abandoned or uncontrolled waste disposal sites. It gave the EPA the authority to determine what contaminants were being released from an abandoned site and assess the associated risk. If the risks were high enough to justify making cleanup of the site a national priority, the site could be added to the National Priorities List (NPL). Eight sites within the NWC are currently on the NPL. The Superfund Amendments and Reauthorization Act of 1986 (SARA) was an amendment to CERCLA and made some changes that reflected 6 years of experience administering the 1980 law. These changes included emphasizing the importance of using innovative cleanup technologies, striving for permanent solutions to contamination, and increasing the involvement of state authorities, citizens, and other stakeholders. DOE laboratories have developed or adapted several cleanup technologies and demonstrated them at NWC sites.

The National Environmental Policy Act requires an environmental impact statement (EIS) for every major federal action "significantly affecting the quality of the human environment."[6] The EIS must describe the environmental impacts of the proposed action and alternatives to that action. Several DOE activities, such as development of waste disposal facilities, may require an EIS.

Several other federal laws that may apply to environmental restoration activities at DOE's Nuclear Weapons Complex sites are listed in Table 10.6.

Table 10.6 Federal laws that may apply to environmental restoration projects at DOE sites

Law	Year passed/amended	Enforcing agency[a]
Atomic Energy Act	1946/1954	DOE
Clean Air Act	1963/1990	EPA
National Historic Preservation Act	1966	DOI
National Environmental Policy Act	1969	EPA
Occupational Safety and Health Act	1970	DOL
Clean Water Act	1972	EPA
Endangered Species Act	1973	DOI
Noise Control Act	1973	EPA
Safe Drinking Water Act	1974	EPA
Hazardous Materials Transportation Act	1975	DOT
Federal Land Policy and Management Act	1976	EPA
Toxic Substances Control Act	1976	EPA
Uranium Mill Tailings Radiation Control Act	1978	DOE
Archeological Resource Protection Act	1979	DOI
Low-Level Radioactive Waste Policy Act	1980/1986	NRC
Nuclear Waste Policy Act	1982/1987	DOE
Oil Pollution Act	1990	EPA

[a]DOE, Department of Energy; DOI, Department of the Interior; DOL, Department of Labor; DOT, Department of Transportation; EPA, Environmental Protection Agency; NRC, Nuclear Regulatory Commission. Compiled from information in Ref. 7.

DOE orders serve as an internal regulatory system for the Department. Several of these orders apply to site remediation activities. Three examples are DOE Orders 5820.2A, 5400.5, and 451.1. DOE Order 5820.2A, Radioactive Waste Management (9/26/88), provides guidelines for managing high-level, low-level, and transuranic wastes and decommissioning of contaminated facilities. It requires DOE waste management activities to comply with relevant federal, state, and local laws. DOE Order 5400.5, Radiation Protection of the Public and the Environment (2/8/90), established an environmental monitoring program, standards for the release of property with residual contamination, and dose limits for members of the public resulting from radioactive material at or from DOE sites. DOE Order 451.1, National Environmental Policy Act Compliance Program (9/11/95), spells out how DOE will comply with NEPA and related regulations. It also sets requirements for DOE's environmental impact statements.[8]

10.5 DEPARTMENT OF ENERGY PLANS FOR ENVIRONMENTAL RESTORATION

In 1989, the Department of Energy planned to bring its sites into compliance with applicable regulations within 30 years. Unofficial cost estimates for the 30-year program ranged from $100 billion to $500 billion or more. By 1996, DOE's Office of Environmental Management (EM) had committed to an accelerated cleanup program, planning to clean approximately 90% of the sites by 2006 and then to focus its attention on the few remaining sites with large amounts of waste or a wide variety of contaminants.[9]

EM's current plans include 353 cleanup projects at 53 sites and are expected to cost $147 billion (in constant 1998 dollars). While most of the projects are scheduled to be completed by 2006, EM expects to spend more than $3 billion per year for several decades after 2006 to finish the most difficult projects. About three-fourths of the money spent after 2006 will be used for cleanup at three sites: the Hanford Site, the Savannah River Site, and the Idaho National Engineering and Environmental Laboratory. Work at all sites is predicted to be complete by 2070.[10]

To complete the cleanup by 2006 and within the projected budget, EM expects to invest in the development of innovative technologies to meet the most critical needs. Top priority will be given to those technologies that can significantly reduce the costs of the most expensive cleanup projects and those expected to perform tasks along the critical path, that is, tasks that are currently impossible and which must be completed before subsequent tasks can begin.[10] Detailed plans for cleanup activities at the major NWC sites are presented in reference 10.

10.6 ENVIRONMENTAL RESTORATION TECHNOLOGIES

When the DOE committed to cleaning up its contaminated sites in 1989, administrators recognized a need to develop safer and more efficient technologies in the following areas:

1. Site characterization and remediation
2. Facility deactivation and decommissioning
3. Waste treatment and disposal

The program established to oversee the development of those technologies was the Characterization, Monitoring, and Sensor Technology (CMST) Program. The technologies themselves were developed by private companies, universities, and federal government laboratories, but CMST was needed to coordinate and integrate those development projects to avoid costly and time-consuming duplication of effort. Between 1990 and 1994, the CMST program focused on technologies useful in characterizing contamination at DOE sites. Some of the technologies developed were as follows:[11]

1. Field-deployable, rapid-turnaround chemical characterization instrumentation
2. Large-area imaging sensor system for surface contamination mapping
3. Advanced technology continuous emission monitors
4. Nondestructive assay/nondestructive evaluation technologies to determine the content of waste drums
5. Integration and application of chemical and radioactivity sensors in robotic characterization, retrieval, decontamination, and waste processing systems

In 1995, the CMST program was reorganized to support DOE's effort to accelerate cleanup of its sites. Five focus areas were established, along with three cross-cutting programs under which technologies that applied to more than one focus area were developed. The initial focus areas and cross-cutting programs are briefly described in the following paragraphs. From time to time, the focus areas are redefined in response to changing priorities in the environmental restoration program.

Contaminant Plume Containment and Remediation Focus Area. This area addresses primarily soil and groundwater contaminants. Technologies are being developed to characterize, contain, and treat, in situ, contaminants at DOE sites.

Mixed Waste Characterization, Treatment, and Disposal Focus Area. Mixed waste is a mixture of hazardous and low-level radioactive wastes. Because regulations governing hazardous waste and those applicable to low-level waste sometimes conflict, mixed waste disposal has been nearly impossible. New technologies are needed to facilitate treatment and disposal of mixed wastes.

Radioactive Tank Waste Remediation Focus Area. Hundreds of thousands of cubic meters of liquid high-level waste are stored at sites in the Nuclear Weapons Complex, most of it in underground tanks, many of which have leaked. The contents of the tanks are mixtures of very hazardous materials. New technologies are needed to safely and economically characterize, retrieve, and treat the wastes to put them in a form acceptable for disposal.

Landfill Stabilization Focus Area. In the early years of the NWC, wastes were often buried at the site where they were generated. Some of those wastes are migrating into the surrounding environment. In this focus area, work is being done on technologies to confine, retrieve, or treat the wastes in situ.

Decontamination and Decommissioning Focus Area. Many of the facilities in the Nuclear Weapons Complex were built in the mid 20th century and are becoming obsolete. In addition, since the Cold War has ended, weapons production has nearly ceased. Buildings and equipment from the NWC are being decontaminated, dismantled, and disposed of. New technologies are needed to make this process as safe and efficient as possible.

The three titles of the three cross-cutting programs clearly describe the nature of the work done in those programs:[11]

1. Characterization, monitoring, and sensor technology cross-cutting program
2. Efficient separations and processing cross-cutting program
3. Robotics technology cross-cutting program

10.7 COMPUTER CODES

Scores of computer codes have been written to model fate and transport of contaminants released to the environment. This section briefly describes four codes that have been used to model radioactive contaminants at DOE sites.

GENII. The Hanford Environmental Radiation Dosimetry Software System. This code calculates radiation doses due to radioactive contaminants in the environment. It can accommodate chronic or acute releases and transport by air, water, or animals.[12]

CAP88-PC. This code calculates doses due to radionuclides released to the air. It can be used for maximally exposed individuals or for entire populations.[13]

RESRAD. The EPA and NRC have both used this suite of codes or allowed them to be used for radiation dose calculations in applications for licenses or permits.[14]

COMPLY. This code to model radiation exposure from radionuclides released into the air was written for the EPA with input from the National Council on Radiation Protection and Measurements (NCRP). The NRC also allows the use of COMPLY to demonstrate compliance with some of its regulations.[15]

10.8 DISCUSSION QUESTIONS AND PROBLEMS

1. Name five of the major Nuclear Weapons Complex sites that are still open and the types of activities that have been performed there.

2. Name and describe four of the regulations and/or orders that apply to the Weapons Complex Sites.

3. Describe two areas in which environmental restoration technologies are being developed by Department of Energy.

4. Which site has the most high-level nuclear waste by volume? By radioactivity?

5. What three nuclear weapons-related sites are located in New Mexico?

6. What type of waste is or is planned to be stored at WIPP?

7. What nuclear weapons related activities were performed at Hanford?

8. Name two of the regulations covering the Nuclear Weapons Complex sites that are enforced by the Environmental Protection Agency.

9. Which Nuclear Weapons Complex Sites were involved with component fabrication?

10. Which Nuclear Weapons Complex Sites were involved with fuel and target fabrication?

11. What name was Fernald opened under?

12. In which state is Hanford located?

13. In which state is the Savannah River Site located?

14. What percentage of the contaminated water is ground water?

15. Why do you think the Department of Energy is first working on the numerous less contaminated sites, instead of the few highly contaminated sites?

16. How clean should "clean" be?

17. What do you think is causing the delay in the permanent disposal of high-level waste?

18. Who should decide which sites need to be cleaned and which sites don't?

19. How much money should be allocated for cleanup? And who should make this decision?

20. Could this money be used in a "better" way, such as for cancer research?

21. Should there be a limit on the amount of time spent discussing nuclear waste storage before action is taken to store the waste? If yes, how much time should be allotted for discussion?

REFERENCES

1. Lawrence Livermore: History. Available at http://www.llnl.gov/llnl/02about-llnl/history.html [9 February 2000].
2. U.S. DOE Office of Environmental Management, Linking Legacies: Connecting the Cold War Nuclear Weapons Production Processes to Their Environmental Consequences. DOE/EM-0319, January 1997.
3. ORNL: A Sense of the Past, an Eye to the Future. Available at http://www.ornl.gov/glance/past-future.html [9 February 2000].
4. U.S. Congress, Office of Technology Assessment, Complex Cleanup: The Environmental Legacy of Nuclear Weapons Production OTA-0-484, U.S. Government Printing Office, Washington, D.C., February 1991.
5. Noyes, R., Nuclear Waste Cleanup Technology and Opportunities, Noyes Publications, Park Ridge, N.J., 1995.
6. The National Environmental Policy Act of 1969, Public Law 91-190, 42, U.S.C. 4321–4347, January 1970.
7. U.S. Department of Energy, Additional Statutes Affecting DOE's Environmental Management Program, August 1994, DOE/EM-0039P (Rev. 1). Available at http://www.em.doe.gov/fs/fs2d.html [15 November 1999].
8. U.S. DOE Office of Environmental Management, Final Waste Management Programmatic Environmental Impact Statements for Managing Treatment, Storage, and Disposal of Radioactive and Hazardous Waste. DOE/EIS-0200-F, Washington, D.C., May 1997.
9. U.S. Department of Energy, Environmental Management Research and Development Program Plan: Solution-Based Investments in Science and Technology, U.S. Department of Energy, Washington, D.C., November 1998.
10. U.S. Department of Energy, Office of Environmental Management, Accelerating Cleanup—Paths to Closure, DOE/EM-0362, June 1998.
11. U.S. Department of Energy, Genesis and History of the CMST Program. Available at http://www.cmst.org/cmst/tech—summ—00 [3 March 2001].
12. Napier, B. A., GENII—The Hanford Environmental Radiation Dosimetry Software System, Pacific Northwest National Laboratory, 15 April 1999. Available at http://www.pnl.gov/health/health—prot/genii.html [15 November 1999].
13. U.S. Department of Energy, CAP88-PC, 10 November 1999. Available at http://www.er.doe.gov/production/er-80/capp88.html [15 November 1999].
14. Environmental Assessment Division, Argonne National Laboratory, The RESRAD Family of Computer Codes, 15 July 1999. Available at http://web.ead.anl.gov/resrad/resrad.html [15 November 1999].
15. U.S. Environmental Protection Agency, Radiation Risk Assessment: Software-COMPLY, 5 August 1999. Available at http://www.epa.gov/radiation/assessment/comply.html [15 November 1999].

ADDITIONAL READINGS

Committee to Provide Oversight of the DOE NWC, Nuclear Weapons Complex, National Academy Press, Washington, D.C., 1989.

U.S. Department of Energy, FY99 CMST Rainbow Book, Available at http://www.cmst.org/cmst/Rainbow—99/index.html [4 November 1999].

U.S. Department of Energy, Laws and Regulations. Available at http://www.em.doe.gov/er/laws.html [15 November 1999].

U.S. DOE Office of Environmental Management, Closing the Circle on the Splitting of the Atom, 2nd Printing, DOE/EM-0266, January 1996.

CHAPTER ELEVEN

TRANSPORTATION

11.1 INTRODUCTION

Transportation is an integral component of waste management, and its safety is as much of public concern as the disposal system. A variety of radioactive wastes are transported in the United States and in other countries. Those wastes include low-level waste, spent fuel, high-level waste resulting from the reprocessing of spent fuel, and transuranic waste. This chapter will focus on the transportation of low-level waste and spent fuel in the United States. Transportation of transuranic waste is covered in Chapter 6. High-level waste resulting from reprocessing of spent fuel is currently transported in other countries which have active reprocessing programs. The United States does not reprocess commercial spent fuel. Eventually, high-level waste resulting from the nuclear weapons program and currently stored at U.S. Department of Energy sites within the United States will be transported to a repository. However, DOE plans to transport spent fuel from commercial power plants first.

This chapter briefly describes the current status of radioactive waste transportation programs in the United States and abroad. It explores public concerns related to radwaste transportation and outlines the regulations that govern radwaste transportation. The types of packages used to transport various categories of radwaste are described. Special attention is given to the DOE's plans for a spent fuel/high-level waste transportation system. Responsibilities of the companies that transport waste are discussed briefly along with the approach to risk assessment of radwaste transportation.

11.2 CURRENT STATUS OF RADWASTE TRANSPORTATION WORLDWIDE

Radwastes are produced throughout the world wherever radioactive materials are used or processed. Thus, generators of radwaste include hospitals, industry, educational institutions, power stations, and fuel reprocessing facilities. Estimates of the volume of waste to be transported are subject to some uncertainty as methods of treatment are developed, old facilities are closed, and new ones are opened. So far, over 200 million packages of radmaterial have been transported safely; quantities range from minute amounts of radioactivity (from hospitals and research work) to very large amounts (e.g., spent fuel from nuclear power stations going to reprocessing plants). The annual amounts of LLW produced in Japan, the United Kingdom, France, and the United States are about 15,000, 10,000,[1] 20,000,[2] and 70,000 m^3 (4.9×10^5, 9.8×10^5, 9.8×10^5, and 2.29×10^6 ft^3), respectively.[3] It is estimated that during the next 15 years in the European community, between 50,000 and 100,000 m^3 (1.64×10^6 and 3.28×10^6 ft^3) of LLW will be conditioned, transported, and disposed of each year. Even when LLW has to be transported a considerable distance, it is mainly a bulk movement problem.

The intermediate-level waste* also generated in the European community in significant quantities was estimated to amount to an additional 150,000–300,000 m^3 (4.9–9.8×10^6 ft^3) to be conditioned, transported, and disposed of by the year 2000.

In the United States, the amount of spent fuel to be transported is significant, and all of it will eventually be transported to a federal interim storage and/or a repository. The inventory of light-water reactor (LWR) spent fuel assemblies in pools is expected to be 150,000, corresponding to over 50,000 MTU. In more than 40 years of civilian nuclear power, approximately 6000 spent fuel assemblies have been shipped. Beginning in 2010 (the currently scheduled repository opening date), the DOE would be responsible for shipping the equivalent of 7000 spent fuel assemblies per year to disposal facilities.

In Europe, HLW other than the spent fuel would require transportation, and the amounts involved are relatively small; as of 1 April 1998 the United Kingdom had 1800 m^3 (ft^3) and France about 3000 m^3 (9.8×10^4 ft^3) of vitrified HLW in storage.[1] The only commercial HLW, other than spent fuel, in the United States will be limited to the solidified waste from the West Valley stocks of liquid HLW. Vitrified HLW is expected to be formed in cylinders about the size of spent fuel bundles and transported in packages similar to those used for spent fuel.

The modes of surface transportation (transport by air is very limited) typically include truck, rail, and barge. In addition, seagoing vessels carry spent fuel from Japan to Europe for reprocessing. The return voyages may transport plutonium for use as reactor fuel and the waste from the reprocessing to Japan.

---

*As stated in Chapter 7, the ILW category used in European countries has a lower specific activity and heat output than HLW. No heat generation needs to be taken into account in the design of storage or disposal facilities for ILW, but it still requires shielding during handling and transporting. It is differentiated from the LLW in that LLW is at a specific activity that does not require shielding during normal handling and transportation.

11.3 PUBLIC CONCERNS

When radioactive materials are transported, they attract a great deal of public attention, and there is particular concern about shipments of SNF and radwastes. One of the common fears is that a nuclear shipment might somehow go awry and cause a serious public hazard; for instance, a shipment of SNF or HLW could be involved in a serious rail or highway accident, releasing radioactive material. Some commonly asked questions and their answers are presented in this section.

The League of Women Voters has issued several publications that attempt to address public concerns about radioactive waste. Some questions and answers about spent fuel transportation follow.[4]

How much spent fuel is there?

Approximately 32,000 metric tons of spent fuel from commercial reactors have accumulated since the mid-1950s. According to utility industry reports, the average commercial reactor produces 20 metric tons of spent fuel annually. In their licensed lifetimes, commercial nuclear reactors will generate a total of 86,000 metric tons of spent fuel.

Is spent fuel being transported now?

Spent fuel shipments, primarily from reactor sites to temporary storage facilities and between utilities in the United States, have been shipped by rail and truck in limited quantities over the last 25 years. Spent fuel is shipped routinely in European countries and Japan.

What is the safety record for transporting nuclear fuel?

In the more than 2500 spent fuel shipments made since 1971, seven accidents have occurred. Four of these took place on rail shipments and three during truck shipments. None resulted in a radioactive release.

What risks are associated with transporting spent nuclear fuel?

Risks associated with transporting spent nuclear fuel encompass both the probability that an accident will occur and the seriousness of an accident, should it occur. Experience to date shows that the occurrence of a transportation accident is rare: as noted above, seven accidents resulted from 25,000 shipments, over a period of 25 years.

How is commercial spent fuel packaged for transport?

Commercial spent fuel is placed in transport casks that are designed to provide a barrier to radiological exposure. The cask design includes protective shields for gamma rays and neutrons, a heat transfer surface, a storage cavity, a basket for the fuel assemblies made of stainless steel or borated stainless steel and a lid. In preparing the shipment, an additional protective barrier may be added around the transport cask.

Who regulates transportation of radioactive spent fuel?

Federal, State, Native American and local government agencies all have roles in regulating the transportation of radioactive spent fuel and they do so to varying degrees and according to pertinent laws, guidelines and designated responsibilities.

Who chooses the transport routes?

DOT issued regulations establishing the Interstate Highway System, the shortest routes to those highways, and bypasses around urban centers as the preferred routes for radioactive waste shipments on highways. DOT also established the rule that carriers must use those preferred routes.

The Atomic Energy Commission (the forerunner of DOE) compiled a list of "most often asked" questions and answers including the following[5]:

How do you keep the casks from breaking open or leaking in an accident?

The casks are kept from breaking open or leaking in an accident by being designed to resist the stresses that might occur during accidents. They are independently reviewed and approved, then fabricated, maintained, and prepared for shipment via quality assurance procedures.

Which is safer, truck or rail?
The overall accident rates on a per rail car mile or per truck mile basis are about the same (or there is a small difference). Rail transport is preferred because rail casks are larger and carry more spent fuel bundles so fewer shipments would be required.

What maximum speeds will the vehicle be permitted to travel? Should there be special speed limits on such shipment?
Both rail and truck shipments for the most part will be made in regular scheduled service at normal speeds. Considering that the waste casks are built to withstand high-speed accidents, reduction of the speed limit on shipments of nuclear waste to, say, 30 miles per hour would not measurably reduce the nuclear hazard. Within reason, lower speeds for trucks or trains would reduce serious accidents just as low speeds would reduce serious automobile accidents, independent of what cargo is carried. Several studies were made regarding the special train issue raised by railroads for nuclear material transport.[6,7] Evaluating the environmental impact of radmaterial shipments by both regular and special trains, reference 6 concluded that in all cases (normal and accident conditions) the incremental environmental impact of special trains is very small. A related study further concluded that the special train alternative not only is not cost effective but also does not even appear desirable from a radiological health viewpoint.[7] Through litigations, the issue was finally settled by the Supreme Court decision of October 6, 1980 that since the railroads have acted in the capacity of common carriers for other materials they could not claim private carrier status for nuclear materials.[8]

Will the casks be able to be hit by a tornado, hurricane, lightning, or earthquake without spreading radioactivity?
In the case of tornado, the analysis indicates that there is a possibility that the 100-ton cars could be rolled over as a result of the high wind velocities, but no radioactive material would be released from the cars. About the worst conditions that can be postulated are that the cars would be turned over and the casks dumped on the ground. In any event, no release of radioactivity could take place. Hurricanes or earthquakes would not stress the casks beyond their design integrity either, and the casks would be able to withstand a lightning strike without significant damage.

11.4 REGULATIONS GOVERNING RADWASTE TRANSPORTATION

11.4.1 International Regulations

A number of international bodies deal with the transportation of radioactive materials, and the majority are sanctioned by or affiliated with the United Nations. Regulations promulgated by these agencies are recommended to member states as a basis for national regulations. The primary agency is the International Atomic Energy Agency (IAEA). In the air transport mode, the International Civil Aviation Organization (ICAO) is active in regulating the transport of dangerous materials including radmaterials. The International Air Transport Association (IATA), made up of member air carriers, also publishes regulations for air transport of restricted articles including radmaterials. In the water transport mode, the International Maritime Organization (IMO) publishes regulations that deal with the carriage of radmaterials by vessel. Both IMO and ICAO regulations are based on the regulations of the IAEA but are more explicit in the compliance actions and requirements for shippers and carriers. International regulations can be found in the following documents:

1. Regulations for the Safe Transportation of Radioactive Materials, as amended, Safety Series 6, 1996 Rev. Ed., IAEA, Vienna (available from UNIPUB, 1180 Avenue of the Americas, New York, NY 10038).
2. The Safe Transport of Dangerous Goods by Air, 1999 2nd Ed., incorporating amendments 1–5, ICAO (available from INTEREG, P.O. Box 60105, Chicago, IL 60660).
3. International Maritime Dangerous Goods (IMDG) Code, 1994, including amendments 29–98.
4. Dangerous Goods Regulations, 41st Ed., 2000, IATA, with Supplement and Amendment issued March 1, 1981 (available from International Air Transport Association, 2000 Peel Street, Montreal, Quebec, Canada H3A 2R4).

11.4.2 Federal Regulations

The principal federal regulations pertaining to the transport of radioactive materials are listed in Table 11.1.[9] These regulations are published by three agencies: the U.S. Department of Transportation (DOT), the U.S. Nuclear Regulatory Commission (NRC), and the U.S. Postal Service. DOT has regulatory responsibility for safety in the transportation of all hazardous materials including radmaterials, that is, shipments by all modes of transport in interstate or foreign commerce (rail, highway, air, water) and by all means (truck, bus, automobile, ocean vessel, airplane, river barge, rail car, etc.) except for the Postal Service. The Interstate Commerce Commission (ICC), which formally had

Table 11.1 Sources of federal regulations

Title 49: Department of Transportation's Hazardous Materials Regulations, Parts 100–177 and 178–199
49CFR106 Rulemaking Procedures
49CFR107 Hazardous Materials Program Procedures
49CFR171 General Information, Regulations and Definitions
49CFR172 Hazardous Materials Tables and Hazardous Materials Communications Regulations
49CFR173 Shippers—General Requirements for Shipments and Packagings
49CFR174 Carriage by Rail
49CFR175 Carriage by Aircraft
49CFR176 Carriage by Vessel
49CFR177 Carriage by Public Highway
49CFR178 Shipping Container Specifications
49CFR179 Specifications for Tank Cars
Title 10: Nuclear Regulatory Commission
10CFR71 Packaging of Radmaterials for Transport and Transportation of Radmaterials under Certain Conditions
Title 39: U.S. Postal Service
Domestic Mail Manual, U.S. Postal Service Regulations, Part 124. (Postal regulations for transport of radioactive matter are published in U.S. Postal Service Publication No. 6 and in the U.S. Postal Manual.)

From A Review of the Department of Transportation Regulations for Transportation of Radmaterials, Research and Special Programs Administration, Materials Transportation Bureau, Department of Transportation, Washington, D.C., Rev. 1983. Reprinted by permission.

jurisdiction over both the safety and economic aspects of the transport of radmaterials by surface modes, transferred the safety aspects to DOT in 1967 but still exercises jurisdiction over the economic aspects of radmaterial transport through the issuance of operating authorities' licenses to carriers. The NRC also has responsibility for safety in the possession, use, and transfer (including transport) of most types of radioactive material, and a license from the NRC is required for such possession and use. The NRC also assists and advises DOT in the establishment of both national and international safety standards. Adopting by reference portions of the DOT regulations, NRC inspects its licensees for compliance with DOT regulations applicable to shippers. The primary division of responsibility between DOT and NRC is such that DOT sets packaging and shipping standards for certain LLW and for general labeling, handling, placarding, loading, and unloading requirements, while NRC sets standards only for the packaging and containment of certain high concentrations of radmaterials including large quantities, special nuclear materials, and spent fuel. Several states have entered into formal agreements with the NRC whereby the regulatory authority over by-products, source, and less than critical quantities of special nuclear material has been transferred to the states from the NRC. These "agreement states" have adopted uniform regulations pertaining to interstate transportation of radmaterials; many states have formally adopted the DOT regulations and apply these requirements to both interstate and intrastate transportation. Routing of radmaterials is governed by routing rules, that is, requirements that direct, redirect, restrict, or delay the movement of radmaterials:

1. The first rule is a general set of regulations that require carriers to consider such factors as population, accident rates, and transit time when choosing routes.
2. The second rule applies only to motor vehicles transporting large quantities of radmaterial or SNF and includes the preferred routes, requirements of routing plan, and driver training certification. Also, under this rule, state agencies may designate alternative preferred routes for large quantities of radmaterials.

Warning labels. Each package of radmaterial, unless excepted, must be labeled on two opposite sides with a distinctive warning label bearing the unique trefoil symbol recommended by the International Commission on Radiation Protection (ICRP) and adopted by the American National Standards Institute (ANSI) (Figure 11.1). There are three

RADIOACTIVE-WHITE I   RADIOACTIVE-YELLOW II   RADIOACTIVE-YELLOW III

Figure 11.1 Package warning labels. From A Review of the Department of Transportation Regulations for Transportation of Radmaterials, Research and Special Programs Administration, Materials Transportation Bureau, Department of Transportation, Washington, D.C., Rev. 1983. Reprinted by permission.

label categories: (1) all-white background (indicating low external radiation level), (2) upper half of the label, yellow background with two red stripes (indicating that an external radiation level or fissile properties may require consideration during transportation), and (3) upper half, yellow label with three red stripes (indicating that the shipping vehicle must be placarded RADIOACTIVE). The criteria that the shipper must consider in choosing the appropriate label are listed in Table 11.2.

11.4.3 State Regulations

By 2000, over 30 states had formally become "agreement states" with NRC regulatory authority over shipments of most radioactive materials being formally transferred to those states. Generally, agreement states formally incorporate the NRC and DOT regulations governing the transportation of radioactive material into their state laws and then designate a state agency to be responsible for enforcing these laws.

Each state is also responsible for designating preferred routes for shipments of spent nuclear fuel and other large quantities of highly radioactive materials through the state.

Issues arising from conflicts between federal and state or local agencies. Transport of hazardous and radioactive materials is a controversial subject, but much of the controversy is focused on the transport of SNF and other highly radioactive materials. Significant conflicts between the DOT and states or local agencies exist over the federal presumption of state and local routing regulations. For instance, the Hazardous Materials Act of 1975[10] preempts any state or local regulations inconsistent with federal rules, unless DOT determines that the local requirements afford public protection equal to or greater than that provided by the federal rules and *do not unnecessarily burden commerce*. Over 200 states and local jurisdictions have adopted bans or special regulations.

Table 11.2 Radioactive materials packages labeling criteria[a]
(CFR Title 49, Section 172.403)

Transport index (TI)	Radiation level at package surface (RL)	Fissile criteria	Label category[a]
NA[b]	RL ≤ 0.5 mrem/hr	Fissile class I only, no fissile class II or III	White-I
≤1.0	0.5 mrem/hr < RL ≤ 50	Fissile class I, fissile class II with TI ≤ 1.0, no fissile class III	Yellow-II
<TI	50 mrem/hr < RL	Fissile class II with 1.0 < TI, fissile class III	Yellow-III

[a] Any package containing a highway-route-controlled quantity (Section 173.403) must be labeled as Radioactive Yellow-III.

[b] Not available.

From A Review of the Department of Transportation Regulations for Transportation of Radmaterials, Research and Special Programs Administration, Materials Transportation Bureau, Department of Transportation, Washington, D.C., Rev. 1983. Reprinted by permission.

11.5 RADIOACTIVE WASTE PACKAGES FOR TRANSPORTATION

11.5.1 Factors Determining Types of Package

Several factors determine the type of packaging:

Specific activity of the waste. For purposes of transportation, materials that are regulated as radioactive are those which spontaneously emit ionizing radiation and have a specific activity in excess of 0.002 μCi/g of material. Those with a specific activity lower than 0.002 μCi/g are not regulated by DOT or IAEA; however, they may be subject to use or transfer regulations issued by NRC or EPA. Over 250 specific radionuclides are listed in reference 11.

Quantity of the radionuclides. The packaging requirements are related to the total quantity in a package (in curies).

Forms of the radionuclides. There are two forms, namely special forms and normal forms. *Special form materials* are those that, if released from a package, might present a hazard of direct external radiation. However, due to their high physical integrity, they are very unlikely to be dispersed. The key characteristic of a special form material may be natural (e.g., massive solid metal with high physical integrity) or an acquired (e.g., sealed in a very durable capsule). *Normal form radmaterials* are any radmaterials that do not qualify as special form; they can be solid waste material in a plastic bag, liquid material in a bottle within a metal container, powder in glass, or waste gas in cylinders.[11]

11.5.2 Types of Radwaste Packages

Since radioactive wastes are produced in many different forms and volumes and with a range of specific activities, several different types of packages are used to transport the wastes. The main types of packages are referred to as *limited-quantity, low-specific activity* (LSA), *type A*, and *type B*. In current regulations, limiting values A_1 (for radionuclides in special form) and A_2 (for normal form) specify the maximum activity of the radionuclide that may be transported in a type A package. Table 11.3 gives examples of A_1 and A_2 values for common radionuclides (other nuclides are listed in reference 11). Quantities

Table 11.3 Type A package quantity limits for selected radionuclides

Radionuclide	Element (atomic number)	A_1 (Ci) (special form)	A_2 (Ci) (normal form)
^{14}C	Carbon (6)	1000	60
^{137}Cs	Cesium (55)	30	10
^{99}Mo	Molybdenum (42)	100	20
^{235}U	Uranium (92)	100	0.2
^{226}Ra	Radium (88)	10	0.05
^{201}Pb	Lead (82)	20	20

From A Review of the Department of Transportation Regulations for Transportation of Radmaterials, Research and Special Programs Administration, Materials Transportation Bureau, Department of Transportation, Washington, D.C., Rev. 1983. Reprinted by permission.

exceeding these limits for type A packages require type B packaging. Quantities greater than 3000 times A_1 or A_2 are called *highway-route-controlled quantities* and are subject to additional regulations. Some limited-quantity and LSA materials below the A_1 and A_2 values may be shipped in packages exempted from some requirements for type A packages.

Since the special form materials are generally much less likely to spread contamination in the event of package failure, the A_1 values are generally substantially larger than A_2 values in a given package. For mixtures of radionuclides, rules are specified for determining whether the type A quantity has been exceeded. The rule applied in most cases is the ratio rule, which involves dividing the activity of each radionuclide present by its A_1 or A_2 value and summing the resulting ratios. If the sum is 1.0 or less, the mixture does not exceed a type A quantity. Provisions for the limited-quantity, LSA, type A, type B, and highway-route-controlled quantity in the regulations are all related to A_1 and A_2 values.

Packages for limited quantities, instruments, and articles. For limited-quantity packages and instruments and articles (for both the item limit and package limit), the A_1 and A_2 values are used as a basis for defining the package quantity limits and are excepted from some of the requirements that apply to type A packages, provided they also meet the following conditions:

1. The material must be packed in strong, tight packages that will not leak any of the radmaterial under normal transportation conditions.
2. The radiation level at any point on the external surface of the package cannot exceed 0.5 mrem/hr.
3. The external surface of the package must be free of significant removable contamination.
4. For instruments or articles, the radiation level at 4 in. from any point on the surface of the unpacked instrument or article may not exceed 10 mrem/hr.
5. A prescribed description of the contents is given on a document that is in or on the package or forwarded with it.

The U.S. Postal Service specifies that mailable amounts of radioactive materials can be no more than 1/10 of the values listed in the DOT regulations.

Packages for LSA materials. LSA materials are those that present a relatively low hazard as a result of their radioactive concentration. Some of these materials are listed by name, e.g., uranium ore and concentrates. Other radmaterials must meet certain limitations related to their radioactive concentration; e.g., tritium oxide in aqueous solutions (trititated water) cannot exceed 5.0 μCi/ml. The allowable radioactive concentration for various LSA materials is related to the A_2 values of the radionuclides present as follows:

A_2 (Ci)	Maximum allowable activity per gram of material (μCi)
$A_2 < 0.05$	0.0001
$0.05 < A_2 < 1.0$	0.005
$A_2 > 1.0$	0.3

LSA materials can be transported by nonexclusive shipments in essentially type A packages (i.e., must survive the physical tests such as the drop and compression tests for type A

FIBERBOARD BOX   WOODEN BOX   STEEL DRUM

Figure 11.2   Typical type A packages. Package must withstand normal conditions (173.465) of transport only without loss or disposal of the radioactive contents. From A Review of the Department of Transportation Regulations for Transportation of Radmaterials, Research and Special Programs Administration, Materials Transportation Bureau, Department of Transportation, Washington, D.C., Rev. 1983. Reprinted by permission.

packages) or by exclusive use of strong, tight packages (i.e., no specific test requirements but a performance criterion of no release of radioactive content during transportation must be met).

Type A packaging. Figure 11.2 illustrates the typical type A packaging, which must be designed in accordance with the applicable general packaging requirements as prescribed in the DOT regulations[11] and must be adequate to prevent the loss or dispersal of its radioactive contents and to maintain its shielding properties under normal conditions of transport. Typically, the type A packaging prescribed in the regulations is the performance-based DOT Specification 7A,[11] for which individual shippers must make their own assessment and certification of the package design against the performance requirements. Prior specific approval by DOT is not required. Foreign-made type A packages are acceptable internationally, provided they are marked as type A and comply with the requirements of the country of origin.

Type B packaging. In addition to meeting the general packaging requirements and all the performance standards for type A packaging, type B packaging (Figure 11.3) must

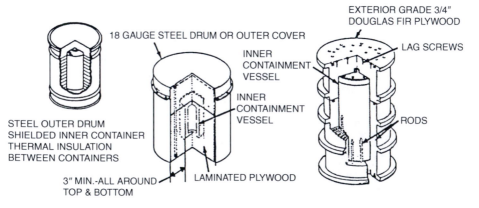

Figure 11.3   TN-8 overweight truck spent fuel cask. From Transportation of radioactive waste by R. M. Burgoyne, in *Radioactive Waste Technology*, edited by A. A. Moghissi et al., p. 423, ASME/ANS, Engineering Center, New York, 1986. Copyright 1986 by American Society of Mechanical Engineers. Reprinted by permission.

withstand certain serious accident damage test conditions; that is, there must be only limited loss of shielding capability and essentially no loss of containment. The performance criteria for type B packages against empirically established, hypothetical accident test conditions of the transport are prescribed in NRC regulations, which include the following:

1. A 30-ft free drop onto an unyielding surface.
2. A puncture test, which is a free drop (over 40 in.) onto a 6-in.-diameter steel pin.
3. Thermal exposure at 1475°F for 30 min.
4. Water immersion for 8 hr immediately after the thermal test (for fissile material packaging only).

Except for a limited number of specification type B packages, as described in the regulation, all type B package designs require prior approval of NRC or DOE.[11]

11.5.3 Packages for Specific Types of Radioactive Wastes

Low-level radioactive waste. Since low-level radioactive waste (LLW) can be in many forms and have a wide range of concentrations of radioactive material, it can be shipped in a variety of packages. LLW is typically shipped in LSA or type A packages, although it is sometimes shipped in type B packages.

LSA packaging Typical radwaste from nuclear power plants shipped in LSA packaging includes contaminated clothes, cleaning cloths, and hardware and has the least stringent packaging requirements of any waste type. The primary packaging requirement is that the package must be a strong, tight container as stated in Section 11.4.2. Figure 11.4 depicts a plywood box for LSA packaging, which is constructed of exterior-grade plywood, joined with resin-coated nails, and reinforced with horizontal and vertical steel banding after loading.[12]

Type A packaging As shown in Figure 11.2, this type of packaging must meet radiation containment and shielding limits after being subjected to the normal conditions of transportation. Type A nuclear power plant waste includes dewatered filter resins, irradiated hardware, and highly contaminated clothing and cleaning cloths.

Type B packaging Type B packaging is used for the shipment of type B solid, nonfissile, irradiated, and contaminated hardware and neutron source components, provided that the applicable certificate of compliance permits the use of the cask for this purpose. Some other type B casks are available for shipping radwastes (Figure 11.3).

High-level radioactive waste. High-level waste (HLW) and spent nuclear fuel are typically shipped in type B packages. Shipping casks for spent nuclear fuel are being used frequently in other countries, and the DOE is developing a fleet of spent fuel shipping casks for use in the United States. These casks will be discussed in Section 11.5.

Fissile radioactive materials Shippers of fissile radioactive material must take into account packaging and shipping requirements to ensure the absence of nuclear criticality. The design of such packaging, the transport index (TI) to be assigned, and any special

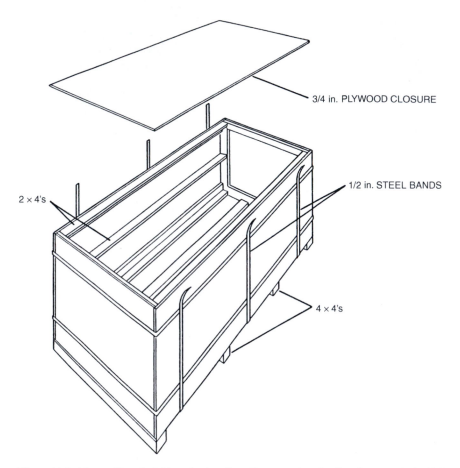

Figure 11.4   Plywood box for LSA packaging. From Transportation of radioactive waste by R. M. Burgoyne, in *Radioactive Waste Technology*, edited by A. A. Moghissi et al., p. 423, ASME/ANS, Engineering Center, New York, 1986. Copyright 1986 by American Society of Mechanical Engineers. Reprinted by permission.

procedures for packaging are all covered in both DOT and NRC regulations, which specify that the packaging must be ensured against nuclear criticality under both normal and hypothetical accident test conditions and prevented from loss of contents in transportation. (The TI is designed to ensure criticality safety by limiting the amount of fissile material in one location or limiting the radiation level, in which case TI equals the highest dose rate at 1 m from any accessible exterior surface of the package.) Fissile materials are classified into three fissile classes with different transport indexes and shipment controls.

Highway-route-controlled quantities Certain quantities of radmaterials, called highway-route-controlled quantities, are subject to additional controls during transportation. The quantity is defined as an amount of material in a single package that exceeds either (1) 3000 times the A_1 quantity for special form material, (2) 3000 times the A_2 quantity for normal form material, or (3) 30,000 Ci, whichever is least. Such packages are subject to specific routing controls that apply to the highway carrier. The carrier must operate on preferred routes that are in conformance with regulations and must report to the shipper the route used in making the shipment. The shipper, in turn, is required to report the routing information to the Materials Transportation Bureau.

11.6 HIGH-LEVEL WASTE AND SPENT FUEL TRANSPORTATION

The Nuclear Waste Policy Act (NWPA) of 1982 created the Office of Civilian Radioactive Waste Management (OCRWM) to oversee management and disposal of spent fuel and high-level waste. OCRWM's responsibilities include transportation of spent fuel and HLW. Some spent fuel was transported in the United States prior to 1982 and spent fuel is routinely transported in many countries around the world. Several type B packages for spent fuel, called shipping casks, already exist. This section describes OCRWM's transportation program, some typical existing shipping casks, and plans for future shipping cask development.

11.6.1 OCRWM Transportation Plan

The Office of Civilian Radioactive Waste Management (OCRWM) was created within the Department of Energy by the Nuclear Waste Policy Act of 1982 to carry out the activities assigned by the Act, including the transport of SNF and HLW to the repository. Initially, the transportation plan consisted of two parts, and each part was addressed in a separate document. One document, The Transportation Institutional Plan, dealt with institutional issues such as interfacing with various state and federal agencies, and the other, The Transportation Business Plan, dealt with the design, purchase, and operation of a fleet of casks for transporting SNF and HLW. In 1986, these two documents were combined, and OCRWM began to focus on design and development of shipping casks.

Contracts were put in place to develop five different high-capacity casks in 1988. In 1991, two cask designs were selected for additional work. There was a 100-ton rail/barge cask built by Babcock & Wilcox and a 80,000-lb legal weight truck cask built by General Atomics. By 1994, OCRWM had refocused its transportation program. It canceled work on the rail/barge cask and began to support development of a multipurpose canister (MPC) to be used with the truck cask. The MPC is a strong metal container in which SNF will be sealed. The sealed MPC is then placed in the appropriate overpack (or cask) for storage, transportation, or disposal. MPCs help to standardize the waste handling process and reduce the number of times spent fuel assemblies must be handled. In 1998, OCRWM suspended work on the MPC.

In accordance with the Nuclear Waste Policy Act, OCRWM has always emphasized the use of private contractors to carry out its transportation responsibilities. In 1998, OCRWM posted a draft Request for Proposals (RFP) for services and equipment related to transportation of spent nuclear fuel. Shortly after the posting, work on the RFP was suspended until a repository siting decision is made. That decision was expected to be made in 2001.

A short excerpt from the draft Request for Proposals gives some insight into DOE's plans for a transportation system in 1998:[13,14]

Acquisition Details
- Competitive, fixed-price type or fixed-rate type contracting is planned.
- Multiple awards will be achieved by dividing the country into four regions and contracting in the initial phase for one or more contractor(s) to service each of the four regions that co-incide with the four NRC regions.
- After the initial contract phase, no contractor, or Regional Servicing Contractor (RSC), would be awarded more than two regional servicing contracts.
- RSCs would:
  — Work with utilities to determine the best way to service a site and integrate site planning into a regional servicing plan
  — Identify proposed transportation routes and submit approved routes to NRC
  — Provide all hardware, including transportation cases, canisters, and ancillary lifting equipment
  — Comply with applicable NRC, DOT, State, local and tribal regulations
  — Interact with those State, local and tribal governments as appropriate
  — Provide all waste acceptance and transportation services necessary to move spent nuclear fuel from the Purchaser's sites to the Federal facility.
- Contracts will be accomplished in three phases:
  Phase A: development of site specific and regional servicing plans, followed by authorization of one RSC per region to continue work into Phase B.
  Phase B: mobilization of transportation services, finalization of transportation routes and training, acquisition or transportation hardware (through either lease or purchase)
  Phase C: actual performance of waste acceptance activities and movement of spent fuel once a Federal facility becomes operational.

11.6.2 Existing Shipping Casks

The shipping cask is the packaging* used for the transport of spent fuel from nuclear power reactors, which requires type B packaging for transport. These shipping casks are massive and reusable vessels which are manufactured in size classes according to their primary mode of transportation, truck or rail. Table 11.4 shows the current and near-term spent fuel shipping casks, which include truck and rail casks.[12] The cask parameters are listed in Table 11.5.

The *NLI-1/2 shipping cask* is a legal-weight truck cask that has a payload capacity of one PWR or two BWR fuel assemblies and was licensed in 1973 (see Figure 3.22 in Chapter 3). It is designed for double containment using an inner container and gasketed closure independent of the main cask body and its closure.[12,15]

The *TN-8* and *TN-9 spent fuel casks* were initially designed for use in Europe and subsequently licensed for use in the United States in 1975. Essentially the same casks, their difference is in their fuel capacity. The TN-8 cask has three compartments in the inner cavity for PWR fuel assemblies (Figure 11.5), while TN-9 has a capacity for seven BWR fuel assemblies. They can be used for rail, water, or truck transport, although an overweight permit is required if used for truck transport.[12,16]

---

*Packaging means the assembly of components necessary to ensure compliance with the packaging requirements as set forth by the federal regulations and is distinguished from the package of radmaterials, which contains the radioactive contents.

Table 11.4 Current and near-term U.S. spent fuel shipping casks

Shipping cask identity	Licensed use	Number available	Owner/operator developer
Truck casks			
NLI-1/2[a]	LWR SF	5	Nuclear Assurance Corp.
TN-8/TN-9[a]	LWR SF	2/2[b]	Transnuclear, Inc.
NAC-1/NFS-4[c]	LWR SF	6	Nuclear Assurance Corp.
FSV-1	HTGR SF	3	GA Technologies Inc.
Rail casks			
IF-300 (4)	LWR SF	4	General Electric Co.
NLI-10/24 (4)	LWR SF	2	Nuclear Assurance Corp.
TN-12[d]	—	0	Transnuclear, Inc.

[a] Casks presently certified by the NRC and listed by DOE as present-generation spent fuel casks.[12]
[b] There are two each of the TN-8 and TN-9 casks in the United States.
[c] The NAC-1 cask was previously designated NFS-4.
[d] The TN-12 cask operates in Europe but is not yet licensed for use in the United States.
From Transportation of radioactive waste by R. M. Burgoyne, in *Radioactive Waste Technology*, edited by A. A. Moghissi et al., ASME/ANS, Engineering Center, New York, 1986, p. 423. Copyright 1986 by American Society of Mechanical Engineers. Reprinted by permission.

The *NAC-1/NFS-4 cask* also has a design capacity of one PWR or two BWR fuel assemblies; it was licensed in 1972 (Figure 11.6). As shown in Table 11.5, this cask uses water as the cavity coolant, but uses air if the fuel has cooled more than 2.5 years. This cask is the only one without a redundant lift.[12,17]

The *IF 300 rail shipping cask* was designed for normal shipment by railroad flatcar and licensed in 1973. The cask may be shipped for short distances by a special over-weight truck to service reactors that do not have rail facilities. It is unique in that it exclusively uses metal fins for impact protection (Figure 11.7).[12,18]

The *NL-10/24 cask* for rail shipment of 10 PWR or 24 BWR fuel assemblies was licensed in 1976 (Figure 11.8). Two materials are used for gamma shielding (lead along the length and depleted uranium at the ends) and for neutron shielding (water jacket along the length and Ricorad PPV-C, a silver-based alloy, at the ends).[12,19]

11.6.3 Plans for Future Shipping Cask Development

Advanced spent fuel casks. Spent fuel storage/transport casks. Section 11.5.2 provides brief descriptions of some casks that have been licensed for transportation of spent nuclear fuel in the United States. Most of those casks were designed and licensed in the 1970s when numerous shipments of spent nuclear fuel were thought to be imminent. But neither an operating repository nor a central spent fuel storage facility has been opened, and most spent fuel is being kept at the reactors where it was generated. Many of the utilities' spent fuel storage pools are full, and the fuel is now being stored in dry casks at the reactor site. As a result, vendors are focusing on design and production of dry storage casks. Vendors and utilities alike recognize that the spent fuel will eventually be transported to a central facility for treatment, storage, or

Table 11.5 Spent fuel shipping cask parameters

Cask model	Weight (tons)[b]		Length (in.)		Diameter (in.)		Capacity (intact assemblies)		Cavity coolant	Shielding material		Transport mode
	Empty	Loaded	Overall[b]	Cavity	Overall	Cavity	PWR	BWR		Gamma	Neutron	
NLI-1/2	22	23	193	180	45	13.375	1	2	Helium	SS–B–DU[a]	Water	LWT[a]
TN-8	37	39	188	168	67	3×9.0^d	3	—	Air	SS–Pb	Resin	OWT[a]
TN-9	37	39	198	178	67	7×5.9^d	—	7	Air	SS–Pb	Resin	OWT[a]
NAC-1/ NFS-4	24	25	200	178	50	13.5	1	2	Water[e]	SS–Pb	Water	LWT[a]
FSV-1	22–23	23.5–25.5	212	190	31[d]	17.76	1	3	Air	SS–DU[a]	Resin[f]	LWT[a]
IF-300	63	68	184 + 15 in. lid	169	64[g]	37.5	7	—	Water[e]	SS–DU[a]	Water	Rail, intermodal
	65	70	184 + 15¾ in. lid	180	64[g]	37.5	—	18	Water[e]	SS–DU[a]	Water	Skid (OWT)
NLI-10/24	90	97.5	204	179.5	88[h]	45	10	24	Helium	SS–Pb–DU	Water	Rail
TN-12	87	97	208	181	90	48	12	32	Air	Steel	Resin	Rail

[a]SS, Stainless steel; DU, Depleted Uranium; LWT, Legal Weight Truck; OWT, Overweight Truck.

[b]Cask weight excludes impact limiters, includes neutron shielding and intact fuel.

[c]Length without impact limiters.

[d]36 in. with neutron shield in place.

[e]Air, if fuel has cooled >2.5 years.

[f]Removable shield. Use only if neutron dose rate is significant.

[g]Add 13 in. to one sector for expansion tanks.

[h]Add 4 in. to one radius (by 10 in. wide) for tie–down lugs.

From Transportation of radioactive waste by R. M. Burgoyne, in *Radioactive Waste Technology*, edited by A. A. Moghissi et al., ASME/ANS, Engineering Center, New York, 1986, p. 423. Copyright 1986 by American Society of Mechanical Engineers. Reprinted by permission.

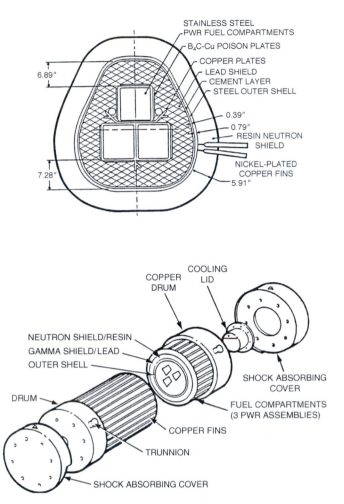

Figure 11.5 Typical type B packages. Package must withstand both normal (173.465) and accident (10CFR Part 71) test conditions without loss of contents. From A Review of the Department of Transportation Regulations for Transportation of Radmaterials, Research and Special Programs Administration, Materials Transportation Bureau, Department of Transportation, Washington, D.C., Rev. 1983. Reprinted by permission.

disposal, and it would be very convenient if the storage casks could also be used for transportation.

Obviously, this approach would require the cask to satisfy both sets of regulations, one for storage[20] and one for transportation,[11,21] even though a cask designed to satisfy the transportation regulations would generally satisfy the storage requirements, too. Since type B transportation packaging is licensed for a period of only 5 years, there is some question whether a cask used to store spent fuel for a number of years would still be licensable for transportation in the presence of potential regulatory changes.[22,23] Several casks belong to the group that are licensed for storage but may also be usable for transportation later: REA 2023,[24] the CASTOR cask,[25] and the TN-1300 cask.[26] Figures 3.24 and 3.15 depict

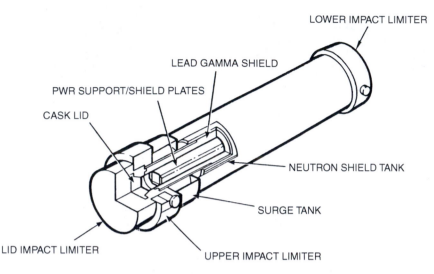

Figure 11.6 NAC-1/NFS-4 spent fuel cask. From Transportation of radioactive waste by R. M. Burgo, in *Radioactive Waste Technology*, edited by A. A. Moghissi et al., p. 423, ASME/ANS, Engineering Center, New York, 1986. Copyright 1986 by American Society of Mechanical Engineers. Reprinted by permission.

these casks. On the REA 2023 cask, lead and steel are used for gamma shielding and a 6.0-in.-thick 48/52% mixture of borated water/ethylene glycol is provided for neutron shielding. For transportation, neutron shielding at the top and bottom of the cask must be added, and a solid-type shielding is preferable to liquid. Impact limiters must be added to the cask package to meet the requirements imposed by the hypothetical accident conditions. The CASTOR cask is a family of dual-purpose casks being developed in Germany.

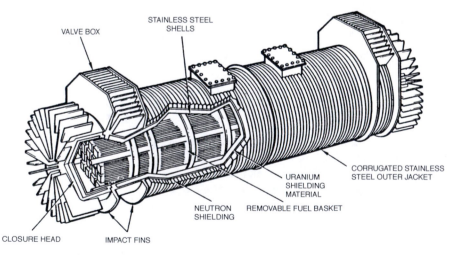

Figure 11.7 IF-300 spent fuel rail shipping cask. From Transportation of radioactive waste by R. M. Burgoyne, in *Radioactive Waste Technology*, edited by A. A. Moghissi et al., p. 423, ASME/ANS, Engineering Center, New York, 1986. Copyright 1986 by American Society of Mechanical Engineers. Reprinted by permission.

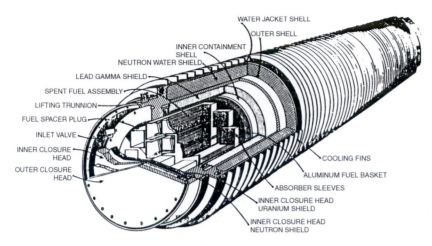

WATER JACKET SHELL
OUTER SHELL
INNER CONTAINMENT SHELL
NEUTRON WATER SHIELD
LEAD GAMMA SHIELD
SPENT FUEL ASSEMBLY
LIFTING TRUNNION
FUEL SPACER PLUG
INLET VALVE
INNER CLOSURE HEAD
OUTER CLOSURE HEAD
COOLING FINS
ALUMINUM FUEL BASKET
ABSORBER SLEEVES
INNER CLOSURE HEAD URANIUM SHIELD
INNER CLOSURE HEAD NEUTRON SHIELD

Figure 11.8 NLI-10/24 rail cask. From Transportation of radioactive waste by R. M. Burgoyne, in *Radioactive Waste Technology*, edited by A. A. Moghissi et al., p. 423, ASME/ANS, Engineering Center, New York, 1986. Copyright 1986 by American Society of Mechanical Engineers. Reprinted by permission.

11.6.4 Other Equipment in Cask Systems and Transportation Operations Systems

In addition to casks, other equipment requirements to support transportation operations are substantial:[27]

1. Transport vehicles: tractor/trailer, engine/rail car, tug/barge, as well as escort vehicles.
2. Ancillary equipment: fuel baskets/spacers/sleeves; impact limiters/structures; tie-down equipment/turning fixture; personnel barrier; cooling systems (as applicable); instrumentation for measuring temperature, gamma/neutron dose rates, and so forth; special tools and closurehead stands.
3. Cask-handling equipment: In-transit lifting/uprighting equipment in the event of an incident, slings, strong-backs and lifting yokes, crane hook adapators/spacers, off-site intermodal transfer equipment, emergency response equipment.
4. Other systems to support transportation operations: facility communications equipment, tracking system equipment, radiologic safety and security equipment, cask maintenance and fleet maintenance equipment, radwaste treatment/packaging equipment, etc.

11.7 SHIPPER AND CARRIER RESPONSIBILITIES

The "shipper" is the company or government agency that owns the radioactive material being transported. The "carrier" is the company that owns and operates the vehicle on which the material is being carried. Together, the shipper and carrier are responsible for the safe transportation of the radioactive material. Specific responsibilities of the shipper and the carrier are prescribed in the regulations and outlined in the following sections.

11.7.1 Shipper Responsibilities[9]

The shipper must (1) select the proper packaging for the specific contents, (2) consider the radiation level limits, (3) consider the contamination limits, and (4) label correctly. In addition, the shipper must ensure compliance with the following:

Package markings: The outside of the package must be marked with the (1) proper shipping name, (2) identification number as shown in the list of hazardous materials, and (3) appropriate specification number or certificate number, as applicable.

Shipping papers: As with other hazardous materials shipments, the following information must be included on the shipping papers:

1. Requirements: (a) Proper shipping name, (b) hazard class (unless contained in the shipping name); (c) identification number; (d) net quantity of material by weight or volume, or measured in curies, and the TI; (e) radionuclide(s) contained in package; (f) physical and chemical form of material, or special form; (g) activity in curies, millicuries, or microcuries; (h) category of radioactive labels, TI if labeled Radioactive Yellow-II or Yellow-III; (i) special information for fissile radmaterial; (j) identification markings shown on the package; and (k) other information as required by the mode of transport or subsidiary hazard of the material.
2. Other information. Other descriptive information is allowed, such as the functional descriptions of the product.
3. Exceptions for limited quantity packages, instruments or articles, and articles manufactured from natural or depleted uranium or natural thorium. These items must be documented for transport by including a notice in, on, or forwarded with the package that includes the name and address of the cosigner or cosignee and a specific statement selected on the basis of the proper shipping name for the package (e.g., "This package conforms to the conditions and limitations specified in 49CFR173.422 for excepted radioactive material, instrument, UN2911.").

Shipper's certification: This certification, signed by the shipper, must appear on the paper that lists the required shipping description.

Security seal: The outside of each type A or B radmaterials package must incorporate a feature, such as a seal, that is not readily breakable and that, while intact, will be evidence that the package has not been illicitly opened.

Minimum dimension: The smallest outside dimension of any radmaterial package (other than excepted quantities) must be 4 in.

Liquid packaging provision: Liquid radmaterial must be packaged in a leak-resistant inner container, which must be adequate to prevent loss or dispersal of the radioactive contents from the inner container if the package is subjected to the 30-foot drop test, and enough absorbent material must be provided to absorb at least twice the volume of the radioactive liquid contents.

Surface temperature of package: Maximum surface temperature limits on packages, resulting from radioactive thermal decay energy of the contents, is either 122°F or, in the case of exclusive use shipments, 180°F.[28]

Quality control requirements: Certain quality control requirements are prescribed in the regulations for the construction of radmaterials packaging and before each shipment of a package.[9]

11.7.2 Carrier Responsibilities

While most regulatory requirements for the assurance of safety in the transport of radmaterials are directed toward safety through proper packaging, and thus apply to the shipper, the following responsibilities belong to the carrier.

Shipping papers and certification: Carriers may not knowingly accept transport packages of radmaterials that have not been described and certified by the shipper pursuant to the regulation. Carriers may prepare and carry with the shipments appropriate bills of lading, waybills, etc., based on the information derived from the shippers' shipping papers. For shipments by vessel, a dangerous cargo manifest or storage plan is also required.

Placarding: The RADIOACTIVE placard must be applied to the transport vehicle (rail or highway) if any radmaterial package on board bears a Radioactive Yellow-III label. Vehicles transporting any package that contains a highway-route-controlled quantity must also display the square white background.

Radiation exposure control by maximum total transportation index (TI) versus distance: For any group of yellow-labeled packages, the carrier must assure that the total TI does not exceed 50 and that such groups of yellow-labeled packages are kept separated from undeveloped film shipments and areas normally occupied by persons.

Reporting of incidents: The carrier must assure that DOT and the shipper are notified in the event of fire, breakage, spillage, or suspected radioactive contamination involving a shipment of radmaterial. To obtain technical assistance in radiologic monitoring, carriers may call on the services of local or state radiologic authorities. Federal assistance in resolving a radiologic emergency may be provided if requested by state or local authorities. For this and other security reasons, the carrier should establish a communication network and coordinate with the local law enforcement.

Routing control: The carrier is responsible for giving advance notification on the routing and receiving inspection and monitoring by the NRC or designated state agency.

11.7.3 Emergency Response

As spent nuclear fuel is transported across the country, it will pass through many states, counties, cities, and townships. Public safety officials in each of those jurisdictions will need to be prepared to respond to any accidents or emergencies related to the shipment of spent nuclear fuel. In addition, they will probably be asked to provide information to the public about the normal transportation of SNF. Under Section 180 (c) of the Nuclear Waste Policy Act of 1982, the Department of Energy is required to provide technical assistance and funds to train public safety officials in states and Indian reservations through which high-level nuclear waste is expected to be transported. The assistance is to prepare officials for both routine transportation and emergency response. DOE expects to begin the assistance 3–5 years before the first HLW shipments are made. However, in 1992, OCRWM prepared two documents outlining its approach to implementing Section 180 (c): Strategy for OCRWM to Provide Training Assistance to State, Tribal, and Local Governments (November 1992, DOE/RW-0374P), and Preliminary Draft Options for Providing Technical Assistance and Funding under Section 180 (c) of the Nuclear Waste Policy Act, as Amended (November 1992).

The Department of Energy consults with several government agencies and other organizations of emergency response. Government agencies include the Nuclear Regulatory Commission, Department of Transportation, and Federal Emergency Management Agency. Other organizations with which OCRWM plans to communicate about emergency response include the Commercial Vehicle Safety Alliance, the Conference of Radiation Control Program Directors, the League of Women Voters Education Fund, the National Association of Regulatory Utility Commissioners, and the National Conference of State Legislatures.[29]

11.8 RISK ANALYSES FOR TRANSPORTATION

All activities have some associated risks, including the transport of spent fuel and radwaste. To conduct a sensible transport operation and to be able to deal reasonably with questions from the public (see Section 11.2), it is necessary to be fully aware of the magnitude of these risks.[3] Risks from transportation can be considered under two conditions: normal operations and accident conditions.

11.8.1 Risks under Normal Transportation Operations

Normal operations are those that do not involve accidents; hence the only hazard arising from these operations is the radiation exposure resulting from the contents and from any contamination on the outside of the package. Those exposed are the transport workers and the public along the route. Because of the stringent shielding requirements for the packages, the radiation exposures are very small. Calculations were made by Rogers and Associates Engineering corporation (RAE) to provide rapid and generic estimates of risks to individuals and populations from transportation of spent fuel to an HLW repository or to temporary storage. The projected maximum individual exposures from normal spent fuel transport by truck cask and rail cask are shown in Tables 11.6 and 11.7.[30] A preliminary analysis of the risks arising from the transportation of spent fuel and HLW to five potential repository sites (for 26 years) was reported to yield probably upper limits of latent cancer fatalities (LCF) between 13 and 26 (if all by rail, corresponding to about 15,000 shipments) or between 9 and 13 (if all by truck, with 90,000 shipments). In the same period of 26 years, the LCF from the background radiation in the United States could be as high as 120,000.[3,31] A survey in the United Kingdom by the National Radiological Protection Board (NRPB) indicated that the collective radiation exposure to the public in the United Kingdom from gamma radiation due to the transport of Magnox fuel (1000 MTU/year) amounts to about 2 person-rem/year,[32] and the annual collective

Table 11.6 Projected maximum individual exposures from normal spent fuel transport by truck cask[a]

Description (service or activity)	Distance to center of cask (m)	Exposure time	Maximum dose rate and total dose
Caravan			
Passengers in vehicles traveling in adjacent lanes in the same direction as cask vehicle	10	30 min	40 μrem/min, 1 mrem
Traffic obstruction			
Passengers in stopped vehicles in lanes adjacent to the cask vehicle which have stopped due to traffic obstruction	5	30 min	100 μrem/min, 3 mrem
Residents and pedestrians			
Slow transit (due to traffic control devices through area with pedestrians)	6	6 min	70 μrem/min, 0.4 mrem
Truck stop for driver's rest Exposures to residents and passers-by	40	8 hr[b]	6 μrem/min, 3 mrem
Slow transit through area with residents (homes, businesses, etc.)	15	6 min	20 μrem/min, 0.1 mrem
Truck servicing			
Refueling (100-gal capacity)	7[c]		60 μrem/min
One nozzle from 1 pump		40 min	2 mrem
Two nozzles from 1 pump		20 min	1 mrem
Load inspection/enforcement	3[d]	12 min	160 μrem/min, 2 mrem
Tire change or repair to cask trailer	5[e]	50 min	100 μrem/min, 5 mrem
State weight scales	5	2 min	80 μrem/min, 0.2 mrem

[a]These exposures should not be multiplied by the expected number of shipments to a repository in an attempt to calculate total exposures to an individual; the same person would probably not be exposed for every shipment, nor would these maximum exposure circumstances necessarily arise during every shipment.

[b]Assumes overnight stop.

[c]At tank.

[d]Near personnel barrier.

[e]Inside tire nearest cask.

From Exposures and Health Effects from Spent Fuel Transportation by G. M. Sandquist, V. C. Rogers, A. A. Sutherland, and G. B. Merrell, RAE-8339/12-1, Rogers & Associates Engineering Corp., Salt Lake City, Utah, 1985. Copyright 1985 Rogers & Associates Engineering Corp. Reprinted by permission.

dose equivalent to all railway workers involved in the transport of spent fuel in the United Kingdom is about 0.5 person-rem,[33] approximately equivalent to the annual collective dose to two people from natural radiation. The corresponding dose to all transport workers involved in the movement of spent fuel *and* LLW by road was less than 9 person-rem. Similar low exposure estimates have been reported from France.[3] In ships, the most realistic assessment of the radiation dose accumulated by the most highly exposed individual is less than 30 mrem and the average dose is 8 mrem, while the average annual natural radiation dose per person in the United Kingdom is about 200 mrem.[3]

Table 11.7 Projected maximum individual exposures from normal spent fuel transport by rail cask[a]

Description (service or activity)	Distance to center of cask (m)	Exposure time	Maximum dose rate and total dose
Caravan			
Passengers in rail cars or highway, vehicles traveling in same direction and vicinity as cask vehicle	20	10 min	30 μrem/min, 0.3 mrem
Traffic obstruction			
Exposures to persons in vicinity of stopped/slowed cask vehicle due to rail traffic obstruction	6	25 min	100 μrem/min, 2 mrem
Residents and pedestrians			
Slow transit (through station or due to traffic control devices) through area with pedestrians	8	10 min	70 μrem/min, 0.7 mrem
Slow transit through area with residents (homes, businesses, etc.)	20	10 min	30 μrem/min, 0.3 mrem
Train stop for crew's personal needs (food, crew change, first aid, etc.)	50	2 hr	50 μrem/min, 0.6 mrem
Train servicing			
Engine refueling, car changes, train maintenance, etc.	10	2 hr	50 μrem/min, 6 mrem
Cask inspection/enforcement by train, state, or federal officials	3	10 min	200 μrem/min, 2 mrem
Cask car coupler inspection/maintenance	9	20 min	70 μrem/min, 1 mrem
Axle, wheel, or brake inspection/lubrication/maintenance on cask car	7	30 min	90 μrem/min, 3 mrem

[a] These exposures should not be multiplied by the expected number of shipments to a repository in an attempt to calculate total exposures to an individual; the same person would probably not be exposed for every shipment, nor would these maximum exposure circumstances necessarily arise during every shipment.

From Exposures and Health Effects from Spent Fuel Transportation by G. M. Sandquist, V. C. Rogers, A. A. Sutherland, and G. B. Merrell, RAE-8339/12-1, Rogers & Associates Engineering Corp., Salt Lake City, Utah, 1985. Copyright 1985 by Rogers & Associates Engineering Corp. Reprinted by permission.

11.8.2 Risks under Transportation Accidents

The events usually regarded as the precursors to serious accidents to packages are impact, fire, and immersion in water or some combination of these events. As noted before, no accident has been reported worldwide that has resulted in a release of radioactive material from a cask transporting spent fuel.[3,30] Furthermore, no release of radmaterial has occurred from any package designed as an accident-resistant package, as stipulated in both the U.S. and IAEA regulations. These standards provide a higher degree of safety to the public than most, if not all, other transport of hazardous materials.[3] This excellent safety record, however, provides no historical data to confirm theoretical models and controlled field and laboratory experiments, but the record does demonstrate that

Table 11.8 Maximum individual radiation dose estimates for rail cask accidents

	Dose (mrem)[a]			
Accident class	Inhalation	Plume gamma	Ground gamma	Dust inhalation
Impact	179	10.7	12.3	0.0001
Impact and burst	6130	71.1	90.9	0.004
Impact, burst, and oxidation	8950	547	707	0.0006

[a] Maximum individual dose occurs about 70 m downwind of the release point and assumes that the individual remains at this location for the duration of the passage of the plume of nuclides that are released.

From Exposures and Health Effects from Spent Fuel Transportation by G. M. Sandquist, V. C. Rogers, A. A. Sutherland, and G. B. Merrell, RAE-8339/12-1, Rogers & Associates Engineering Corp., Salt Lake City, Utah, 1985. Copyright 1985 by Rogers & Associates Engineering Corp. Reprinted by permission.

the probability of a cask failure and radmaterial release is very small. This probability is estimated to be no greater than two occurrences in 1 million rail transport accidents.[34] In the referenced study,[30] the radiation exposures and health effects from the worst-case accidents were analyzed, and the following conclusions were reached:

• A person responding to the emergency caused by the severe but credible rail car accident—a severe impact followed by a massive fire fed by large quantities of fuel— could receive a dose of up to 10 rem in a few hours if no protective equipment is worn and no attempt is made to avoid inhalation of radionuclides in the atmosphere. This dose is not unreasonable, considering the circumstances and small probability of occurrence (Table 11.8).

• For the highest population assumed (3860 persons/km^2 in an urban area) to be exposed from such an accident, up to 22 latent health effects (LHE)* might be expected over the succeeding 50 years (Table 11.9). This is compared with 470,000 cancer fatalities that the same population would experience over 50 years from all causes.

• If such an accident occurs beside a reservoir of about 100 acres (containing 3.8 × 10^6 m^3 or 1.34 × 10^8 ft^3 of water), up to 13 LHE could result among the general population that the water quantity would service (about 37 million people) (Table 11.10). This number is based on the assumption that no measure is taken to remove radionuclides resulting from the accident from the water consumed by the population. The same population would experience about 72,000 cancer deaths per year from all causes, using a cancer rate of 0.00194 fatal cancer/year from all other sources.[35] The study by Marsden[31] also considered probable accidents resulting from the transportation of spent fuel and HLW to the various potential repositories. During the 26 years of operations, it was estimated that if all the movements were by rail, there would be a maximum of two fatalities; if all were by truck, there would be 15–38 accident fatalities, depending on the location of the repository. These numbers can be compared with present general accident rates in the United States: about 32,000 would die from train accidents and 65,000 would die from truck accidents in a

*LHE estimates are based on 1 person-rem = 2 × 10^{-4} LHE. It is defined as an early cancer death of an exposed person or a serious genetic health problem in two generations after those exposed. About half of the LHE would be cancers to the exposed generation and the other half would be genetic health problems.

Table 11.9 Fifty-year population dose estimates for spent fuel rail cask accidents[a] with no cleanup of deposited nuclides

Accident class	Urban area (3860 people/km^2)				Rural area (6 people/km^2)			
	Inhalation	Plume gamma	Ground gamma	Total	Inhalation	Plume gamma	Ground gamma	Total
Impact								
Dose (person-rem)	3.09	0.33	936	939	0.005	0.0005	1.45	1.45
Latent health effects[b]				0.19				0.00029
Impact and burst								
Dose (person-rem)	106	2.23	13,400	13,500	0.16	0.0034	20.8	21
Latent health effects[b]				2.7				0.0042
Impact, burst, and oxidation								
Dose (person-rem)	154	17.2	112,000	112,000	0.24	0.027	174	174
Latent health effects[b]				22				0.035

[a]The ground gamma dose is what would be received if each member of the population stayed at the same location for 50 years. The inhalation dose is a 50-year dose commitment from inhalation of the passing plume. Doses are for the population within 80 km of the release point. It is assumed that there is no cleanup of deposited nuclides and that no other measures are used to reduce radiation exposures.

[b]Based on 1 person-rem = 2×10^{-4} latent health effects. A latent health effect here is defined as an early cancer death of an exposed person or a serious genetic health problem in the two generations after those exposed. About half of the latent health effects are expected to be cancers and the rest genetic health problems.

From Exposures and Health Effects from Spent Fuel Transportation by G. M. Sandquist, V. C. Rogers, A. A. Sutherland, and G. B. Merrell, RAE-8339/12-1, Rogers & Associates Engineering Corp., Salt Lake City, Utah, 1985. Copyright 1985 by Rogers & Associates Engineering Corp. Reprinted by permission.

Table 11.10 Population radiation exposure from water ingestion for severe but credible spent fuel rail cask accidents

Accident class	Total release from rail cask (Ci)[a]	Population dose effects from water ingestion
Impact	8.07	182 person-rem 0.036 LHE[b]
Impact and burst	153	6,870 person-rem 1.4 LHE[b]
Impact, burst, and oxidation	1379	63,000 person-rem 12.6 LHE[b]

[a] The noble gas ^{85}Kr is omitted because of its negligible uptake by a surface water body.
[b] Latent health effect (LHE) estimates are based on 1 person-rem = 2×10^{-4} LHE.
   From Exposures and Health Effects from Spent Fuel Transportation by G. M. Sandquist, V. C. Rogers, A. A. Sutherland, and G. B. Merrell, RAE-8339/12-1, Rogers & Associates Engineering Corp., Salt Lake City, Utah, 1985. Copyright 1985 by Rogers & Associates Engineering Corp. Reprinted by permission.

26-year period. The Womack Study in the United Kingdom[36] of rail transport of fuel at 30 MT/year from the proposed PWR at Sizewell to Sallafield (635 km) estimated that the frequencies of severe impact and of severe fire are 1.5×10^{-9}/year and 10^{-9}/year, respectively. These very low frequencies are not those of a major breach in the containment of the cask, but of a minor breach such as seal leakage. Considering the consequences of a postulated severe fire (2 hr at 1000°C, or 1832°F) with the accident happening in London,[37] the NRPB indicated that, even if no countermeasures were taken, the probability of fatal cancer to the closest individuals is about 2.5×10^{-4}, which is about 1/1000 of the natural probability of fatal cancer, with two fatal cancers resulting. Risk analyses of the transport of HLW by train have also been made in Germany[38] for 1500 MTHM in glass transported some 360 km each year. The exposure amounted to 2×10^{-7} rem/year for 41-year-old HLW or 6×10^{-7} rem/year for 5-year-old HLW. The most effective method of reducing the risk was by an adequate emergency response involving decontamination of the area involved. In sea transport the most severe hazard is a ship collision, with a fire on board. The frequency of such an accident has been calculated to be about 3×10^{-5}/year. The maximum collective dose, in the event of such an accident near a major city, has been estimated to be 3×10^4 person-rem. Thus the risk is about 1 person-rem/year. Even in the event of such an improbable accident, the major immediate danger is likely to be from fire or drowning and to involve the ship's crew.[3] Based on the studies mentioned above, it was concluded justifiably in reference 3 that there is no technical reason why the industry cannot expand the volume and scope of its radwaste transport activities to meet the increased needs of the future, and that it would be necessary to maintain a significant public relations effort so that the public and decision makers support its future development.

11.8.3 Response to Hypothetical Accident Conditions

Despite the safety record and the strict regulations to protect public health and safety by subjecting the packages used for spent fuel or HLW shipments to the hypothetical accident conditions test, the adequacy of the tests remains a subject of public concern. The specified tests[21] are adequate to ensure high-integrity packaging and are generally recognized to be rigorous. The extent to which the tests simulate real accidents, however, had not been fully developed. To address these concerns, the NRC initiated a study, which became known as the Modal Study, to evaluate the safety of spent fuel shipments in terms of severe accidents that actually occurred in nonnuclear shipments in surface transport modes.[39,40] A generic reference package design was used in this study. Its response to the range of forces encountered in real accidents was expressed in measured strain on the inner steel shell, and its response to thermal input was measured by the temperature of the midline of the lead shielding. The degree of package response was determined by computer models, which defined "package response states" for the various combinations of strain and temperature. The radiologic hazard of each road or rail package (i.e., the amount of radmaterial released or increase in external radiation level, if any) was estimated for each response state. Results of the study indicate that if the representative (reference) truck casks were involved in 1000 accidents, the forces involved in 994 of the accidents would not exceed either the strain or thermal responses corresponding to the hypothetical accident conditions. For these 99.4% of accident cases damage would be superficial and any release of radmaterial or increase in radiation exposure levels would be well within NRC acceptance criteria. Of the 6 remaining accident cases, 4 would cause minor functional damage to the cask; however, the analyses indicate that the radiologic hazard of these accidents would be small, still well within NRC's acceptance criteria. In 2 of the 1000 accidents, cask structural damage could be significant, although rupture of the cask's containment shield would not be expected, and the thermal damage could include some lead shield melting. The analysis predicts that the radiologic hazard of these 2 accidents would only slightly exceed the regulatory acceptance values. Thus the Modal Study clarifies the level of safety provided under real accident conditions by spent fuel packages designed to current standards and practices.[39]

Calculations of potential health risks resulting from the accidents and releases postulated by the NRC Modal Study provide further capabilities for route-specific risk analyses.[41] State-level data bases developed for accident rates, farmland utilization (i.e., percentage land in farms), and agricultural productivity are incorporated in the risk analysis. To obtain the overall radiologic health risk from spent fuel transportation for a specific route, the risk is integrated by the segments of the route that traverse population zones in the affected states. The results also indicate that risk from rail transport can be greater than that from truck transport, although truck transport has consistently higher accident rates. This is primarily attributed to the fact that the spent fuel inventory is about six times greater per rail cask than per truck cask.[41]

In another study,[42] the risks were compared in terms of years before a serious accident (i.e., with a potential to release radioactive material) in the total system might occur, using statistical information from the Transportation Technology Center (TTC) of Sandia National Laboratories[43] as shown in Table 11.11. The data indicated that only 1 in 1000 of the accidents specified in this table would be severe enough to cause sufficient deformation of the cask that there would be potential for some release of radioactive material.

Table 11.11 Transportation risk data

400,000 truck miles per accident resulting in more than $250 damage
660,000 rail car miles per accident resulting in more than $1500 damage
One in 100 of these are severe enough to approach design conditions
One in 10 of the above could deform the cask so that there was potential for some release
All analyses show that this release would be small
To date there has never been an accident causing release from a type B package

From The Probability of Spent Fuel Transportation Accidents by J. D. McClure, Report SAND 80-1721, Sandia National Laboratories, Albuquerque, N.M., 1981.

11.9 COMPUTER PROGRAMS

RADTRAN. This risk assessment program was developed at Sandia National Laboratory to calculate the radiologic consequences associated with radmaterial transportation. It uses population, weather, packaging, transportation, and health physics data to calculate the accidental dose risks from groundshine, inhalation and resuspension, cloudshine, and ingestion.[44]

RISKIND. This program was developed by Argonne National Laboratory to investigate the risks associated with the transportation of spent nuclear fuel. It estimates the radiologic health consequences and risks during normal transport and accident conditions.[45]

TRAGIS. The Transportation Routing Analysis Geographic Information System (TRAGIS) is sponsored by Oak Ridge National Laboratory. The advantage of this program is its ability to display graphics of the routes in addition to calculating the optimal routes.[46]

GRAIL. This program estimates the risk associated with the transport of hazardous cargo, including radioactive materials. It is able to calculate the projected health and economic consequences due to an accident.[47]

11.10 DISCUSSION QUESTIONS AND PROBLEMS

1. What are four of the major public concerns about the transportation of radioactive waste? Do you share those concerns? Why or why not?

2. Which three titles of the Code of Federal Regulations cover radwaste transportation, and which agency is responsible for each title?

3. Describe the dimensions and amounts and type of waste carried in two of the shipping casks listed in this chapter.

4. What criteria do you think should be considered when determining radwaste shipping routes?

5. Compare the risks associated with normal transportation conditions of spent nuclear fuel with those associated with accident conditions.

6. Packaging for low-level radioactive waste is divided into three categories. What are these categories, and what are the requirements for each type of packaging?

7. What are the two categories of high-level radioactive waste and the packaging requirements for each category?

8. What does the acronym OCRWM stand for? Which law created this government office and when was the law passed?

9. Compare the responsibilities of shippers and carriers. List at least three responsibilities for each party.

10. What are the three warning label categories for radmaterial packaging?

REFERENCES

1. NIREX, Radioactive Wastes in the UK: A Summary of the 1998 Inventory. Available at http://www.nirex.co.uk/reports/Inventory/Summary.pdf [29 February 2000].
2. ANDRA, Types of Radioactive Waste 7. Available at http://www. Andra.fr/en/savoir/dechets/classdech—en.htm [7 March 2000].
3. Salmon, A., The Transportation of radwaste—A review, in *Waste Management '87*, vol. I, p. 387, University of Arizona Press, Tucson, Ariz., 1987.
4. League of Women Voters Education Fund, Transporting Radioactive Spent Fuel: An Issue Brief, Publication # 1052, July 1996.
5. U.S. AEC, Everything You Always Wanted to Know about Shipping HLW, Wash-1264 Rev., 1974.
6. U.S. ICC, Final Environmental Impact Statement Transportation of Radmaterials by Rail, Interstate Commerce Commission Office of Proceedings, Section of Energy and Environment, Washington, D.C., August 1977.
7. Smith, D. R., and J. M. Taylor, Analysis of the Radiological Risks of Transporting Spent Fuel and Radwastes by Truck and by Ordinary and Special Trains, SAND 77-1257, Sandia National Laboratories, Albuquerque, N.M., 1978.
8. Klassen, D., Transportation of Radmaterial by Rail: Special Train Issue, SAND 81-1447, TTC-0226, Sandia National Laboratories, Albuquerque, N.M., 1982.
9. U.S. DOT, A Review of the Department of Transportation Regulations for Transportation of Radmaterials, Research and Special Programs Administration, Materials Transportation Bureau, Department of Transportation, Washington, D.C., Rev. 1983.
10. U.S. Congress, 93rd, Hazardous Materials Transportation Act of 1975, Public Law 93-633, 1975.
11. U.S. Code of Federal Regulations, Title 49, Part 173, Requirements for Transportation of Radmaterials, 1983.
12. Burgoyne, R. M., Transportation of radioactive waste, in *Radioactive Waste Technology*, A. A. Moghissi et al., eds., p. 423, ASME/ANS, Engineering Center, New York, 1986, Also Callaghan, E. F., and W. H. Lake, DOE procurement activities for spent fuel shipping casks, *J. Inst. Nucl. Mater. Manage*, vol. 16, no. 3, p. 20, 1988.
13. U.S. Department of Energy, Office of Civilian Radioactive Waste Management, OCRWM Transportation Report, June 1995.
14. U.S. Department of Energy, Waste Acceptance and Transportation: Competitive, Private Sector Acquisition Process. Available at http://www.rw.doe.gov/wasteaccept/acqprogress/acqprogress.htm [16 August 2000].
15. U.S. NRC, NLI-1/2 Spent Fuel Shipping Cask Safety Report, Amend. No. 1, December 1972.
16. U.S. NRC, Safety Analysis Report for Transnuclear, Inc., Spent Fuel Shipping Cask Model No. TN-8 and TN-9, Rev. 7, Dockets 71-9015 and 71-9016, 1976; also TN-8 and TN-9 Safety Analysis Report, Rev. 8, Transnuclear, Inc., 1980.
17. U.S. NRC, Safety Analysis Report for Nuclear Fuel Services, Inc., Spent Fuel Shipping Cask Model No. NFS-4, Docket 71-6698, 1972.
18. U.S. NRC, Design and Analysis Report IF-300 Shipping Cask, NEDO 10084-1, Docket 71-9001, 1972.
19. U.S. NRC, Safety Analysis Report NLI-10/24 Spent Fuel Shipping Cask, Rev. 5, Docket 71-9023, 1976.
20. U.S. Code of Federal Regulations, Title 10, Part 72, Licensing Requirements for the Storage of Spent Fuel in an Independent Spent Fuel Storage Installation (ISFSI), March 26, 1984.
21. U.S. Code of Federal Regulations, Title 10, Part 71, Packaging of Radmaterials for Transport and Transportation of Radmaterial under Certain Conditions, September 6, 1983.

22. Allen, G. C., Advanced Transportation System Option for Spent Fuel and HLW, SAND 84-0826C, Sandia National Laboratories, Albuquerque, N.M., CONF 8404127, 1984.

23. Saling, J. H., and Y. S. Tang, Spent nuclear fuel casks development, in *Proc. Southeastern Symposium on In-Situ Treatment and Immobilization of Hazardous and Radwastes*, Knoxville, Tenn., June 1986.

24. Best, R. E., Status of development of the REA 2023 dry storage cask for LWR spent fuel, presented at the Spent Fuel Storage Seminar, Institute of Nuclear Materials Management, Washington, D.C., January 1985.

25. Droste, B., H. W. Hubner, and U. Probst, Qualification and certification criteria of casks for the intermediate dry storage of spent fuel in the FRG, in *PATRAM '83, Proc. 7th International Symposium on Packaging and Transportation of Radmaterials*, vol. I, p. 224, New Orleans, May 1983; also Wakeman, B. H., Status of Metal Cask Storage Program at the Surry Power Station, presented at the INMM Spent Fuel Storage Seminar, Washington, D.C., January 1987.

26. Keese, H., W. Anspack, and R. Christ, The TN-1300 transport/storage cask system, in *PATRAM '83, Proc. 7th International Symposium on Packaging and Transportation of Radmaterials*, vol. I, p. 242, New Orleans, La., May 1983.

27. ORNL, Preliminary Description of the Transportation Operations Systems, ORNL/Sub/86-02217/1, Oak Ridge National Laboratory, Oak Ridge, Tenn., 1988.

28. Goldman, K., and W. C. Gekler, Safety Criteria for Spent Fuel Transport, EPRI NP-4573, Electric Power Research Institute, Palo Alto, Calif., 1986.

29. Safe Transportation and Emergency Response Training; Technical Assistance and Funding, *Fed. Reg.*, vol. 60, no. 1, pp. 99–100, 1995.

30. Sandquist, G. M., V. C. Rogers, A. A. Sutherland, and G. B. Merrell, Exposures and Health Effects from Spent Fuel Transportation, RAE-8339/12-1, Rogers & Associates Engineering Corp., Salt Lake City, Utah, 1985.

31. Marsden, M. M., A preliminary analysis of the risk of transporting nuclear waste to potential candidate commercial repository sites, in *Waste Management '84*, University of Arizona Press, Tucson, Ariz., 1984.

32. Gelder, R., et al., Radiation Exposure Resulting from the Normal Transport of Radmaterials within the U.K., NRPB-R155, National Radiological Protection Board, U.K., 1984.

33. Shaw, K. B., Worker Exposure: How Much in the U.K.? IAEA Bulletin, Vienna, Spring 1985.

34. Wilmot, E. L., Transportation Accidents Scenarios for Commercial Spent Fuel, SAND 80-2124, TTC-0156, Sandia National Laboratories, Albuquerque, N.M., 1981.

35. ACS, *1985 Cancer Facts and Figures*, American Cancer Society, 1985.

36. Womack, C. J., *Proof of Evidence to Sizewell Public Inquiry*, Central Electricity Generating Board, London, 1982.

37. Clark, R., Potential consequences of accidents, in *The Urban Transportation of Irradiated Fuel*, Macmillan, London, 1984.

38. Schneider, K. A., and E. Merz, Risk analysis of transporting vitrified HLW by train, in *PATRAM '86, Proc. 8th International Symposium on Packaging and Transportation of Radmaterials*, West Germany, 1986.

39. Cook, J. R., W. R. Lahs, and W. H. Lake, The modal study: The response of spent fuel packages to severe transportation accidents, in *Waste Management '87*, vol. 1, p. 437, University of Arizona Press, Tucson, Ariz., 1987.

40. LLNL, Shipping Container Response to Severe Highway and Railway Accident Conditions, NUREG/CR-4829, Lawrence Livermore National Laboratory, 1987.

41. Chen, S. Y., and Y. C. Yuan, Calculation of health risks from spent nuclear fuel transportation accidents, in *Proc. ANS Topical Conference on Population Exposure from the Nuclear Fuel Cycle*, ORNL, Oak Ridge, Tenn., 1987.

42. Saling, J. H., and C. R. Bolmgren, Phase 1 Study of Metallic Cask Systems for Spent Fuel Management from Reactor to Repository, WTSD-TME-085, Westinghouse Advanced Energy Systems Division, Madison, Pa., 1986.

43. McClure, J. D., The Probability of Spent Fuel Transportation Accidents, Report SAND 80-1721, Sandia National Laboratories, Albuquerque, N.M., 1981.

44. Sandia National Laboratories, RADTRAN. 7. Available at http://www.ttd.sandia.gov/risk/rt.htm [3 May 2000].

45. Argonne National Laboratory, RISKIND. Available at http://web6.ead.anl.gov/~hmrt//riskind.html [8 May 2000].
46. Oak Ridge National Laboratory, TRAGIS Routing Model. Available at http://www.ornl.gov/ttg/tragis.htm [8 May 2000].
47. Lawrence Livermore National Laboratory, System Sciences-Transportation Risk Assessment. Available at http://www.llnl.gov/eng/eetd/sysscig/transport.html [8 May 2000].

ADDITIONAL READINGS

Foratom, French Nuclear Forum. Available at http://www.foratom.org/Member/France/france.html [29 February 2000].
Foratom, German Nuclear Forum. Available at http://www.foratom.org/Member/Germany/germany.html [29 February 2000].
Japan Nuclear Fuel Limited, Outline of Facilities. Available at http://www.jnfl.co.jp/gennen/shisetsu-e.html [3 March 2000].
Pacific Northwest National Laboratory, Department of Energy, Japan-Nuclear Fuel Cycle and Material Production. Available at http://etd.pnl.gov:2080/fac/japan/nuclear.html [6 January 2000].
U.S. Department of Energy, Integrated Data Base Report-1996, DOE/RW-0006, Rev. 13. Available at http://gpo.osti.gov:901/dds/advanced.html.
Waste Management in the U.S.A. Available at http://trinity.tamu.edu/COURSES/NU415/waste—management/usa.htm [3 March 2000].

TWELVE

DECONTAMINATION AND DECOMMISSIONING

12.1 INTRODUCTION

Decontamination (decon) and decommissioning of nuclear facilities have received considerable attention in recent years, especially with regard to the technical, financial, and environmental issues. While decontamination is always required for decommissioning, the reverse is not true, as decon is not always followed by decommissioning of the facility. The normal decon processes for LLW-contaminated material have been covered in Chapter 7; thus this chapter will consider decon as a part of decommissioning. The objectives of decommissioning operations can be stated in general as follows:

1. To place the site in a long-term radiologically safe condition and available for unrestricted usage.
2. To dismantle the facility safely and cost-effectively.

However, it sometimes may not be economically feasible to clean a site well enough to release it for "unrestricted" use. Decon work essentially involves the external plant structural and equipment surfaces preceding dismantling work and internal components such as tank internal surfaces, internally contaminated piping embedded in concrete, and the reactor purification system. Decon may allow unrestricted release of components for use on site or off site and reduces the occupational exposure of the dismantling personnel.

This chapter will cover the expected wastes from reactor decommissioning, the decommissioning decision and alternatives, decommissioning engineering and techniques, environmental impacts and regulatory guides, decommissioning experience, and cost estimates for decommissioning.

12.1.1 Surplus Facilities Management

DOE's Surplus Facilities Management Program (SFMP) has the objectives of surveillance and maintenance of surplus contaminated facilities awaiting decommissioning as well as implementation of a structured decommissioning program. In meeting these objectives, it also performs the following:

1. Cost estimating, budgetary management, planning and scheduling, engineering and technology development to support decommissioning.
2. Identifying and making available for potential reuse materials, equipment, facilities, and surplus property.
3. Assisting in planning and technical implementation of facility decommissioning through technology transfer to civilian nuclear industry.
4. Providing a program-level information center and literature reference system, disseminated technology, and program standards for record retention.

After a 1999 reorganization in DOE, these functions were assigned to the Nuclear Material and Facility Stabilization Program under DOE's Office of Environmental Management. DOE participates in international activities in support of DOE agreements with other countries for technology exchange. One example of such exchange was the American Nuclear Society (ANS) International Topical Meeting in Niagara Falls, New York,[1] where experience in decommissioning was described and technology exchanged among 26 countries. In 1994, the ANS held another topical meeting,[2] in Washington, D.C., on the same subject. Locations of DOE radioactively contaminated surplus facilities are shown in Figure 12.1.

12.1.2 Waste from Decommissioning of Fuel Cycle Facilities

Different amounts of wastes are generated from decommissioning of commercial fuel cycles, that is, facilities for uranium conversion, uranium enrichment, fuel fabrication, and reactors and reprocessing plants. Table 12.1 indicates relative wastes generated by decommissioning of these facilities.[3] The facility size varies as shown, using one 1000-Mwe LWR as the basis. For example, the uranium conversion plant is large enough to support 60 such reactors. All facilities are assumed to have a 40-year lifetime, and reactors are assumed to have a 70% capacity factor. The waste volumes represent the volume (in cubic meters) of waste with its packaging for disposal, and the relative decommissioning waste is compared with that from one 1000-MWe PWR. Thus it becomes obvious that decommissioning waste from reactors is two or three orders of magnitude greater than that from other facilities. Other than reprocessing plants, all wastes from decommissioning these facilities are LLW except for a small amount of high-activity waste from certain reactor core internal parts, which may be treated in part as RH-TRU wastes and in part as a special class of LLW requiring

Figure 12.1 Locations of DOE radioactively contaminated surplus facility, Surplus Facilities Management Program.

special disposal methods beyond those needed for normal LLW. Wastes from decommissioning the reprocessing plants includes hulls and structural material that are TRU waste.[4]

12.1.3 Radioactive Inventory

The radioactive inventory data play a major role not only in deciding on the mode of decommissioning, but also in planning and scheduling. The degree of decontamination and the shipping and disposal of radmaterial generated from decommissioning will be a direct function of the type and magnitude of the contamination source. An accurate radioactive inventory estimate is necessary to determine the radiation exposure and to prepare the environmental impact assessments.

The radioactive inventory of a nuclear reactor or facility to be decommissioned can be divided into two categories: (1) the radioactivity induced by neutron activation of certain elements in a reactor vessel, reactor components, and adjacent structures and (2) the radioactive material deposited on the structures and external surfaces of various systems.[5] Table 12.2, which is taken from reference 5, shows typical radionuclides of concern in neutron-activated materials. Cobalt-60, a significant gamma emitter, is of prime concern in decommissioning the reactor and determines the requirements for remote operation and worker shielding. Other radionuclides of concern are ^{63}Ni, ^{59}Ni, ^{94}Nb, and ^{14}C because of their long half-lives. Nickel-59 has not been of major importance in previous reactor decommissioning because the reactors had not operated long enough to create significant quantities of this radionuclide. Carbon-14 is of concern for high-temperature, gas-cooled reactors. In decommissioning, a radionuclide half-life of less than 1 month is considered short.

Table 12.1 Wastes from decommissioning of commercial fuel cycle facilities and power reactors

Fuel cycle facility[a]	Capacity of typical plant	Number of 1000-MWe reactors supported by typical facility	Wastes generated during operating lifetime[b] (m³)	Decommissioning wastes[c] (m³)	Relative decommissioning wastes per 100-MWe PWR[d]
Uranium conversion	10,000 MTHM/yr	60	LLW: 44,000[e]	LLW: 1,260[c] (1,200 kg uranium)	0.001
Uranium enrichment	8.75×10^6 SWU/yr	90	LLW: 11,200	LLW[f]: 25,470	0.019
Fuel fabrication	1,000 MTHM/yr	33	LLW: 99,000	LLW: 1,100 (150 kg uranium)	0.002
Pressurized water reactor	1,000 MWe	1	LLW: 26,000	LLW: 15,200 (<50,000 Ci); activated waste: 75 (4.1×10^6 Ci)	1.000
Boiling water reactor	1,000 MWe	1	LLW: 48,000	LLW: 16,300 (<15,000 Ci); activated waste: 77 (5.7×10^6 Ci)	1.072
Reprocessing plant	1,500 MTHM/yr	50	LLW: 54,000; TRU[g]: 91,000	LLW: 3,100 (4,000 Ci); TRU: 4,600 (2.5×10^7 Ci)	0.011

[a] Assuming 40-year lifetime for all facilities, 70% capacity factor for reactors.

[b] Facility size varies—for example, 60 reactors are supported by one uranium conversion plant.

[c] Volume of waste to disposal site, including packaging.

[d] Relative decommissioning wastes based on fuel cycle facilities to support one PWR reactor.

[e] Decommissioning data on solvent extraction type facility. Operating data on direct fluoridation facility.

[f] Assume density of waste 1 g/cm³ to convert mass of waste to volume.

[g] Includes hulls and structural material.

From *Residual Radionuclide Contamination within and around Commercial Nuclear Power Plants: Assessment of Origin, Distribution, Inventory, and Decommissioning* by K. H. Abel et al., NUREG/CR-4289, 1986. Reprinted by permission.

Table 12.2 Typical radionuclides of concern in neutron-activated materials

Isotope	Half-life (yr)	Means of production	Emission	Energy (MeV)
		Base material: carbon and stainless steel		
^{14}C	5730.0	^{14}N(n,p)	β^-	0.156
^{49}V	0.906	^{52}Cr(p,α)	γ, β^-	0.6[a]
^{54}Mn	0.856	^{56}Fe(d,α)	γ	0.835
^{55}Fe	2.6	^{54}Fe(n,γ)	γ	0.23[a]
^{59}Ni	8×10^4	^{58}Ni(n,γ)	ε	1.06[a]
^{63}Ni	100.0	^{62}Ni(n,γ)	β^-	0.066
^{65}Zn	0.667	^{64}Zn(n,γ)	$\gamma, \varepsilon, \beta^+$	1.115, 1.352, 0.325
^{58}Co	0.194	^{55}Mn(α,n)	β^+, γ	0.474, 0.810
^{60}Co	5.263	^{59}Co(n,γ)	β^-, γ, γ	0.314, 1.17, 1.33
^{93}Mo	3.5×10^3	^{92}Mo(n,γ)	ε	Nb x rays
^{94}Nb	2×10^4	^{93}Nb(n,γ)	β^-, γ, γ	0.49, 0.702, 0.871
^{95}Nb	0.096	^{95}Zr decay	β^-, γ	0.16, 0.765
^{95}Zr	0.175	^{94}Zr(n,γ)	β^-, γ, γ	0.396, 0.724, 0.756
		Base material: concrete		
^{14}C	5730.0	^{14}N(n,p)	β^-	0.156
^{35}S	0.238	^{34}S(n,γ)	β^-	0.167
^{36}Cl	3.01×10^5	^{35}Cl(n,γ)	β^-, ε	0.714, 1.18[a]
^{37}Ar	0.0953	^{36}Ar(n,γ)	ε	0.81[a]
^{39}Ar	269.0	^{38}Ar(n,γ)	β^-	0.565
^{40}K	1.28×10^9		β^-, γ	1.314, 1.46
^{41}Ca	8×10^4	^{40}Ca(n,γ)	γ	K x rays
^{45}Ca	0.446	^{44}Ca(n,γ)	β^-	0.257
^{46}Sc	0.229	^{45}Sc(n,γ)	$\beta^-, \beta^-, \gamma, \gamma$	1.48, 0.357, 0.889, 1.12
^{54}Mn	0.856	^{56}Fe(d,α)	γ, ε	0.835, 0.829, 1.379
^{55}Fe	2.6	^{54}Fe(n,γ)	γ	0.23
^{59}Fe	0.122	^{58}Fe(n,γ)	β^-, γ, γ	1.57, 1.1, 1.29
^{58}Co	0.194	^{55}Mn(α,n)	β^+, γ	0.474, 0.81
^{60}Co	5.263	^{59}Co(n,γ)	β^-, γ, γ	0.314, 1.17, 1.33
^{59}Ni	8×10^4	^{58}Ni(n,γ)	ε	1.06[a]
^{63}Ni	100.0	^{62}Ni(n,γ)	β	0.067
^{65}Zn	0.667	^{64}Zn(n,γ)	$\gamma, \varepsilon, \beta^+$	1.115, 1.352, 0.325
^{94}Nb	2×10^4	^{93}Nb(n,γ)	β^-, γ, γ	0.49, 0.702, 0.871
^{95}Nb	0.096	^{95}Zr decay	β^-, γ	0.16, 0.765
^{93}Mo	3.5×10^3	^{92}Mo(n,γ)	γ	Nb x rays
		Base material: aluminum		
^{46}Sc	0.229	^{45}Sc(n,γ)	$\beta^-, \beta^-, \gamma, \gamma$	1.48, 0.357, 0.889, 1.12
^{54}Mn	0.856	^{56}Fe(d,α)	γ	0.835
^{55}Fe	2.6	^{54}Fe(n,γ)	γ	0.23[a]
^{54}Fe	0.122	^{52}Fe(n,γ)	β^-, γ, γ	1.57, 1.1, 1.29
^{60}Co	5.263	^{59}Co(n,γ)	β^-, γ, γ	0.314, 1.17, 1.33
^{65}Zn	0.667	^{64}Zn(n,γ)	$\gamma, \varepsilon, \beta^+$	1.115, 1.352, 0.325
^{110m}Ag	0.69	^{109}Ag(n,γ)	β^-, γ^b	0.087, 0.6577

[a] Continuous spectrum of x-ray energies below this number, due to bremsstrahlung.

[b] Energy of most probable energy β^- and most probable energy γ given.

From *Decommissioning* by T. S. LaGuardia, in *Radioactive Waste Technology*, edited by A. A. Moghissi et al., New York, 1986. p. 499, ASME/ANS, Copyright 1986 by American Society of Mechanical Engineers. Reprinted by permission.

12.2 DECOMMISSIONING COMMERCIAL NUCLEAR POWER PLANTS

Decommissioning of commercial nuclear power plants is governed by Nuclear Regulatory Commission (NRC) regulations 10CFR2, 10CFR50, and 10CFR51, the last two of which were first issued in 1988.[6] After utilities had decommissioned a few commercial reactors and the NRC had an opportunity to gather information on the process, the Commission revised all three regulations. The new regulations became effective in August 1996.[7] To help utilities comply with the regulations, the NRC issues Regulatory Guides.[6–8]

When power reactors are permanently shut down, they must be decommissioned within 60 years. That is, within 60 years, radioactive material on site must be reduced to a level that will allow the NRC license to be terminated. Three decommissioning methods were initially allowed by the NRC: (1) immediate dismantlement (DECON), (2) safe storage for a time followed by dismantlement (SAFSTOR), and (3) entombment (ENTOMB). However, under the 1996 revised regulations, ENTOMB may not be acceptable.

To date, 22 reactors in the United States have been decommissioned. Table 12.3 gives the name, location, type, size, startup date, shutdown date, and decommissioning option selected for each reactor. Section 12.2.1 describes in detail the decommissioning alternatives for an experimental power plant, Shippingport, decommissioned by the U.S. Department of Energy to demonstrate the decommissioning process.

12.2.1 Decommissioning Management Decision and Alternatives

After reaching the decision to decommission, the mode of decommissioning has to be decided. For example, for the Shippingport atomic power station, DOE, in accordance with public law,[11] published an environmental impact statement (EIS) for the decommissioning of the facility.[12] All of the alternatives that were considered in reaching the decision must be discussed in the EIS. Such alternatives include (1) no action, (2) immediate dismantlement, (3) safe storage followed by deferred dismantlement, and (4) entombment. A record of decision to decommission was then provided on the decommissioning mode along with background information and the rationale used in the selection, which completes the National Environmental Policy Act (NEPA) process for decommissioning of the facility. The four alternatives were evaluated, considering the work plan and schedule, the radiation doses expected for occupational workers and the general public, waste disposal, and economic considerations.

No action. Three options are considered under the no-action alternative: to continue operation of the station to produce electricity, to close the station while continuing existing maintenance and surveillance, or to close the station and do nothing further. To continue commercial operation of the station by a utility, an NRC operating license would be required; this would likely require extensive analyses and modifications, because the reactor is over 20 years old. No interest in this has been expressed by the owner of the site and the turbine generator (Duquesne Light Company).

Table 12.3 Commercial power plants decommissioned in the United States

Name	Plant type	Location	Capacity (MWe)	Startup date	Shutdown date	Decommissioning option
Big Rock Point	BWR	Charlevoix, MI	67	12/08/62	08/29/97	DECON
Bonus	BWR	Puerto Rico	17	08/14/64	06/01/68	ENTOMB
CVTR	PHWR	Parr, SC	17	12/18/63	01/01/67	SAFSTOR
Dresden-1	BWR	Morris, IL	197	04/15/60	10/31/78	SAFSTOR
Elk River	BWR	Elk River, MN	22	08/24/63	02/01/68	DECON
Fermi-1	FBR	Monroe Co., MI	65	08/05/66	09/22/72	SAFSTOR
Fort St. Vrain	HTGR	Platteville, CO	330	12/11/76	08/18/89	DECON
Haddam Neck	PWR	Haddam, CT	560	08/07/67	07/22/96	DECON
Humboldt Bay	BWR	Eureka, CA	63	04/18/63	07/02/76	SAFSTOR
Indian Point-1	PWR	Buchanan, NY	257	09/16/62	10/31/74	SAFSTOR
LaCrosse	BWR	LaCrosse, WI	48	04/26/68	04/30/87	SAFSTOR
Maine Yankee	PWR	Wiscasset, ME	860	11/08/72	12/06/96	DECON
Millstone-1	BWR	New London, CT	641	11/29/70	07/17/98	Undecided
Pathfinder	BWR	Sioux Falls, SD	59	07/25/66	09/16/67	DECON
Peach Bottom-1	HTGR	York Co., PA	40	01/27/67	10/31/74	SAFSTOR
Rancho Seco-1	PWR	Sacramento, CA	873	10/13/74	06/07/89	SAFSTOR
San Onofre-1	PWR	San Clemente, CA	436	07/16/67	11/30/92	SAFSTOR
Three Mile Island-2	PWR	Middletown, PA	880	04/21/78	03/28/79	PDMS
Trojan	PWR	Rainier, OR	1095	12/23/75	11/09/92	DECON
Yankee	PWR	Franklin Co., MA	167	11/10/60	10/01/91	DECON
Zion-1	PWR	Zion, IL	1040	06/28/73	—	—
Zion-2	PWR	Zion, IL	1040	12/26/73	09/19/96	SAFSTOR

Compiled from information in references 9 and 10.

For the second option, following fuel removal, all of the Shippingport station's systems and components could be left intact while continuing existing surveillance, maintenance, and monitoring. Liquids would not be removed (removal of liquids is considered to be part of safe storage). Cost of 24-hr surveillance, full-time maintenance, and routine monitoring would amount to at least $200,000 annually. Occupational radiation doses would be as high as 25 person-rem per year, principally because of maintenance on the liquid handling systems.

The last option of shutting down the facility and doing nothing further following defueling is not feasible because of the radioactivity left inside the facility and the potential environmental impacts.

Immediate dismantlement. Immediate dismantlement is removal from the site, within a few years after shutdown, of all fluids, piping, equipment, components, structures, and wastes having radioactivity levels greater than those permitted for unrestricted use of the property. Removal of nuclear fuel, the blanket, and the reflector assemblies will have been accomplished immediately after shutdown, as fuel removal is considered to be part of final reactor shutdown and not part of decommissioning. Dismantlement activities would begin with draining, decon, and removal of nonessential systems and continue to completion of the final radiation survey. Dismantlement of the plant would require about 5 years. The reactor vessel can be removed and shipped intact together with appropriate shielding to a disposal site by barge or rail car. Furthermore, the internals, the reactor vessel, and the neutron shield tank can be shipped as one package to a disposal site by barge, in which case the internals would be left in the vessel and the neutron shield tank would be filled with concrete to provide shielding. Coolant piping and other systems and equipment would be decontaminated on a selective basis to reduce radiation dose rates to ALARA levels.

Dose rates used in estimating the occupational radiation dose for immediate dismantlement are based on measured dose rates in the Shippingport station, assuming exposure occurs 2 years after reactor shutdown. The estimated occupational dose for this mode of decommissioning is 1275 person-rem. This may be compared to a maximum dose of 2500 person-rem if each of the 100 (average number) workers received a maximum allowed dose of 5 rem per year over a 5-year period. A public radiation dose results from transportation of the radwaste from the site to the disposal site, assuming it is hauled to DOE burial sites at Hanford, Washington, or at Savannah River Plant, South Carolina, which are 2380 and 715 miles from Shippingport, respectively. The dose estimate (shown in Table 12.4, pages 358–359) is based on transporting all solidified decon wastes and radioactive, cask-contained, low-specific-activity material (as defined in 49CFR173) according to DOT regulations. A public radiation dose also results from airborne releases from routine decommissioning activities. Calculation for a much larger PWR (1175 MWe) shows that the radiation dose to the public from these sources is trivial.

Liquid radwastes from the immediate dismantlement are from two sources: (1) the existing liquid inventory in the piping, components, and fuel handling canal and (2) the water used in the decon procedures. All radioactive liquids would be filtered, demineralized, and evaporated in the existing liquid waste processing system, and liquid waste would not be transported off site. Solid radmaterials that must be removed from the site during immediate dismantlement are of three types: neutron-activated material, contaminated material, and radwaste. Most of the neutron-activated materials must be shipped in shielded containers to meet the allowable surface dose rate limits for transport. The bulk of the neutron-activated material is contained in the metal in the pressure vessel and its internals. Contaminated material would be handled as low-specific-activity material as defined in 49CFR173; for instance, spent ion-exchange resins and evaporator concentrates would be solidified with concrete in 208-L drums. Combustibles and other dry active waste (DAW) would be volume reduced in the waste compactor and place in 208-L drums for disposal.

Safe storage, SAFSTOR (mothballing). Safe storage is defined as the activities required to place and maintain a nuclear facility in such a condition that risk from the facility to public safety is within acceptable bounds and the facility can be safely stored for as long as desired. Safe storage consists of a period of facility and site preparation followed by a period of continuing care that involves security, surveillance, monitoring, and maintenance. During the storage period, the reactor facility is put in a hardened safe storage condition. The deferred dismantlement of the facility is started at the end of safe storage, when the structure is reopened and all materials that still have radioactivity greater than the levels permitted for unrestricted use are removed and shipped to a disposal site.

Preparation for safe storage would include disposing of all radioactive fluids and wastes and of some selected components. Highly radioactive reactor vessel and internals and the primary coolant system piping and components would be sealed within a safe storage barrier to prevent unauthorized access during the storage period. The hardened safe storage (and entombment, as discussed in the next section) boundaries for Shippingport might consist primarily of reinforced concrete and steel containment structures. Top and side views of such boundaries are shown in Figures 12.2 and 12.3, respectively. Preparation for safe storage should be completed in about 31 months. Security, surveillance, periodic inspections, radiation surveys, and maintenance of the storage boundary are provided. A 98-year storage period would permit the radioactivity to decay to a level such that some material and equipment could be released to unrestricted use and most other components could be dismantled manually rather than remotely within the occupational dose rate limit (300 mrem/hr). The neutron-activated reactor vessel and vessel internals would still have to be removed and shipped to a controlled burial ground for disposal, with less of the shielding requirement than before. The key factor for selecting the length of the safe storage period is the dose rate from ^{60}Co, which decays with a half-life of 5.27 years. The deferred dismantlement would start when the storage period ends, when the radiation dose rates would be significantly lower than that for the immediate dismantlement mode, most originally contaminated piping and equipment would present little radiation hazard to the decommissioning workers, and the neutron-activated reactor vessel and internals would require less sophisticated tooling for cutting and removal.

Occupational radiation doses would be received by the decommissioning workers during preparation for safe storage, during the storage period, and during deferred dismantlement. Based on actual measurements at the Shippingport station during reactor

Table 12.4 Summary of decommissioning alternatives—Shippingport station

Parameters	1. No action	2. Immediate dismantlement	3. Safe storage followed by deferred dismantlement	4. Entombment
Activities	Continue operation generating electricity Defuel and close the station; continue maintenance and surveillance Defuel, close the station, and do nothing further	Removal of all fluids, piping, equipment, components, structures, and wastes	Preparation period Continuing care Deferred dismantlement	Same as in 3
Time required	—	5 years	Preparation complete in about 31 months	42 months
Radiation doses[a] Occupational	—	1275 person-rem	505 person-rem	617 person-rem
General public Truck to SRP (drivers/public)		24 person-rem/ 4 person-rem	18 person-rem/ 3 person-rem	14 person-rem/2 person-rem
Truck to Hanford (drivers/public)		67 person-rem/ 12 person-rem	53 person-rem/ 9 person-rem	40 person-rem/6 person-rem
Waste disposal Liquid wastes Existing	—	In piping, components fuel handling canals	Same as in 2	Same as in 2
Decon water		Decon generated, high	Less than in 2	Less than in 2

Solid waste:				
Neutron activated		Shipped in shielded container	Less due to decay	Remove reactor internals to reduce time for release
Contaminated material		Low specific activity	Only outside boundary treated	Same as in 3
Radwastes		Solid, with concrete	Same as in 2	Same as in 2
Cost	Maintenance and surveillance cost 0.2×10^6/yr	100×10^6	Most costly	Between 2 and 3
Remarks	—	Prohibits unrestricted use of facility until dismantlement		May not be viable due to low subbasement floor relative to Ohio River level

[a]Two years after shutdown.

From Final Environment Impact Statement, Decommissioning of the Shippingport Atomic Power Station by U.S. DOE, DOE/EIS0080F, Nuclear Waste Management and Fuel Programs Remedial Action Program Office, Washington, D.C., 1982. Reprinted by permission.

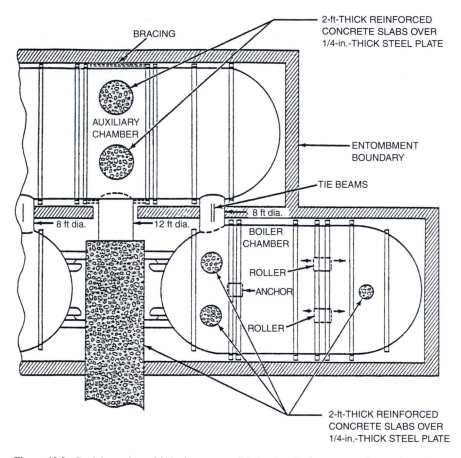

Figure 12.2 Partial top view of Shippingport possible hardened safe storage and entombment boundaries. From Final Environment Impact Statement, Decommissioning of the Shippingport Atomic Power Station by U.S. DOE, DOE/EIS0080F, Nuclear Waste Management and Fuel Programs Remedial Action Program Office, Washington, D.C., 1982. Reprinted by permission.

shutdown, the occupational radiation dose is estimated to be 505 person-rem. Estimates of the public radiation dose resulting from safe storage followed by deferred dismantlement are based on the same assumptions as used for estimating that from immediate dismantlement. These values are shown in Table 12.4. The inventory of radioactive liquids present at final shutdown of the Shippingport station would be the same irrespective of the decommissioning alternative selected. For SAFSTOR, the volume of liquid waste from decon operations at the time of safe storage preparation would be less than the volume generated in immediate dismantlement because only loose contamination in safe storage is removed. Reliance in this case is placed on radioactive decay rather than on decon as a mechanism for reducing the radiation dose to workers. Solid radmaterials that have to be removed from the site are of the same three types as for immediate dismantlement. During the preparation period, the bulk of the solid radmaterial removed would be radioactive waste (liquid concentrate,

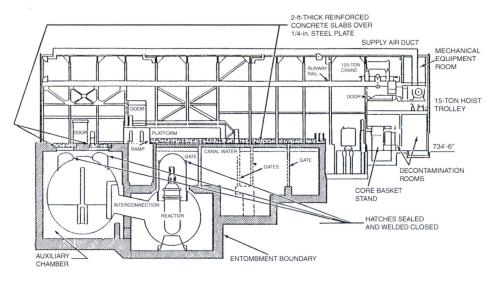

Figure 12.3 Side view of Shippingport possible hardened safe storage and entombment boundaries. From Final Environment Impact Statement, Decommissioning of the Shippingport Atomic Power Station by U.S. DOE, DOE/EIS0080F, Nuclear Waste Management and Fuel Programs Remedial Action Program Office, Washington, D.C., 1982. Reprinted by permission.

combustibles, and other DAW). Only the contaminated piping and equipment located outside the storage boundary would be decontaminated or removed. When deferred dismantlement is accomplished, the quantity of neutron-activated material is reduced because of radioactive decay; thus, more material can be reclaimed and less material must be sent to a burial ground than in the case of immediate dismantlement. Spent ion-exchange resins and evaporator concentrates are solidified with concrete in drums, and combustible and other DAW are compacted in drums as in the immediate dismantlement case.

Entombment (ENTOMB). Entombment is the encasement of radmaterials and components in a massive structure of concrete and steel, which must be sufficiently strong and long-lived to ensure retention of the radioactivity until it has decayed to levels that permit unrestricted use of the site. Unlike the SAFSTOR mode, where the structure will be reopened at the end of safe storage, the ENTOMB mode requires a very careful and complete inventory of the radmaterial to be entombed. The structural lifetime of the entombing facility is estimated to be 200 years, based on the effects of freeze–thaw cycles on concrete and on corrosion rates of the steel chambers inside the entombment boundary. A long period of continuing care, consisting of security, surveillance, and maintenance, follows the entombment of the radmaterials and lasts until the radioactivity has decayed to levels that permit unrestricted use of the site. No access is available to the interior of the entombed structure and no utilities are available within the entombed structure, such as the use of interior sump pumps. Thus there must be no possibility of groundwater or flood water seeping into the structure, which may cause seepage of contaminated water out of the structure later. It should be noted that the lowest level of the subbasement is 203 m (666 ft) above sea level, while the normal elevation of the Ohio River is 202.7 (665 ft) and the project design flood level is 214.9 m (705 ft). Seepage into the subbasement has been observed in the past. The entombment alternative is similar to the SAFSTOR alternative in that the entombment structure is the same as the hardened safe storage boundary (Figures 12.2 and 12.3). All of the radionuclides in the neutron-activated materials except ^{94}Nb and ^{59}Ni will decay sufficiently in 125 years to reduce the dose rate to 0.1 mrem/hr at 2.5 cm (1 in.) from the surface. The reactor internals and the reactor vessel stainless steel cladding will contain these long-lived radionuclides. To reduce the time until the entombed reactor can be released for unrestricted use, the pressure vessel internals and the pressure vessel cladding that received the highest radiation exposure would have to be removed and shipped off site to a controlled burial ground. Entombment could be accomplished in about 42 months after shutdown.

The estimate of the occupational radiation dose of entombment, on the same basis used for immediate dismantlement, is 617 person-rem. The reduction in radiation dose as compared with immediate dismantlement stems from a less thorough initial decon, which results in less contaminated waste to process, and from the avoidance of having to remove the pressure vessel and the reactor coolant system. The estimates of the public radiation dose from entombment, which again are based on the same assumptions as those used in estimating the public radiation dose from immediate dismantlement, are shown in Table 12.4.

Although the inventory of radioactive liquid at reactor shutdown that must be disposed of would be the same as for immediate dismantlement, the volume of liquid waste generated by flushing and decon would be less than the volume from immediate dismantlement because of the removal of only loose contamination. Neutron-activated material, contaminated piping and equipment outside the entombment boundary, and radwaste (liquid concentrates, combustibles, and other DAW) would constitute the solid radmaterials that must be removed from the site. As mentioned before, the important neutron-activated material to be removed consists of the reactor internals and cladding from the inside of the reactor vessel, but some contaminated material would be placed within the entombment structure.

Summary of alternatives. Table 12.4 shows the summary of decommissioning alternatives for the Shippingport station in terms of the activities involved, time required, estimated radiation doses, waste disposal requirements, and costs. Immediate dismantlement results in the highest radiation dose, the least cost, and the largest amount of land committed to the disposal of solid radwastes. However, it also results in the complete removal of radioactivity and the release of the facility and/or site for unrestricted use just a few years after the facility ceases operation. SAFSTOR results in less radiation dose and less land committed to the disposal of radwastes than immediate dismantlement and also costs more than the latter. It prohibits unrestricted use of the facility and site until deferred dismantlement is completed. Entombment is intermediate between immediate dismantlement and SAFSTOR in cost and radiation dose as well as in the amount of land committed to the disposal of radwaste if the land area occupied by the entombed structure itself is included. Entombment may

Table 12.5 Factors contributing to alternative selection

Public health and safety	Cost

Radiation exposure
  During decommissioning
  Transportation
  Accident consequences
 	
Occupational safety
  Radiation exposure and personnel safety during
    decommissioning
  Accident consequences
 	
Environmental impact
  Site dedication
  Protected storage facility form
  Program accomplishment impact
  End product, site/facility use
  Waste type and volumes
  Repository availability
	Program costs: labor, materials, equipment rental,
	  services, waste containers, transportation,
	  burial, etc.
	
	Safe storage costs: duration of storage period
	  facility operation, security/surveillance,
	  environmental monitoring, etc.
	
	Value of site/facility for future use
	
	Impact of alternatives on financing methods,
	  regulatory interaction
	
	Other influences
	  Federal/state/local regulations
	  Decommissioning process
	  Required safe storage period and condition
	  after that period

From Shippingport Station Decommissioning Project (SSDP)—A progress report by G. R. Mullee and J. M. Usher, in *Spectrum '86, Proc. ANS International Meeting on Waste Management and D/D*, Niagara Falls, New York, vol. II, p. 1213, September 1986. Copyright 1986 by American Nuclear Society. Reprinted by permission.

not be a viable alternative for decommissioning the Shippingport station because of the low elevation of the containment building subbasement relative to the normal elevation of the Ohio River and the possibility of seepage into the entombed structure.

12.2.2 Factors Contributing to Alternative Selection

From the example in the previous sections, it can be shown that a number of factors will contribute to alternative selection. These factors are listed in Table 12.5. The assessment of alternatives and decision on the decommissioning mode can be considered as a preconceptual engineering assessment.[12] As indicated in Table 12.5, the environmental impacts of decommissioning alternatives, including public and occupational safety, are important factors. A review of environmental considerations is called for by the NEPA, and these impacts are described in Section 12.3.

12.3 DECOMMISSIONING ENVIRONMENTAL IMPACTS AND REGULATORY GUIDES

12.3.1 Environmental Impact Statement

NEPA requires government agencies to determine the need for an EIS for proposals or actions significantly affecting the quality of the human environment. The statement must report the following:[14]

1. The environment impact.
2. Any adverse environmental effects that cannot be avoided.
3. Alternatives to the proposed action.
4. Relationship between the local short-term uses of the human environment and the maintenance and enhancement of long-term productivity.
5. Any irreversible and irretrievable commitments of resources involved.

12.3.2 Regulatory Guides

Regulatory guides[7] define the key elements to be addressed in the licensing phase related to decommissioning and dismantling as follows:

1. Long-term use of the land and postdecommission site condition.
2. Amount of land irretrievably committed.
3. Environment consequences.
4. Cost of decommissioning.

The environmental assessment of a specific decommissioning program should include:

1. Impact on land resources.
2. Occupational and nonoccupational radiation exposures.
3. Industrial safety consideration.
4. Nonradiologic effluent releases.
5. Sociological/economic impacts.
6. Program-related resource commitments.

The environmental effects of decommissioning alternatives include direct and indirect effects such as occupational and public radiation dose (including that from transportation), aesthetic impact of the protected storage facility, and the effects of program accomplishment on financing, the labor force, other socioeconomic respects (housing, school, traffic), and the use of materials and natural resources (e.g., energy requirement and depletable material such as concrete, chemicals, gases, and water). The effects on urban quality and historic and cultural resources also should be considered, as these are the effects of end-product use interaction with the environment. Possible conflicts with government actions or regulations and mitigation of adverse environmental effects should be considered as potential environmental impacts of decommissioning as well.

12.3.3 Evaluation of Impacts

There are no formal regulations or guidelines for evaluating the environmental impacts of decommissioning. However 10CFR50.82 (a)(4)(i) states that a post-shutdown decommissioning activities report (PSDAR) include "a discussion that provides the reasons for concluding that the environmental impacts associated with site-specific decommissioning activities will be bounded by appropriate previously issued environmental impact statements."[16] Evaluations may be conducted using methods consistent with current licensing practices. Thus the following methods may be used in a decommissioning assessment:[5]

1. Comparison of environmental impacts of decommissioning with the impacts caused by the construction or operation of the facility.
2. Comparison of decommissioning environmental impacts with federal, state, and local environmental regulations.
3. Pathway analysis of effluent releases to the environment.

12.4 DECON AND DECOMMISSIONING TECHNIQUES AND TECHNOLOGY DEVELOPMENT

12.4.1 Decon Plan and Specifications

A decon plan is prepared to provide a basis for the project participants in organizing the decon project activities and an impetus for making program decisions as well as to obtain concurrence from the licensing agency for the decon of the facility. The plan focuses on the cleanup criteria for decon of the facility, including limits for surface contamination and dispersed activity. Examples of radiologic cleanup criteria are those provided in the decon plan of Aerojet's California Depleted Uranium Manufacturing Facility, which was based on State of California guides and supplemented by EPA and NRC guides and standards.[15] These criteria are as follows:

1. Removable contamination, determined by smearing with a dry filter: 1000 dpm/100 cm^2 (1 dpm $=$ 1 disintegration per minute, or 1/60 Bq, or 0.4 pCi).
2. Average total contamination, based on a maximum area of 1 m^2: 5000 dpm/100 cm^2.
3. Maximum total contamination, based on an area of not more than 100 cm^2: 15,000 dpm/100 cm^2.
4. External radiation exposure to less than 0.10 mSv/year above natural background (net residual external gamma exposure rate).
5. Dispersed activity in soil, 1.3 Bq/g (33 pCi/g).

Note that the external whole-body gamma dose from surface contamination of 5000 dpm/100 cm^2 of depleted uranium is about 0.01 mSv/year, based on 100% occupancy time and dose rate conversion factors of references 17 and 18. Often, decon work may be performed by a contractor, and specifications for the decon work must be included in the contract, broken down into detailed tasks. The performance of the specifications will require decon not only to the release criteria, but also to ALARA levels. The example of such specifications from reference 15 includes such items as

1. Clean a section of pipe.
2. Remove jib crane foundations.

3. Remove roof ventilators.
4. Remove tanks.
5. Clean/chip out a section of floor control joints.
6. Clean lights.
7. Remove insulation.

12.4.2 Decon Operations

Some activities performed at the Babcock & Wilcox (B&W) Lynchburg site research reactor and fuels development facility exemplify such decon operations,[19] as described in this section.

Removal of paint, floor tile, and tile cement from walls, ceilings, and floors was necessary because many of the walls and ceilings had been repainted and new floor tile had been installed as laboratories were converted to other uses. A potential existed for alpha contamination to be hidden under the new paint and floor tile. Where contaminated areas were found, walls were decontaminated by removing portions of cinder block and floors were decontaminated by chipping up portions of the concrete floor.

Removal of hot drain lines and of the floor from above each hot drain line that was installed under the floor was done after the base concrete floor had been surveyed and released. This uncontaminated concrete rubble was disposed of as clean landfill. Each hot drain line was cut into sections, removed, and surveyed. If the surface activity was less than the specified limits, the pipe was released as clean scrap. Otherwise, the pipe was decontaminated or disposed of as contaminated waste.

Samples were taken of the soil lying under the drain line after the pipe was removed, to examine its radioactivity. A special soil assay program was developed using gamma spectroscopy instead of conventional radiochemistry, which was judged to be too slow, tedious, and expensive.

Extensive decon of masonry and steel surfaces was needed, which meant removing paint as well as floor mastic from concrete and block surfaces. Through extensive testing of various methods, the following preferred methods were selected: chipping to remove paint, chemicals to remove tile mastic, and scabbling and wet scrubbing for the removal of concrete.

Decon of 5000-gal underground, concrete retention tanks posed some different problems. Tank sludge had to be removed, solidified, and disposed of. Since the tanks constituted enclosed spaces, special ventilation and respiratory protection were needed. Descaling, scabbling, and wet scrubbing were all successful in cleaning retention tank surfaces.

12.4.3 Chemical Decon[20]

Chemical decon of reactor systems has been successfully demonstrated in decommissioning programs as well as at operating commercial power plants for maintenance and refurbishment. In most cases, the objective has been to reduce radiation levels so that workers may have a longer access time to segment piping and components. In general, the processes now in use for maintenance of operating plants do not decontaminate surfaces to unrestricted access levels since care must be taken to avoid deterioration of a system that must be subsequently operated. For decommissioning this concern does not exist and thus more vigorous agents (e.g., stronger acids) may be used. Since most of the materials in the facilities are not activated but are contaminated on the surface, they can, in principle, be decontaminated to unrestricted release levels. Note that complete decon could significantly decrease the amount of LLW requiring disposal.

12.4.4 Segmentation and Demolition

Segmentation processes. Segmenting of piping, tanks, and ancillary components is a major activity in a dismantling program, especially when radioactive contamination or activation is present. Consequently, remote removal may be necessary, in which case remote cutting will be required. Various segmenting processes are described in reference 5, including the use of plasma arc, oxygen burner, thermite reaction lance, explosive cutting, hacksaws and guillotine saws, circular cutters, abrasive cutter, and arc saw. A summary of the application characteristics of each process is given in Table 12.6.[5]

Plasma arc process A dc arc between a tungsten electrode and any conducting metal is established in a gas (e.g., argon) stream at a temperature of 10,000–24,000°C

Table 12.6 Application characteristics of segmenting processes

Process	Application	Relative cost	Notes[a]
Plasma arc	All metals ≤6 in. (15 cm)	High	P, R, S
Oxygen burner	Mild steels, all thicknesses	Low	P, R, S
Thermite reaction lance	All metals, all thicknesses	Low	P
Explosive cutting	All metals ≤6 in. (15 cm)	High to very high	R
Hacksaws and guillotine saws	All metals, piping ≤18 in. (45 cm) diameter	Low	P, R
	All metals, piping or stock ≤24 in. (60 cm)	Low	S
Circular cutter	All metals, piping ≤6 in. (15 cm) diameter with wall thickness ≤3 in. (7.5 cm)	Low	P, R
Abrasive cutter	All metals, piping or stock ≤2 in. (5 cm) chord	Low	P
	All metals, piping or stock ≤8 in. (20 cm) chord	Low	S
Arc saw	All metals ≤36 in. (91 cm) chord	High	S

[a]Recommended operating modes for the cutting processes: P, portable application where personnel bring the process equipment to components being disassembled; R, remote application where remotely operated mechanisms are required to segment components; S, stationary application where material is brought to a permanently established workstation for segmenting.

From Decommissioning by T. S. LaGuardia, in *Radioactive Waste Technology*, edited by A. A. Moghissi et al., p. 499, ASME/ANS, New York, 1986. Copyright 1986 by American Society of Mechanical Engineers. Reprinted by permission.

(17,500–42,700°F). Figure 12.4 shows a schematic representation of a torch system in position for segmenting a reactor vessel for disassembly in the Elk River Reactor and Sodium Reactor Experiment.[5,21,22]

Oxygen burner A flowing mixture of a fuel gas and oxygen is ignited at the orifice of a torch. This is sometimes referred to as oxyacetylene cutting. An oxygen-burning torch ordinarily cannot cut stainless steel, aluminum, other nonferrous metals, or ferrous-high percentage alloy metals.

Thermite reaction lance An iron pipe is packed with a combination of steel, aluminum, and magnesium wires and a flow of oxygen gas is maintained, with a temperature at the tip in the range 2250–5500°C (4000–10,000°F).

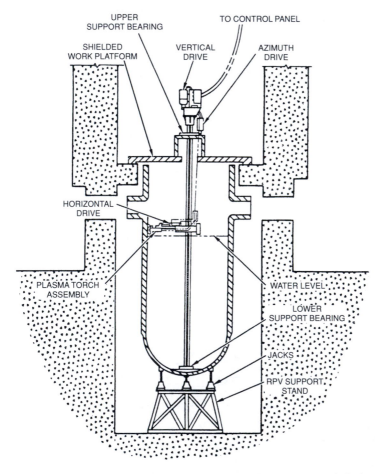

Figure 12.4 Plasma torch system for reactor vessel. From Decommissioning by T. S. LaGuardia, in *Radioactive Waste Technology*, edited by A. A. Moghissi et al., p. 499, ASME/ANS, New York, 1986. Copyright 1986 by American Society of Mechanical Engineers. Reprinted by permission.

Explosive cutting An explosive core, surrounded by a casing of lead, aluminum, copper, or silver, causes a high-explosive jet of detonation products of combustion through deformed casing metal. Explosive cutting is normally used when the geometry of the object is too complex or when several cuts must be made simultaneously (e.g., removal of a large prestressed beam).

Hacksaws and guillotine saws These are relatively common industrial tools for cutting metals with a reciprocating-action, hardened steel saw blade.

Circular cutter This is a self-propelled, circular saw that cuts as it moves around the outside circumference of a pipe on a track. Historically, circular machines have been used primarily for pipe weld preparation, but they are effective decommissioning tools for segmenting pipe and round vessels.

Abrasive cutter This is an electrically, hydraulically, or pneumatically powered wheel formed of resin-bonded particles of alumina or silicon carbide. It cuts through the workpiece by grinding the metal away. Cutting rates for stationary abrasive wheels are approximately 6.5 cm^2 (1 in.^{2}) of cut area every 7 sec.

Arc saw This is a circular toothless saw blade that cuts conducting metal without physical contact with the workpiece. The rotation of the blade is essential to its operation.

Concrete demolition. Nearly every decommissioning program involves either the demolition or surface decon of a concrete structure, which was described in Section 12.4.2. Activated concrete in the region immediately surrounding the core belt line represents the most difficult concrete removal activity because of the relatively high radiation dose and potential for release of radioactive particles during demolition. The concrete removal processes and application characteristics of each process are listed in Table 12.7.[5]

Controlled blasting This is recommended for demolition of massive or heavily reinforced thick concrete sections. The direction of material movement is controlled by a delayed firing technique.

Wrecking ball Typically, this is used on nonreinforced or lightly reinforced concrete structures less than 1 m (3 ft) thick. A 2- to 5-ton ball or flat slab suspended from a crane boom is used.

Air and hydraulic rams Rams are used for concrete structures less than 0.6 m (2 ft) thick with light reinforcement and for interior demolition in confined areas. They cause low noise and low vibration. With the ram head mounted on a backhoe, the operator has approximately a 6- to 7.6-m (20–25 ft) reach and the ability to position the ram in structures with limited access.

Flame cutting A thermite reaction process in which a powdered mixture of iron and aluminum oxidizes in a pure oxygen jet is used. The temperature in the jet is

Table 12.7 Application characteristics for concrete removal processes

Process	Application	Feasibility	Relative equipment cost
Controlled blasting	All concrete ≥ 2 ft (0.6 m)	Excellent	High
Wrecking ball	All concrete ≤ 3 ft (1 m)	Excellent for nonradioactive concrete; not recommended for radioactive concrete	Low
Air and hydraulic rams	Concrete ≤ 2 ft (0.6 m)	Good	Low
Flame cutting	Concrete ≤ 5 ft (1.5 m)	Fair	Low
Rock splitter	Concrete ≤ 12 ft (3.7 m)	Good	Low
Bristar demolition compound	All concrete ≥ 1 ft (0.3 m)	Fair	Low
Wall and floor sawing	All concrete ≤ 3 ft (1 m)	Good	Low
Core stitch drilling	Concrete ≥ 2 ft (0.6 m)	Poor	High
Explosive cutting	Concrete ≥ 2 ft (0.6 m)	Good	High
Paving breaker	Concrete ≤ 1 ft (0.3 m)	Poor	Low
Drill and spall	Concrete ≤ 2 in. (5 cm)	Excellent	Low
Scarifier	Concrete ≤ 1 in. (2.5 cm)	Excellent	Low

From Decommissioning by T. S. LaGuardia, in *Radioactive Waste Technology*, edited by A. A. Moghissi et al., p. 499, ASME/ANS, New York, 1986. Copyright 1986 by American Society of Mechanical Engineers. Reprinted by permission.

approximately 8900°C (16,000°F) and causes rapid decomposition of the concrete in contact.

Rock splitter A splitter is used for fracturing concrete by hydraulically expanding a wedge into a predrilled hole until tensile stresses are large enough to cause fracture. The unit is powered by a hydraulic supply system and operates at a pressure of 500 kg/cm^2 (7100 psi). It may be operated by air pressure, gasoline engine, or electric motor sources. Figure 12.5 shows a schematic of the splitter operating principle.

Bristar demolition A chemically expanding compound is poured into predrilled holes and causes tensile fractures in the concrete on hardening. Bristar is a proprietary compound of limestone, siliceous material, gypsum, and slag, which will develop to over 300 kg/cm^2 (4300 psi) within 10–20 hr.

Core stitch drilling A diamond- or carbide-tipped drill bit is used in close-pitched drilling of holes in concrete. The hole pitch is such that there is very little concrete left between adjoining holes (less than half of the radius of the holes). The method is not recommended for reinforced concrete.

Explosive cutting Described in the preceding section.

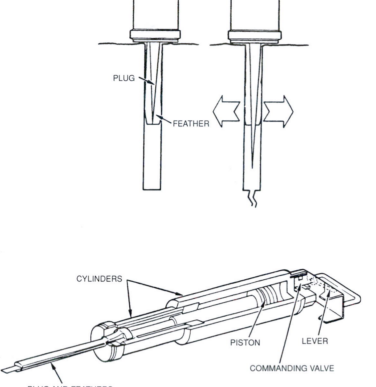

Figure 12.5 Schematic of rock splitter. From Decommissioning by T. S. LaGuardia, in *Radioactive Waste Technology*, edited by A. A. Moghissi et al., ASME/ANS, New York, 1986. Copyright 1986 by American Society of Mechanical Engineers. Reprinted by permission.

Paving breakers (jackhammer) Concrete and asphalt are removed by mechanically fracturing localized sections of the surface by the impact of a hardened tool steel bit of either a chisel or moil point shape.

Drill and spall Contaminated surfaces of concrete are removed without demolishing the entire structure by drilling holes 2.5–3 cm (1–1.5 in.) in diameter and 7.6 cm (3 in.) deep and inserting a hydraulically operated spalling tool (an expandable tube).

Scarifier (trade name Scabbler) This is best suited for the removal of thin layers (up to 2.5 cm or 1 in. thick) of contaminated concrete, using pneumatically operated piston heads that strike the surface to chip off the concrete.

12.4.5 Asbestos Insulation Removal[23]

Because of the specialized nature of the work, the unique hazards involved, and the special training and precautions required, asbestos removal from diverse areas of components and piping in the Shippingport Station Decommissioning Project (SSDP) was assigned to a single subcontractor. Development of an asbestos removal specification for a nuclear decommissioning project requires consideration of asbestos regulations, disposal, ALARA principles, wastewater generation, and the dismantling plan. The cost–benefit study for asbestos disposal showed that it was cost-effective at SSDP to ship all asbestos generated in radiologically controlled areas as radwaste and to ship it for burial in a plywood overpack box. ALARA considerations can influence asbestos removal in a highly contaminated nuclear installation, but SSDP was not significantly affected since it is very clean from a radiologic standpoint. Water is frequently used to assist in asbestos removal work for cleanup and control of airborne fibers, but the SSDP subcontractor was instructed not to generate wastewater in the radiologically controlled parts of the plant. Project dismantling plans also can influence the scope of asbestos removal. For instance, the SSDP plan called for the removal of steam generator pressurizer, blowoff, and flash tanks and gas stripper in one piece, as opposed to the in situ segmentation. An engineering evaluation resulted in a procedure in which the pressurizer, blowoff, and flash tanks were stripped of asbestos prior to removal. The precautions of concern were those related to the asbestos worker and the work environment—for instance, respiratory protection, immediate analysis of air sample filters each evening, protective clothing, training in safe asbestos removal techniques, and provision of air lock entry/exit of asbestos work area. Primary plant protection features include work area containments, bagging, radioactive asbestos storage area, and egress routes for asbestos-filled bags.

12.4.6 Decommissioning Technology Development

Additional technology development in four primary areas to reduce radiation exposure, waste volume, and costs has been recommended by an NEA expert group:[20]

1. In situ chemical decon of piping and components. Methods for accomplishing complete decon while generating a minimum volume of waste are being developed. They include electropolishing (a process that uses an electric current and a solvent to remove microscopic layers of metals) and decontaminant gels (gel-based decontaminant compounds that are applied to the area to be decontaminated).[24]

2. Dismantling operations (including remote-operated equipment and tools for segmenting piping, components, and reactor vessel and internals). Remote segmentation of the reactor vessel and internals was successfully demonstrated in decommissioning the Elk River Reactor and the Sodium Reactor Experiment (Figure 12.4). Additional development work is being done in several countries to further automate the process by using microcomputer control of cutter location, cutting parameters, and verification of cut completion. Improvements are desirable for equipment setup and relocation as well as for removal of segmented sections of the vessel with reduced worker exposure. Development work includes both mechanical and thermal cutting methods. Research to develop remotely operated disassembly of piping and components to combine crimping and arc sawing or shearing is also under way in the United Kingdom to reduce the worker exposure. Much work has been done in the United Kingdom to develop a large-scale (2.5-m) diamond saw cutting machine. Further work is desirable to develop methods for the quick removal of cut

sections. The process for cleaning out debris following blasting or ram-breaking might be improved by adopting mining machinery for more rapid handling and removal of materials. Research to develop safer and more rapid techniques for removing asbestos insulation from pipe is being done in West Germany and the United States.

    3. Methods and equipment for waste management, volume reduction (VR), and efficient packaging. The disposal of decommissioned waste can be facilitated by active development of VR techniques, including VR for packaging piping and components. Methods that are being used to reduce the volume of LLW from operating reactors can also be used to reduce the volume of LLW from decommissioning. Further attention to waste management techniques and procedures was needed in decommissioning waste types such as tritiated wastes from HWRs, mobile cesium waste from decon, and high-radioactivity waste from metallic reactor internals.

    4. Improved measuring techniques to facilitate the segregation of different categories of waste, thereby reducing waste volume for disposal. Better technology is needed for rapid measurement of radioactivity over large areas of materials in applying the exemption level of contamination. The exemption, or unrestricted release, level of contamination is a level at which materials may be released for general use or disposal without further concern for residual radioactivity; it can be defined in terms of the potential dose rate to individuals who may use these materials or in terms of the risk of adverse health effects. In the United States and some other countries release of materials with residual surface contamination is done in conformity with the U.S. NRC Guide 1.86. Because scrap materials are used in international commerce, there is a need for international agreement on acceptable release levels of a residual surface contamination.

12.5 DECOMMISSIONING EXPERIENCE

This section describes the decommissioning of five different types of nuclear facilities. These are not necessarily typical decommissioning projects. However, they do illustrate the range of factors that might be considered when planning and conducting a decommissioning project. The facilities are (1) Shippingport, decommissioned by DOE as a demonstration project, (2) West Valley commercial fuel reprocessing plant, (3) a uranium manufacturing facility, (4) Three Mile Island-2, a commercial nuclear power plant that suffered a partial meltdown, and (5) a test reactor and a power plant in Europe.

12.5.1 Shippingport Station

The Shippingport power station was constructed during the mid-1950s as a joint project of the U.S. Government and the Duquesne Light Company for the purpose of developing and demonstrating PWR technology and for generating electricity. The station consists of a PWR last rated at 72 MWe, a turbine generator, and associated facilities. Located on the south bank of the Ohio River at Shippingport, Pennsylvania, on land owned by Duquesne Light Company, the reactor and steam-generating portions of the station are owned by the DOE and the electrical generating portion of the station is owned by Duquesne Light Company. From December 1957 through December 1980, the station produced more than 6.6 billion kW-hr of electricity from three cores of reactor fuel (the first two being PWR cores and the last one a light-water breeder reactor that was shut down after end-of-life testing and removed in 1985).

The station consists of a 275 × 60 ft fuel handling building containing the reactor containment chamber, service building, turbine building, radwaste processing building, administrative building, and other support buildings. Figure 12.6a shows the SSDP site plan and Figure 12.6b shows the systems, components, and structures of the station, with four primary system loops inside the containment chambers. Each loop consists of two main coolant isolation valves located next to the PV nozzles, a second set of isolation valves, the loop check valve, horizontal steam generators, and the main coolant pump. A concrete shield wall separates the loops to reduce radiation

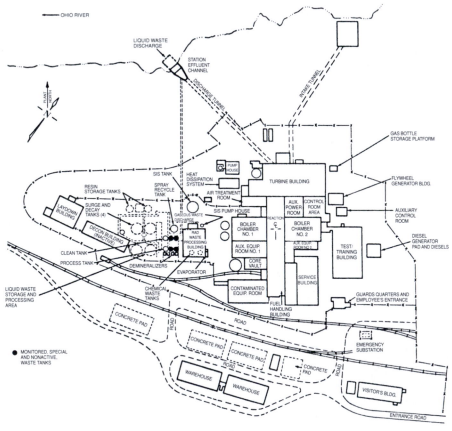

(a)

Figure 12.6 (a) SSPP site plan. (b) SSDP systems, components, and structures prior to removal. From *Decommissioning U.S. Reactors: Current Status and Development Issues* by K. E. Schwartztrauber, EPRI NP-5494, Final Report, 1988. Copyright 1988 by Electric Power Research Institute. Reprinted by permission.

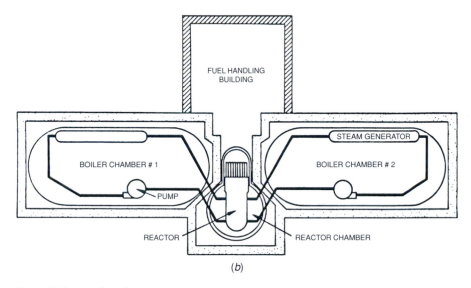

(b)

Figure 12.6 (**continued**)

levels when one of the loops requires maintenance. The steam and feedwater lines to and from the balance of the plant pass through the auxiliary equipment chamber. The radioactivity and waste material inventory are listed in Table 12.8.[25] Of residual radioactivity on the site, 99% is confined to the interior of the reactor pressure vessel. Typical readings are <100 mR/hr, with "hot spots" ranging from 100 to 500 mR/hr. The total exposure for the entire project is estimated as 1000 person-rem. There were up to 30 subcontractors, with a total work force averaging 200 for the three peak activity years (1986–1988). According to DOE guidelines for allowable levels of residual radioactivity at remote sites in the Surplus Facilities Management Program (SFMP),[26,27] doses to the public from this residual activity must not exceed 500 mrem/year (for up to 5 years) or 100 mrem/year for a lifetime. Furthermore, the concentrations must be low enough so that additional action that might be taken to reduce them cannot cost-effectively decrease the dose to the public (ALARA). Thus, a cost–benefit analysis based on actual conditions at Shippingport would identify these "low enough" concentrations to be release criteria for the site.[26]

Prior to the final phase—physical decommissioning operations—there were phases of preconceptual engineering decommissioning assessment (as described in Section 12.2.2), conceptual engineering baselines, and a detailed engineering decommissioning plan. The plant was permanently shut down in October 1982. Defueling of the station was completed during the summer of 1984. The last spent fuel shipment left the station on September 6, 1984, after which the responsibility for the station was transferred from DOE Naval Reactors to DOE Richland Operations Office, and the operational responsibility for station surveillance, maintenance, and operations was transferred on schedule from the Duquesne Light Company to the decommissioning operations contractor. With 1 year of caretaker site preparation, the decommissioning was initiated in September 1985. The removal of the 40-ton pressurizer, flash tank, and blowoff tank was completed in November 1985; the irradiated components loading was completed in March 1987,

Table 12.8 Radioactivity and waste material inventory of SSDP

Radioactive inventory	
Reactor vessel package	13,300 Ci
Activated core components	19 Ci
Resins	120 Ci
Large vessels/steam generators	20 Ci
Piping/components/surfaces	6 Ci
Liquids	1 Ci
Total activity	13,466 Ci
Reactor vessel package	770 tons
Radwaste volume	3,000 cubic yards
Chamber steel	22,400 tons
Contaminated concrete	50 cubic yards
Noncontaminated rubble	15,000 cubic yards
Contaminated pipe	56,000 linear feet
Noncontaminated pipe	55,000 linear feet
Asbestos waste	400 cubic yards
Waste quantities	
Solids	
Reactor vessel package	10,200 ft^3
Resins/solidified wastes	4,700 ft^3
Solid wastes	63,260 ft^3
Total volume	78,160 ft^3
Liquids	
Reactor coolant	14,000 gal
Canal pool water	320,000 gal
Waste processing tanks	52,200 gal
Neutron shield (cromates)	24,000 gal
Total volume	411,200 gal

From Shippingport Station Decommissioning Project Technology Transfer Program by L. A. Pasquini, in *Proc. 1986 Joint ASME/ANS Nuclear Power Conference*, p. 89, Philadelphia, Penn., 1986. Copyright 1986 by American Society of Mechanical Engineers. Reprinted by permission.

structural removal was completed in 1989, and reactor pressure vessel (RPV) package preparation and removal was scheduled for later.

An important feature of the SSDP was the one-piece removal of the RPV. It was estimated that this approach would save about $7 million, reduce personnel radiation exposure from 250 to 140 person-rem, and reduce the total decommissioning schedule by about 1 year compared with segmentation of the vessel and internals. Figure 12.7 shows how the pressure vessel is shielded and prepared for removal. It weighed about 820 tons. Figure 12.8 shows the lifting tower to be used to remove the vessel. The final design, calculations, specifications, and drawings for the RPV/neutron shield tank (NST) lifting skirt and beam were completed in 1985 and reviewed and approved by an independent design review team. An order was placed for recommending an RPV fill material (grout and concrete), and low-specific-activity (LSA) shipments of the package were planned by

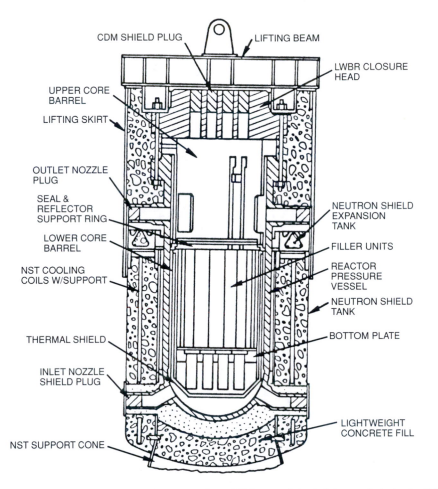

Figure 12.7 SSDP reactor vessel package. From SSDP start of physical decommissioning by F. P. Crimi, in *Proc. 1986 Joint ASME/ANS Nuclear Power Conference*, p. 85, Philadelphia, 1986. Copyright 1986 by American Society of Mechanical Engineers. Reprinted by permission.

barge, truck, or rail, as appropriate, in compliance with federal regulations, DOE orders, and state and local regulations. The estimated cost for SSDP is $98.3 million (1987 dollars) and is broken down as follows:[28]

Engineering	$6.0 million
Technical management support	7.8
Station operator support	1.2
Site management and support	37.5
Decommissioning activities	35.0
Subtotal	87.5
Contingency	10.8
Total	$98.3 million

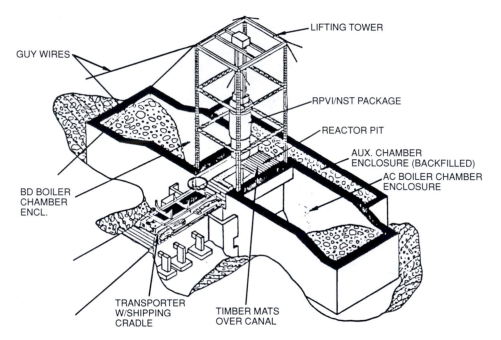

Figure 12.8 SSDP lifting of RPV/NST package. From SSDP start of physical decommissioning by F. P. Crimi, in *Proc. 1986 Joint ASME/ANS Nuclear Power Conference*, p. 85, Philadelphia, 1986. Copyright 1986 by American Society of Mechanical Engineers. Reprinted by permission.

Finally, as one of the program's objectives is to provide planning data for future decommissioning projects, there is a technology transfer program to collect and archive the decommissioning data base and make it available to the nuclear industry. These data include the following:[28]

1. Project management information
2. Dismantling and demolition data
3. Health safety and environmental protection data
4. Radiologic controls data
5. Decon data
6. Special tools, services, and materials handling data
7. Quality assurance data
8. Training programs

Shippingport decommissioning was completed in December 1989 and cost $91.3. Thus the project finished 4 months ahead of schedule and $7 million under budget.[29]

12.5.2 Decontamination and Decommissioning (D/D) of West Valley Demonstration Project (WVDP)[30]

D/D at the West Valley site is a good example of the radioactive cleanup and dismantling operation associated with a nuclear fuel reprocessing plant. The objective of the previtrification D/D operation is to make the various shielded cells and other plant facilities suitable for installation and operation of the HLW processing and vitrification systems. The reprocessing plant, which is the largest building on the 3300-acre West Valley site (Figure 12.9), has six elevations. There are 235 area divisions in this building, including a fuel storage pool, 24 shielded cells of various sizes, several laboratories, and several auxiliary areas. Contamination levels up to 1800 rem/hr were present in some of the cells and radiation fields in excess of 1 rem/hr existed in many other cells.

Before starting the D/D work, the radiation and contamination levels were characterized on a cell-by-cell basis by lowering thermal luminescence dosimeters (TLDs) into the cells and taking temperature measurements and airborne samples as well as smear samples from the floors and walls. Detailed procedures with signoff tasks were prepared, reviewed, and approved for each step of the D/D work. These procedures were practiced by the radiation worker teams before entering the contaminated areas through a double-chambered contamination control tent erected and sealed around the entry to the cell. The D/D work activities included the following:

1. *Master/slave manipulator repair shop.* The shop has been decontaminated from loose surface contamination of the 105 dpm per 100 cm^2 and radiation fields up to 20 mR/hr to an uncontrolled-area level. All of the tools, equipment, and furnishings were removed, walls were stripped and repainted, and the contaminated concrete floor was removed and disposed of as LLW. A new concrete floor was poured with embedded stainless steel anchor strips to which stainless steel sheets were welded extending 18 in. up the walls from the floor. This provides WVDP with an easily decontaminated D/D facility.

2. *Plutonium product storage area.* This area has been cleaned out and decontaminated, refurbished, and converted for radioactive waste handling, packaging, and compaction. A segmented gamma scanner for surveying waste drums has been installed in this area. The equipment is providing volume reduction of LLW, which results in improved burial ground utilization.

3. *Laboratory areas.* The standards and quality control laboratory and the hot laboratory were decontaminated, dismantled, and refurbished as an analytical chemistry laboratory and radiochemistry laboratory, and the old mass spectroscopy laboratory was decontaminated, dismantled, and refurbished as a quality assurance instrumentation laboratory.

4. *Chemical crane room.* The room previously had radiation fields of 50 mR/hr and hot spots of 10 R/hr. It has been decontaminated by stripping the paint off the concrete floor, grinding the floor surface, repainting and applying strippable coating to the floor, and then covering the floor with Herculite sheeting.

5. *Other D/D work* completed including the equipment decon room, extraction chemical room, and fuel receiving and storage area.

Optimum D/D techniques for each particular application are constantly being researched and developed. An example is the adaptation of standard tooling (concrete scarifier) to automate decon operation and thereby save time and reduce personnel exposure. Volume reduction of compactible LLW was accomplished with a 445-kg radwaste compactor. Utilization of the existing on-site burial ground was improved by installing a 2-m-diameter by 17-m-long carbon steel vertical caisson in the burial ground for monitored retrievable storage or permanent disposal while maximizing the

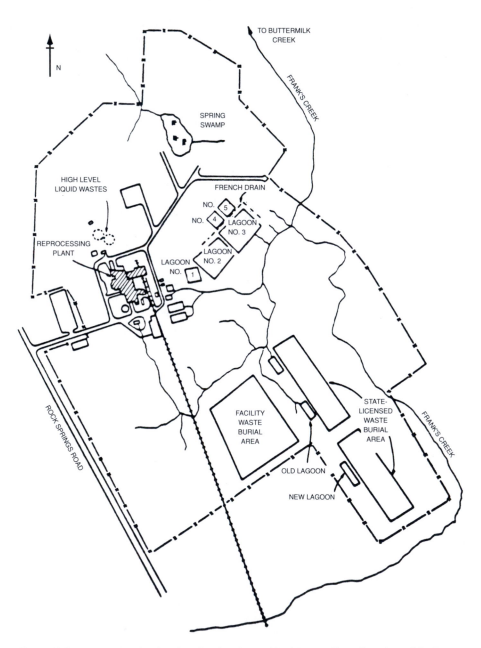

Figure 12.9 West Valley site showing plant location and burial areas. From Overview of the West Valley Demonstration Project by J. L. Knabenschuh and W. H. Hannum, in *Advances in Ceramics*, vol. 8, *Proc. 2nd International Symposium on Nuclear Waste Management*, p. 93, April 1983. Copyright 1983 by American Ceramic Society. Reprinted by permission.

volume of waste buried per unit surface area. Arrays of seven 208-L drums were remotely emplaced in the caisson using a radio-controlled Vac-U-Lift handling device.

12.5.3 Decommissioning of Depleted Uranium Manufacturing Facility[15]

Aerojet Ordnance Tennessee, Inc. (Aerojet), decommissioned its depleted uranium manufacturing facility in California, where manufacturing and research and development activities had been conducted since 1977 under a State of California Source Material License. Its principal activity had been production of GAU-8 penetrators from depleted uranium rod stock, and the building was a commercial warehouse-type structure. Manufacturing activities ceased in mid-May 1986 and manufacturing machines and support equipment were removed in early 1987. The planning and initiation of decontamination of the facility included characterization of the facility and the associated contamination, preparation of a decommissioning plan, and development of detailed technical specifications.

Since the facility was to be released for unrestricted use, it was decontaminated to ALARA levels. The general layout of the property and building is shown in Figure 12.10. The 5800-m² masonry commercial structure is located in a commercial zone area. The primary structural items of interest on the outside of the plant building are:

1. Four-stage clarifier prior to the main sanitary sewer connection on the east side of the facility.
2. Surface drain in the east parking area, which discharges to the street.
3. Afterburner, for the evaporator effluent, located on the east side of the facility.
4. Air compressor shed, which is isolated from the main facility and to the northwest of the facility.

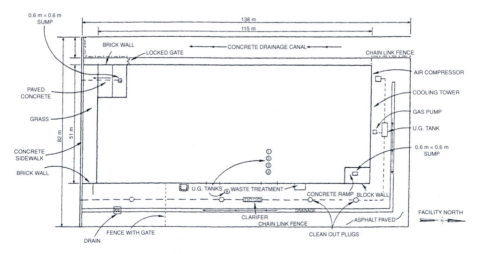

Figure 12.10 General layout of Aerojet Compton site (depleted uranium manufacturing facility). From Decommissioning plan depleted uranium manufacturing facility by D. E. Bernhardt, J. D. Pittman, and S. V. Prewett, in *Waste Management '87*, vol. 1, p. 533, edited by R. G. Post, University of Arizona Press, Tucson, Ariz., 1987. Copyright 1987 by University of Arizona. Reprinted by permission.

5. Wastewater tank under the east parking lot.
6. Gasoline tank and pump, not used by Aerojet.

There were piping and four wastewater tanks under the slab in the building. The tanks were used to store and process contaminated wastewater and were to be removed. There was over 1000 m of piping and electrical conduit in the ceiling area; the piping included compressed air, fire sprinkler, and machine cooling-water systems. Figure 12.11 shows the subgrade sewers of the facility. Much of the inner structure of the building was contaminated with dust containing depleted uranium oxide. The contamination ranged from small amounts (below unrestricted release limits) in the unrestricted area to more significant amounts in the ceiling area of the shop and the shop concrete floor (10-cm slab), which had depleted uranium oxide in pits, cracks, and control joints. There were two primary cracks in the floor, both associated with the operation of swaggers, and a number of joints associated with subslab piping, tanks, and machine footings. Two subgrade concrete pits in the central area of the shop contained steel-lined oil sumps and were part of the vacuum heat treat furnaces.

Routine monitoring was performed throughout the facility for contamination control, worker protection, and compliance with the operating license. Additional monitoring was done to provide a background baseline and assist in the development of specifications, as well as to estimate the required level of effort and costs for decon of the facility. Measurements were taken on the floors in several areas of the shop. Recently cleaned floors measured around 3000 dpm per 100 cm^2, but small pits in the concrete surface, even after normal floor scrubbing, had levels of contamination above 15,000 dpm per 100 cm^2. The radiologic cleanup criteria for the facility were established as described in Section 12.4.1. To reduce contamination levels to ALARA, the

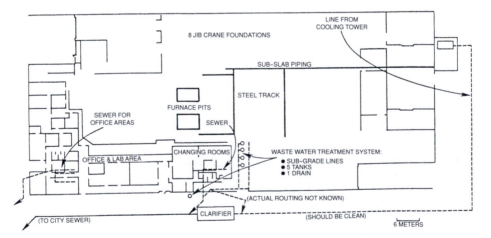

Figure 12.11 Aerojet Compton site subgrade sewers. From Decommissioning plan depleted uranium manufacturing facility by D. E. Bernhardt. J. D. Pittman, and S. V. Prewett, in *Waste Management '87*, vol. 1, p. 533, edited by R. G. Post, University of Arizona Press, Tucson, Ariz., 1987. Copyright 1987 by University of Arizona. Reprinted by permission.

Table 12.9 Decontamination criteria and associated whole-body doses[15]

Criterion	Value	Dose (mSv/yr)
Surface contamination	5,000 dpm	0.01
Dispersed contamination (reduced for shielding)	1.3 Bq/g (35 pCi/g)	0.01
External gamma	0.10 mSv	0.10

From Decommissioning plan depleted uranium manufacturing facility by D. E. Bernhardt, J. D. Pittmann, and S. V. Prewett, in *Waste Management '87*, vol. 1, p. 533, edited by R. G. Post, University of Arizona Press, Tucson, Ariz., 1987. Copyright 1987 by University of Arizona Press. Reprinted by permission.

basic cleaning procedures specified were fully applied, even if the initial contamination was below the limits. The monitoring on the floors and lower walls is based on 1-m grids with at least three recorded measurements and smears per square meter. The monitoring effort for upper walls and ceiling areas was reduced, and attributes-type sampling and statistics was applied. There will be a recorded measurement and smear from each 3-m length of the exterior surface of in-place pipe. The net residual external gamma exposure rate will be less than 0.10 mSv/year, based on State of California criteria (as listed in Section 12.4.1). This will be verified by gamma exposure rate measurements and conversion of soil and surface measurements to gamma exposure rates. The basic decon criteria and associated external gamma whole-body doses are given in Table 12.9. It is evident that the soil and surface contamination criteria are more limiting for this situation than the external gamma dose.

12.5.4 TMI-2 Unique Waste Management Technology

The 1979 accident at the Three Mile Island Unit 2 (TMI-2) commercial reactor at Middletown, Pennsylvania, severely damaged the reactor core and contaminated more than one million gallons of water. Subsequent activities created another million gallons of water. The damaged reactor core represented a new waste form and cleanup of the contaminated water and system components created other new waste forms requiring creative approaches to waste management. Technologies specific to fuel waste management, core debris shipping, processing accident-generated water, and disposing of the resultant waste forms are unique and the experience of the TMI recovery will be valuable to the industry.

Fuel waste management. Fuel and structural core materials from the damaged TMI-2 are loaded in specially designed fuel canisters and shipped by rail in specially designed shipping casks to the DOE's Idaho National Engineering and Environmental Laboratory (INEEL) near Idaho Falls. Loaded casks were carried by rail from Middletown to East St. Louis, Illinois, where the shipments were transferred and shipped to INEEL. This was estimated to take 35–40 shipments during a period of 2 1/2 years. At INEEL, the fuel and core materials will be studied and analyzed as part of the TMI-2 accident evaluation program to provide a complete understanding of the accident sequence and a better understanding of nuclear fuel behavior during severe reactor accidents.

The fuel and core materials will be placed in interim storage at INEEL until a national repository or other alternative becomes available for ultimate disposal. Rail shipment was selected for transporting the fuel and core materials because it was considered safe and economical and would greatly reduce the number of shipments as compared with truck transport (it would take 250 shipments by truck). The shipping carriers were selected because of their extensive expertise in transporting radmaterials. In compliance with regulations, the respective governors' designees in Pennsylvania, Ohio, Indiana, Illinois, Missouri, Kansas, Nebraska, Colorado, Wyoming, and Idaho were notified of the shipping plan.

Canisters and shipping casks. Two specially designed NuPac 125 B rail casks are 280 in. long by 120 in. in diameter and will weigh about 90 tons when fully loaded. Double containment is provided in each cask by the stainless steel inner vessel, which includes a hub-and-spoke arrangement to support tubes holding loaded fuel canisters, and composite-walled (three thick layers of metal) outer vessel. The three layers are an inner shell of 2.54-cm (1-in.)-thick stainless steel and an outer shell of 5.1-cm (2-in.)-thick stainless steel with a 10.2-cm (4-in.)-thick lead layer sandwiched between them for radiation shielding. Attached to each end of the outer vessel are large energy absorbers called overpacks, which are made of stainless steel and filled with foam that crushes on impact and protects the cask body. In addition, to protect the cask in the event of fire there is a thermal shield which consists of wire wrapped around the outer shell and covered by a thin sheet of stainless steel welded over the wire. Thermal shielding is provided by the air gap between the thin sheet and the outer shell. The structural integrity of the cask was demonstrated in drop tests at Sandia National Laboratory simulating severe hypothetical accident conditions and was certified by the NRC. Each cask will hold seven canisters containing fuel and core debris. These canisters, also specially designed form the TMI-2 materials, are made of stainless steel and are 3.8 m (150 in.) long by 35.5 cm (14 in.) in diameter. The integrity of the fuel canisters was demonstrated in a series of full-scale drop tests at Oak Ridge National Laboratory. The canisters experienced no serious damage to the exterior shell or internal structure. Thus, the shipping package of the fuel canisters and the shipping cask provide three separate levels of protection for the radioactive cargo. In addition, each shipment has been thoroughly inspected before leaving TMI by the DOE, NRC, and DOT to ensure that cask, cargo, and rail car meet all necessary federal requirements for safe shipping and the safety requirements of the American Association of Railroads. Radiation surveys are performed prior to the transport, and the track that the shipment would cross is thoroughly inspected prior to the start of shipments by the Federal Railroad Administration.[31]

Processing contaminated water. Water cleanup was accomplished at TMI-2 using two systems: the EPICOR II system, a three-stage ion-exchange cleanup system with prefilters as the first stage and demineralizers as the second and third stages, was used to decontaminate water in the auxiliary fuel handling building. The first stage captured most of the contaminants and resulted in the generation of highly loaded liners that required special handling, while the second and third stages resulted in the generation of demineralizer liners that can be disposed of at a commercial burial site. A solid waste staging facility was constructed for temporary storage of the prefilters at TMI-2. The prefilters were then shipped to INEEL in existing shipping containers and temporarily stored in newly built silos inside the hot cells. High-integrity containers (HICs) were designed and fabricated with a design life of at least 300 years and were to be transported by newly designed shipping containers to a commercial burial site and buried as class C waste. Early processing of contaminated water using the EPICOR II system showed that many hundreds of EPICOR liners would be necessary to decontaminate the total volume of water. To reduce the waste volume being generated, a much more selective SDS was developed. This SDS uses an optimum mixture of inorganic zeolites to remove ions of cesium and strontium, resulting in a smaller number of highly loaded SDS liners that require special handling. The radiolytically generated gases in the liners were managed with a new vacuum outgassing system and hydrogen and oxygen were recombined within the liners. These represent the development of waste management technologies during the TMI-2 program.

12.5.5 Decommissioning Experience in European Countries

As stated in Section 12.1.1, experience in decommissioning was described and technology was exchanged in ANS international meetings. The following two reports on decommissioning work are from references 31 and 32, on the decommissioning of the KEMA Suspension Test Reactor in The Netherlands and the nuclear power plant Niederaichbach in West Germany, respectively.

Decommissioning of the KEMA Suspension Test Reactor.[32] During operation of the KEMA Suspension Test Reactor a powdered fuel of UO_2–ThO_2 was circulated through the primary system of the reactor. Dismantling of such a system, which is contaminated with fission products and alpha emitters, requires a thorough decommissioning plan even if the activity levels are low. Because of the difficulties in measuring any residual alpha activity on the internal surfaces of complicated components, the disposal of decontaminated steel from this reactor posed another problem, which was seldom referred to.

The nearly spherical reactor vessel was connected with piping and such equipment as a heat exchanger, a liquid–gas separator, and the circulating pump through which a suspension of mixed oxide fuel in light water was circulated. The operating conditions were 250°C and 60 bars, the concentration of the 5-μm fuel particles was 400 g/L, and the thermal power of the reactor was 1 MW. The 30-cm-diameter core vessel with 5-mm wall thickness was surrounded by a BeO and graphite reflector and installed inside a pressure vessel. Fission heat from the reactor was transported by the circulating suspension to the primary cooling system in the heat exchanger and then to an air cooler via a secondary cooling system. Hydrogen was injected in the mainstream of the suspension and again extracted in the gas separator to provide for removal and recombination of radiolytic gas, removal of gaseous fission products, and production of particle-free water by condensation.

The nuclear systems (main system, gas purification system, reflector cooling system, sampling system, and instrument sensors) were installed in four big compartments separated by concrete walls. The compartments have been lined with carbon steel plating and will be maintained and reused for other purposes after decommissioning and cleaning. At the top, these compartments were closed by two layers of concrete blocks and made gastight by a flexible, removable sealing sheet. The compartments, which served as a second containment, were brought to lower than ambient pressure by completely burning out the oxygen and then monitored continuously for air inleakage. The reactor hall over these compartments served as a third containment and was also held at underpressure.

The reactor was operated at different power levels during the period 1974–1977. Because the aqueous fuel circulated through the outside loop and small amounts of the mixed oxide fuel remained in the main system and the gas purification system after the bulk of the fuel had been removed, the insides of components and tubing are alpha-contaminated. In addition, the presence of fission products, mainly ^{137}Cs and ^{90}Sr, in the systems mentioned above results in beta and gamma contamination. Many parts of the main system are also activated, especially those inside the pressure vessel. The decommissioning is further influenced by the relatively small dimensions of the components, which makes decontamination of the inside surfaces very time consuming and measurement of the decon results difficult or even impossible. For these special features a compromise had to be found between the effort to prevent spreading of contaminants and the effort needed for volume reduction and decon of the components.

The decommissioning plan called for a collective dose not to exceed 0.2 Sv (20 rem), and no radwaste from the reactor should be left at the KEMA site at the end of the project. The vessel was removed from the system with flanges provided and leak-tightness tested. It was packed in a specially designed container and sent to the United States by air transport, for which special licenses were necessary from the Dutch authorities, Euroatom, and the U.S. DOE. An internal decon of the system was carried out by circulation of water with acids at elevated temperature and subsequent dumping of the liquid into the dump vessel. Internal decon measurements were made to determine the activity levels of the nuclear components and to map the levels in the compartments, which ranged from 0.25 to 5 mSv/hr. The core vessel was internally inspected by using a remotely operated small TV camera, which was brought in at the top of the vessel, and images were collected on videotape. The inside of the vessel was shown to be in perfect condition, without erosion cracks or caking, after 5100 hr of operation. The circulation pump was removed and the pump casing was inspected, which showed erosion of the bottom region, but the casing and the impeller were still acceptable for further use.

To obtain experience in the handling procedures and improve the effectiveness of the working team and tools, the sequence for removal of components was from the conventional secondary systems to the activated and contaminated systems. Special tents were built over the compartments with provision for ventilation, tools, personal protection, and lifting cranes. After removal from the compartments, the components were cut to sufficiently small pieces and packed in 200-L drums, each containing about 300 kg. The empty compartments (cellars) were prepared for decon by installing a steam jet cabin, a glass jet cabin, tanks for chemical decon, water jets, and an electrochemical bath. Special tools were developed to prevent spreading of radmaterial and minimize the exposure dose of the workers; these included hydraulic squeezers for squeezing of tubes, a "water sensing unit" that can penetrate the tube wall and extract active water from a system without spilling, and a remotely operated manipulator. Contaminated steel scrap could be either remelted or stored at an interim location. The former option was preferred because of the unequally distributed contaminants, many of which separated from the molten steel and concentrated in the slag or in the filters of the off-gas system, and because it made possible volume reduction of the waste and a reliable determination of the concentration of radionuclides. However, the remelting of contaminated steel is still restricted to low concentrations of radionuclides. It was estimated that of a total of 80 tons of steel that has to be handled, 60 tons will be remelted. The decommissioning costs were estimated at 3–4% of the construction costs, or U.S. $4 million.

Decommissioning of Niederaichbach nuclear power plant.[33] The Niederaichbach nuclear power plant was permanently shut down in 1974 and brought to safe storage enclosure. The plant, a heavy water-moderated, gas-cooled, pressure tube reactor, was designed for a nominal output of 100 MWe. In the first phase, the safe enclosure, fuel elements, and plant media were removed and the radioactive inventory was enclosed within the safe storage boundary. The follow-up decommissioning plan is divided into five steps:

1. Manual in-place dismantling of the nonradioactive systems.
2. Manual removal of the contaminated material.
3. Remote-controlled dismantling of the activated material.
4. Removal of concrete of the biological shielding by explosives.
5. Conventional demolishing of the building.

Approval for "total dismantlement" in accordance with the German Atomic Law was granted in the summer of 1986.

The dismantling phase for nonradioactive systems and contaminated material will run in parallel because the work can be carried out in two physically separated areas. Contaminated equipment includes all systems—CO_2 systems, He systems, and moderator-related systems—that have been in contact with radioactive media during operation, and the dismantling work will be done with commercial-size industrial tools such as saws, grinders, and thermal cutters. Mechanical cutting tools are mostly used, to avoid aerosols as much as possible. Of the components that do not meet acceptable levels for unrestricted release for recycle or reuse and that must be decontaminated, large parts have to be transported to the crushing house for further cutting. The small parts are collected in special boxes to be decontaminated by melting. The most important contaminated components to be dismantled are the two steam generators, weighing approximately 320 metric tons, including steam generator shells, insulation, upper boiler heads and steam outlets, steam generator pipes, and tubes. The remaining activated parts of the reactor core, the 351 pressure tubes, the moderator tank, and the thermal shielding have to be dismantled, removed, packed, and transported by specially designed remote-controlled manipulators.

A crushing house with recrushing facilities and an adjacent radiation-shielded control station and packaging station were assembled on the reactor platform, where batches of crushed parts are put into containers, concrete-grouted, and closed with lids. A rotary manipulator and manipulator crane were installed in the crushing house. Both manipulating robots are remotely controlled and can operate simultaneously within the cylindrical biological shielding. The dismantling starts with milling off the pressure tube shafts and subsequently cutting the pressure tubes just above the lower neutron shield and removing the steel shielding spheres, followed by mechanical separation of the shield. The pressure tubes are lifted out in batches. The dismantling work will be continued by using the various separating techniques until the entire inside space of the moderator vessel is cleared. Specially designed rotary circular grinding wheel equipment is used for the remaining moderator vessel and the enclosed thermal shielding.

The biological shielding is a concrete cylinder with tight internal and external reinforcements. The inner zone of the cylinder, with a diameter of 8 m (25 ft), has been activated by neutron flux to a depth of about 60 cm (24 in.) within a height of 6 m (19 ft). The maximum specific activity measured is up to 10 Bq/g. The 26-mm (1-in.)-diameter reinforcement wire meshes will be peeled off by precision blasting. For the blasting of the two shells, shot holes must be prepared which are 40 mm (1.6 in.) in diameter and arranged vertically 20 cm (8 in.) and 60 cm (24 in.) in distance to the inner wall in circular patterns. These shot holes are filled with fuse to 90% of the depth, with every 10 of such explosive charges combined to circular sectors. After finishing this crushing, all the equipment is decontaminated, dismantled, and taken out. The remaining building is then decontaminated to far below the release values.

12.6 DECOMMISSIONING COST ESTIMATES

A power reactor licensee must demonstrate that it has sufficient funds to properly decommission the power plant. Minimum amounts required in the decommissioning funds are specified in 10CFR50.75.[16] For PWRs of 3400 MW(thermal) or larger, the amount is $105 million in January 1986 dollars, and for BWRs of the same size, the amount is $135 million. A formula is prescribed for determining the amount required to be set aside for decommissioning smaller reactors. Each year, the licensees must determine the amount required in their decommissioning fund in current-year dollars.

The NRC publishes a document that provides licensees with the information they need to make annual adjustments to the minimum amount required in their decommissioning funds. The document is NUREG-1307.[33] Factors to be considered when calculating decommissioning costs are (1) labor (65% of the total), (2) energy (13%), and (3) waste burial (22%). Labor and energy costs are to be adjusted based on "national producer price indices, national consumer price indices, and local conditions for a given state."[34] Waste burial costs are to be recalculated assuming PWR waste volumes for a reference reactor specified in NUREG/CR-0130, BWR waste volumes specified in NUREG/CR-0672, and burial prices supplied by the burial sites and reported in NUREG-1307.

12.7 COMPUTER CODES

Costs associated with decontamination and decommissioning of a nuclear power plant must be estimated and sufficient funds set aside to cover those costs. The D/D costs vary with the size and type of reactor, typically a PWR or BWR. Pacific Northwest National

Laboratory has developed a pair of Cost Estimating Computer Programs (CECPs) for D/D. These are described below.

CECP (PWR). Estimating pressurized water reactor decommissioning costs. CECP (PWR) is written in Fortran and is designed to run on a PC. It calculates total decommissioning costs incurred to return the site to a condition that would allow the operator's NRC license to be terminated. Some costs included in the calculations are decontamination costs, cost of removing components such as pipes, labor costs, and waste-related costs such as transportation and burial.

CECP (BWR). Estimating boiling water reactor decommissioning costs. CECP (BWR) is a companion to CECP (PWR) and performs the same types of calculations.

12.8 DISCUSSION QUESTIONS

1. What is the difference between decontamination and decommissioning? What is the purpose of each?

2. Into which two categories can the radioactive inventory from a nuclear reactor or facility to be decommissioned be divided?

3. Which regulations govern the decommissioning of nuclear power plants?

4. Which three decommissioning methods were initially allowed by the NRC? Describe each method.

5. What is an EIS? Which decommissioning alternatives must be discussed in an EIS?

6. List at least four of the factors considered in determining which decommissioning alternative is selected.

7. Describe two processes for segmenting pipes, tanks, etc., during decommissioning and two concrete demolition processes.

8. What are some of the similarities and differences of the decommissioning of the five sites discussed in Section 12.5?

REFERENCES

1. Cregnt-Lurie, M., and R. Lourme, French decommissioning experience, in *Spectrum '86, Proc. ANS International Meeting on Waste Management and D/D*, vol. II, p. 1204, J. M. Pope et al., eds., September 1986; also Morell, W., et al., Decontamination experience in Germany as of 1986, in *Spectrum '86, Proc. ANS International Meeting on Waste Management and D/D*, vol. II, p. 1220, September 1986.

2. White, M. G., ed., *Decommissioning, Decontamination, and Environmental Restoration at Contaminated Nuclear Sites*, American Nuclear Society, Washington, D.C., 1994.

3. Abel, K. H., D. E. Robertson, C. W. Thomas, E. A. Lepel, and J. C. Evans, Residual Radionuclide Contamination within and around Commercial Nuclear Power Plants; Assessment of Origin, Distribution, Inventory and Decommissioning. NUREG/CR-4289, 1986.

4. Elder, H. K., Technology, Safety and Costs of Decommissioning Reference Nuclear Fuel Facilities: Classification of Decommissioning Waste, NUREG/CR4519, or PNL-5589, Battelle Pacific Northwest Laboratory, Hanford, Wash., 1986.

5. LaGuardia, T. S., Decommissioning, in *Radioactive Waste Technology*, A. A. Moghissi, H. W. Godbee, and S. Hobart, eds., p. 499, ASME/ANS, New York, 1986.

6. U.S. Nuclear Regulatory Commission, Regulatory Guide 1. 184, Decommissioning of Nuclear Power Reactors, July 2000.

7. U.S. Regulatory Commission, Regulatory Guide 1. 179, Standard Format and Content of License Termination Plans for Nuclear Power Reactors, January 1999.
8. U.S. NRC, Preparation of Environmental Reports for Nuclear Power Stations, NRC Regulatory Guide 4.2, 1975.
9. Masnik, N., and L. Thonus, NRC Regulatory Framework for the Decommissioning of Power Reactors, in *Decommissioning and Restoration of Nuclear Facilities*, M. J. Slobodien, ed., Health Physics Society, 1999.
10. International Atomic Energy Agency, Power Reactor Information Sheet. Available at http://www.iaea.org [December 21, 2000].
11. U.S. NRC, Termination of Operating Licenses for Nuclear Reactors, Regulatory Guide 1.86, Nuclear Regulatory Commission, Washington, D.C., 1974.
12. U.S. DOE, Final Environment Impact Statement, Decommissioning of the Shippingport Atomic Power Station, DOE/EIS-0080F, Assistant Secretary of Nuclear Energy, Deputy Assistant Secretary for Nuclear Waste Management and Fuel Programs Remedial Action Program Office, Washington, D.C., 1982.
13. Mullee, G. R., and J. M. Usher, Shippingport Station Decommissioning Project (SSDP)—A progress report, in *Spectrum '86, Proc. ANS International Meeting on Waste Management and D/D*, vol. II, p. 1213, Niagara Falls, New York, September 1986.
14. Licensing and Regulatory Policy and Procedures for Environmental Protection, 1982. U.S. Code of Federal Regulations, Title 10, Part 51.
15. Bernhardt, D. E., J. D. Pittmann, and S. V. Prewett, Decommissioning plan depleted uranium manufacturing facility, in *Waste Management '87, Proc. Waste Isolation in the United States*, R. G. Post, ed., vol. 1, p. 533, University of Arizona Press, Tucson, Ariz., 1987.
16. U.S. Nuclear Regulatory Commission, 10CFR50, Domestic Licensing of Production and Utilization Facilities, 1998.
17. Kocher, D. C. Dose-rate conversion factors for external exposure to photons and electrons, *Health Phys.*, vol. 45, p. 665, 1983.
18. Kocher, D. C., and A. L. Sjoreen, Dose-rate conversion factors for external exposure to photon emitter in soil, *Health Phys.*, vol. 48, p. 193, 1985.
19. Hoovler, G. S., P. M. Myers, and C. S. Caldwell, Research reactor and fuel development facility decommissioning experience and technology, in *Waste Management '87, Proc. Waste Isolation in the United States*, R. G. Post, ed., vol. 1, p. 539, University of Arizona Press, Tucson, Ariz., 1987.
20. OECD, Decommissioning of Nuclear Facilities-Feasibility, Needs and Costs, Nuclear Energy Agency, Organization for Economic Co-operation and Development, Paris, 1986.
21. U.S. AEC, AEC Elk River Reactor, Final Report, United Power Association (Report DR-1207), COO-651–93, 1974.
22. U.S. DOE, Sodium Reactor Experiment Decommissioning, Final Report, Rockwell International Energy Systems Group, ESG-DOE-13403, 1983.
23. Bauer, R. G., Asbestos insulation removal at the Shippingport Station Decommissioning Project, in *Spectrum '86, Proc. ANS International Topical Meeting on Waste Management and D/D*, vol. II, p. 1350, Niagara Falls, New York, September 1986.
24. IAEA, Decontamination of Nuclear Facilities to Permit Operation, Inspection, Maintenance, Modification or Plant Decommissioning, Technical Report Series No. 249, International Atomic Energy Agency, Vienna, 1985.
25. Pasquini, L. A., Shippingport Station Decommissioning Project technology transfer program, in *Proc. 1986 Joint ASME/ANS Nuclear Power Conference*, p. 89, Philadelphia, July 1986.
26. Eger, K. J., D. L. Gardner, and R. J. Giordano, Release Criteria for Decommissioning of the Shippingport Atomic Power Station, in *Proc. 1986 Joint ASME/ANS Nuclear Power Conference*, p. 97, Philadelphia, July 1986.
27. U.S. DOE, Guidelines for Residual Radioactivity at Formerly Utilized Sites Remedial Action Program and Remote Surplus Facilities Management Program Sites, Rev. 1, July 1985.
28. Crimi, F. P., SSDP start of physical decommissioning, in *Proc. 1986 Joint ASME/ANS Nuclear Power Conference*, p. 82, Philadelphia, July 1986.
29. U.S. General Accounting Office, Shippingport Decommissioning—How Applicable Are the Lessons Learned? GAO/RCED-90-208, September 1990.

30. Knabenschuh, J. L., and W. H. Hannum, Overview of the West Valley Demonstration Project, in *Advances in Ceramics, Proc. 2nd International Symposium on Ceramics in Nuclear Waste Management*, Chicago, April 1983.

31. Bixby, W. W., W. R. Young, and P. J. Grant; TMI-2: Unique waste management technology, in *Waste Management '87*, R. G. Post, ed., vol. 1, p. 17, University of Arizona Press, Tucson, Ariz., 1987.

32. Kuypers, J., and A. Spruyt, Decommissioning of the KEMA suspension test reactor, in *Spectrum '86, Proc. ANS International Topical Meeting on Waste Management and D/D*, vol. II, p. 1502, Niagara Falls, New York, September 1986.

33. Loschhorn, U., U. Birkhold, J. Obst, and W. Stasch, Nuclear Power Plant Niederaichbach—Dismantling and concrete removal, in *Spectrum '86, Proc. ANS International Topical Meeting on Waste Management and D/D*, vol. II, p. 1495, Niagara Falls, New York, September 1986.

34. U.S. Nuclear Regulatory Commission, Report on Waste Disposal Charges: Changes in Decommissioning Waste Disposal Costs at Low-Level Waste Burial Facilities, NUREG-1307, Rev. 9, September 2000.

APPENDIX A

NOMENCLATURE AND UNIT CONVERSIONS

A.1 NOMENCLATURE

a	proportionality constant
a_d	dynamic adhesion term in penetrator equation
A	area
A_c	cross-sectional area of subchannels
b	backfill thickness
B	constant
BOC	beginning of cycle
BOP	balance of the plant
BU	burnup, MW-day/ton
c_p	specific heat at constant pressure, cal/g °C
C_s	solubility, g/m^3
C_U	soil shear strength
D	diameter, cm
D_e	equivalent diameter, cm
EOC	end of cycle
f	friction factor
F_B	buoyant force
F_D	inertial resistance of sediment
F_s	resistance due to soil stresses
FCR	annual fixed charge rate, % per year
g	conversion factor
G	mass flux, kg/sec m^2
h	specific enthalpy, cal/g; or heat transfer coefficient, W/m^2 °C
i	discount rate, % per year
I	capital investment cost, $
k	thermal conductivity, W/m °C
k_{eff}	effective multiplication factor

K	constant, or retardation coefficient
m	mass flow rate, g/sec
M	mass, g
n	neutron density
n	number of events or of years in the evaluation period
N_{ed}	dynamic and heaving capacity factor
O_j	annual operating expenses in year j, \$/year
$\overline{O}$	levelized O & M costs, \$/year
p	pressure, Pa
P	perimeter or pitch, cm
q	heat generation rate, J/sec
q''	heat flux, W/m^2
r	radius, cm
R	thermal resistance, (W/m °C)$^{-1}$
S	residual collective dose commitment
$\overline{S}$	levelized operating cost savings
S_b	surface area of penetrator
S_o	cross-sectional area
t	time, sec
T	temperature, °C
T_o	initial temperature, °C
$T_{1/2}$	half-life
U	overall heat transfer coefficient, W/m^2 °C
V	velocity, m/sec
x	axial length or plate height, cm
X	cost, \$
Z	depth below seabed, m

Greek Symbols

α	radioactive particles
α	ratio of the permeability of backfill to that of rock
β	radioactive particles
γ	radioactive rays
δ	thickness of a plate or peak-to-peak displacement, cm
Δ	change in dimension
ε	porosity
η	frequency, sec^{-1}
λ_i	precursor decay constant of ith group, sec^{-1}
μ	viscosity, g/cm sec
ρ	density, g/cm^3
σ	standard deviation
τ	time increment, sec
ψ	stream function
∇	gradient

Subscripts

a	apparent
av	average
ax	axial
c	coolant, or cold liquid

CL	centerline
conv	convective
eff	effective
gas	gas phase
H	hot liquid
i	index of different types of events
liq	liquid
m	mixture
s	static, or solid
ss	stainless steel
v	vapor
w	wall

A.2 UNIT CONVERSIONS

Acceleration	$1 \text{ m/sec}^2 = 3.280 \text{ ft/sec}^2$
Area	$1 \text{ m}^2 = 10.76 \text{ ft}^2$
Density	$1 \text{ kg/m}^3 = 6.242 \times 10^{-2} \text{ lb/ft}^3$
Force	$1 \text{ N} = 0.2248 \text{ lb}_f$
Heat flow	$1 \text{ W/m}^2 = 0.3173 \text{ Btu/hr ft}^2$
Heat transfer coefficient	$1 \text{ W/m}^2 \, ^\circ\text{C} = 0.1761 \text{ Btu/hr ft}^2 \, ^\circ\text{F}$
Length	$1 \text{ m} = 3.281 \text{ ft}$
Mass	$1 \text{ kg} = 2.205 \text{ lb}$
Mass flow rate	$1 \text{ kg/sec} = 7.938 \times 10^3 \text{ lb/hr}$
Mass flux	$1 \text{ kg/sec m}^2 = 7.374 \times 10^2 \text{ lb/hr ft}^2$
Power	$1 \text{ W} = 3.412 \text{ Btu/hr}$
Pressure	$1 \text{ Pa} = 1.450 \times 10^{-4} \text{ psi}$
Specific heat	$1 \text{ J/kg} \, ^\circ\text{C} = 2.388 \times 10^{-4} \text{ Btu/lb} \, ^\circ\text{F}$
Thermal energy	$1 \text{ J} = 9.479 \times 10^{-4} \text{ Btu}$
Thermal conductivity	$1 \text{ W/m} \, ^\circ\text{C} = 0.5778 \text{ Btu/hr ft} \, ^\circ\text{F}$
Thermal resistance	$1 \text{ m}^2 \, ^\circ\text{C/W} = 5.681 \text{ ft}^2 \, ^\circ\text{F hr/Btu}$
Velocity	$1 \text{ m/sec} = 3.218 \text{ ft/sec}$
Viscosity	$1 \text{ Pa sec} = 0.672 \text{ lb/ft sec}$
Volume	$1 \text{ m}^3 = 3.531 \times 10 \text{ ft}^3$

APPENDIX B

ACRONYMS AND ABBREVIATIONS

AAEC	Australian Atomic Energy Commission	CEC	Commission of European Communities
AEC	Atomic Energy Commission (abolished by the Energy Reorganization Act 1974)	CEN	Belgian Nuclear Research Establishment
ABS	alkali borosilicate	CERCLA	Comprehensive Environmental Response, Compensation, and Liability Act
AECL	Atomic Energy of Canada Limited		
AERE	Atomic Energy Research Establishment	CFCN	calcine-fed ceramic melter
AFR	away from reactor (fuel storage)	CFR	Code of Federal Regulations
AGNS	Allied General Nuclear Fuel Services Company	CH-TRU	contact-handled transuranic
		Ci	curie, a unit of radioactivity
ALARA	as low as reasonably achievable	CMST	Characterization, Monitoring, and Sensor Technology
ANDRA	National Agency for Radwaste Management (France)	COGEMA	Nuclear Fuel Cycle Company (France)
ANL	Argonne National Laboratory	DAW	dry active waste
ANS	American Nuclear Society	D & D	decontamination and decommissioning
ANSI	American National Standards Institute		
		DOE	Department of Energy (U.S.)
BEIR	Committee on Biological Effects of Ionizing Radiation	DOT	Department of Transportation (U.S.)
BNFL	British Nuclear Fuels Limited	DSHS	Department of Social and Health Services (Washington State)
BNFP	Barnwell Nuclear Fuel Plant		
BNL	Brookhaven National Laboratories	DWCF	Demonstrated Waste Calcining Facility
BRH	Bureau of Radiological Health		
BWIP	Basalt Waste Isolation Plant	EA	environment assessment
BWR	boiling-water reactor	EARP	Enhanced Actinide Removal Plant (BNFL)
CEA	(French government) Nuclear Energy Agency	ECI	equivalent capital investment

ECN	Energy Research Center of Netherlands	LET	linear energy transfer; energy loss/distance traveled
EdF	Electricité de France	LLMW	Low-Level Mixed Waste
EIS	environmental impact statement	LLW	low-level waste
EM	Environmental Management	LLWPAA	Low Level Waste Policy Amendments Act
E-MAD	engine maintenance assembly and disassembly building	LRR	levelized revenue requirement
EPA	Environmental Protection Agency	LSA	low specific activity
EPRI	Electric Power Research Institute	LWR	light-water reactor
ERDA	Energy Research and Development Administration	LWT	legal weight truck
		MAW	medium-active waste, $>10^{-10}$ Ci/ml but $<10^{-6}$ Ci/ml
ESTG	Engineering Studies Test Group of NEA	MCC	Material Characterization Center, Pacific Northwest Laboratory
FDA	Food and Drug Administration		
FFTF	Fast Flux Test Facility	MFRP	Midwest Fuel Recovery Plant at Morris, Illinois
GME	Great Meteor East area		
GSF	Gesellschaft für Strahlen- und Umoltforschung mbH	MPBB	maximum permissible body burden
		mrem	millirem (10^{-3} rem)
HEPA	high-efficiency particulate aerosol (filter)	MRS	monitored retrievable storage
		MTR	Material Testing Reactor
HIC	high-intensity container	NCRP	National Council on Radiation Protection and Measurement
HLW	high-level waste		
HMTA	Hazardous Materials Transportation Act	NEA	Nuclear Energy Agency of OECD
		NECO	Nuclear Engineering Company, Inc.
HOCUS	Hole Closure Simulation		
HSWA	Hazardous and Solid Waste Amendments	NEPA	National Environmental Policy Act
		NNWSI	Nevada Nuclear Waste Storage Investigation
IAEA	International Atomic Energy Agency		
		NPL	National Priorities List
IATA	International Air Transport Association	NRC	Nuclear Regulatory Commission
		NRCI	National Research Council
ICAO	International Civil Aviation Organization	NRPB	National Radiological Protection Board (U.K.)
ICPP	Idaho Falls Chemical Processing Plant	NSRA	Japanese Nuclear Safety Research Association
ICRP	International Commission on Radiological Protection	NTS	Nevada Test Site
		NWC	Nuclear Weapons Complex
If T	Institute für Tieflagerung	NWPA	Nuclear Waste Policy Act
ILW	intermediate-level waste; see MAW	NWPAA	Nuclear Waste Policy Amendments Act
IMDG	International Maritime Dangerous Goods Code	NWTS	National Waste Terminal Storage Program
IMO	International Maritime Organization		
INEL	Idaho National Engineering Laboratory	OCRWM	Office of Civilian Radioactive Waste Management
INEEL	Idaho National Engineering and Environmental Laboratory (formerly INEL)	OECD	Organization for Economic Cooperation and Development
		ONWI	Office of Nuclear Waste Isolation
JRC	Joint Research Center	ORNL	Oak Ridge National Laboratory
KBS	(Swedish) Nuclear Fuel Safety Project	OWT	overweight truck
		PL	Public Law
LBL	Lawrence Berkeley Laboratory	PNL	Pacific Northwest Laboratory
LCF	latent cancer fatality	PVRR	present value of revenue requirement
LDC	London Dumping Conference		
LDR	Land Disposal Restrictions	PWR	pressurized water reactor

R	unit of exposure to ionizing radiation (roentgen)	SPDV	site and preliminary design verification
RCCA	reactor core control assemblies	SRP	Savannah River Plant
RCRA	Resource Conservation and Recovery Act	SSDP	Shippingport Station Decommissioning Project
rem	dose equivalent unit, the effective exposure of humans (roentgen equivalent man)	SSP	self-shielded package
		SWG	Seabed Workshop Group
		SYNROC	synthetic rock consisting of small number of titanic mineral phases
RF	Rocky Flats		
RHO	Rockwell Hanford operations	TBP	tributyl phosphate
RH-TRU	remote-handled transuranic	TI	transport index
RLFCM	radioactive liquid feed ceramic melter	TLD	thermoluminescent dosimeter
		TMI	Three Mile Island
RPV	reactor pressure vessel	TOTM	transportation operation and traffic management
SARA	Superfund Amendments and Reauthorization Act		
		TRU	transuranic
SDP	Seabed Disposal Program	TRUPACT	transuranic package for transport
SDS	submerged demineralizer system	TSI	thermal structural interaction
SF, SNF	spent (nuclear) fuel	TTC	Transport Technology Center (Sandia)
SF2	A special spent fuel bundle, form 2		
SFCM	slurry-fed ceramic melter	TVA	Tennessee Valley Authority
SFHPP	Spent Fuel Handling and Packaging Program	TVR	transportable VR and solidification system
SFMP	Surplus Facilities Management Program	URL	(Canadian) Underground Research Laboratory
SIXEP	site ion-exchange plant	USGS	United States Geological Survey
SKBF	Swedish Nuclear Fuel and Waste Management Company	VR	volume reduction
		WIPP	Waste Isolation Pilot Plant
SNAP	Southern Nares Abyssal Plain	WVDP	West Valley Demonstration Project

INDEX